VOLUME NINETY ONE

Advances in MARINE BIOLOGY

Global Knowledge on the Commercial Sea Cucumber *Holothuria scabra*

ADVANCES IN MARINE BIOLOGY

VOLUME NINETY ONE

ADVANCES IN MARINE BIOLOGY

Global Knowledge on the Commercial Sea Cucumber *Holothuria scabra*

Edited by

ANNIE MERCIER
Department of Ocean Sciences, Memorial University, St. John's, Newfoundland & Labrador, Canada

JEAN-FRANÇOIS HAMEL
Society for the Exploration and Valuing of the Environment, St. Philips, Newfoundland & Labrador, Canada

ELSEVIER

ACADEMIC PRESS
An imprint of Elsevier

Academic Press is an imprint of Elsevier
125 London Wall, London, EC2Y 5AS, United Kingdom
The Boulevard, Langford Lane, Kidlington, Oxford OX5 1GB, United Kingdom
525 B Street, Suite 1650, San Diego, CA 92101, United States
50 Hampshire Street, 5th Floor, Cambridge, MA 02139, United States

First edition 2022

Notices
Knowledge and best practice in this field are constantly changing. As new research and experience broaden our understanding, changes in research methods, professional practices, or medical treatment may become necessary.

Practitioners and researchers must always rely on their own experience and knowledge in evaluating and using any information, methods, compounds, or experiments described herein. In using such information or methods they should be mindful of their own safety and the safety of others, including parties for whom they have a professional responsibility.

To the fullest extent of the law, neither the Publisher nor the authors, contributors, or editors, assume any liability for any injury and/or damage to persons or property as a matter of products liability, negligence or otherwise, or from any use or operation of any methods, products, instructions, or ideas contained in the material herein.

ISBN: 978-0-443-18434-5
ISSN: 0065-2881

For information on all Academic Press publications
visit our website at https://www.elsevier.com/books-and-journals

Publisher: Zoe Kruze
Acquisitions Editor: Jason Mitchell
Developmental Editor: Cindy Angelita P. Gardose
Production Project Manager: Vijayaraj Purushothaman
Cover Designer: Matthew Limbert

Typeset by STRAIVE, India

Contributors to Volume 91

Guillaume Caulier
Biology of Marine Organisms and Biomimetics unit, University of Mons, Mons, Belgium

Chantal Conand
Département origines et évolution, Muséum National Histoire Naturelle, Paris, France

Igor Eeckhaut
Biology of Marine Organisms and Biomimetics unit, University of Mons, Mons, Belgium

Jean-François Hamel
Society for the Exploration and Valuing of the Environment (SEVE), St. Philips, Newfoundland & Labrador, Canada

Annie Mercier
Department of Ocean Sciences, Memorial University, St. John's, Newfoundland & Labrador, Canada

Jiamin Sun
Department of Ocean Sciences, Memorial University, St. John's, Newfoundland & Labrador, Canada

Contents

Series Contents for Last Fifteen Years*,†

Volume 44, 2003.

Hirst, A. G., Roff, J. C. and Lampitt, R. S. A synthesis of growth rates in epipelagic invertebrate zooplankton. pp. 3–142.

Boletzky, S. von. Biology of early life stages in cephalopod molluscs. pp. 143–203.

Pittman, S. J. and McAlpine, C. A. Movements of marine fish and decapod crustaceans: process, theory and application. pp. 205–294.

Cutts, C. J. Culture of harpacticoid copepods: potential as live feed for rearing marine fish. pp. 295–315.

Volume 45, 2003.

Cumulative Taxonomic and Subject Index.

Volume 46, 2003.

Gooday, A. J. Benthic foraminifera (Protista) as tools in deep-water palaeoceanography: environmental influences on faunal characteristics. pp. 1–90.

Subramoniam, T. and Gunamalai, V. Breeding biology of the intertidal sand crab, Emerita (Decapoda: Anomura). pp. 91–182.

Coles, S. L. and Brown, B. E. Coral bleaching—capacity for acclimatization and adaptation. pp. 183–223.

Dalsgaard J., St. John M., Kattner G., Müller-Navarra D. and Hagen W. Fatty acid trophic markers in the pelagic marine environment. pp. 225–340.

Volume 47, 2004.

Southward, A. J., Langmead, O., Hardman-Mountford, N. J., Aiken, J., Boalch, G. T., Dando, P. R., Genner, M. J., Joint, I., Kendall, M. A., Halliday, N. C., Harris, R. P., Leaper, R., Mieszkowska, N., Pingree, R. D., Richardson, A. J., Sims, D.W., Smith, T., Walne, A. W. and Hawkins, S. J. Long-term oceanographic and ecological research in the western English Channel. pp. 1–105.

Queiroga, H. and Blanton, J. Interactions between behaviour and physical forcing in the control of horizontal transport of decapod crustacean larvae. pp. 107–214.

*The full list of contents for volumes 1–37 can be found in volume 38

†The full list of contents for volumes 38–43 can be found in volume 78

Braithwaite, R. A. and McEvoy, L. A. Marine biofouling on fish farms and its remediation. pp. 215–252.

Frangoulis, C., Christou, E. D. and Hecq, J. H. Comparison of marine copepod outfluxes: nature, rate, fate and role in the carbon and nitrogen cycles. pp. 253–309.

Volume 48, 2005.

Canfield, D. E., Kristensen, E. and Thamdrup, B. Aquatic Geomicrobiology. pp. 1–599.

Volume 49, 2005.

Bell, J. D., Rothlisberg, P. C., Munro, J. L., Loneragan, N. R., Nash, W. J., Ward, R. D. and Andrew, N. L. Restocking and stock enhancement of marine invertebrate fisheries. pp. 1–358.

Volume 50, 2006.

Lewis, J. B. Biology and ecology of the hydrocoral Millepora on coral reefs. pp. 1–55.

Harborne, A. R., Mumby, P. J., Micheli, F., Perry, C. T., Dahlgren, C. P., Holmes, K. E., and Brumbaugh, D. R. The functional value of Caribbean coral reef, seagrass and mangrove habitats to ecosystem processes. pp. 57–189.

Collins, M. A. and Rodhouse, P. G. K. Southern ocean cephalopods. pp. 191–265.

Tarasov, V. G. Effects of shallow-water hydrothermal venting on biological communities of coastal marine ecosystems of the western Pacific. pp. 267–410.

Volume 51, 2006.

Elena Guijarro Garcia. The fishery for Iceland scallop (Chlamys islandica) in the Northeast Atlantic. pp. 1–55.

Jeffrey, M. Leis. Are larvae of demersal fishes plankton or nekton? pp. 57–141.

John C. Montgomery, Andrew Jeffs, Stephen D. Simpson, Mark Meekan and Chris Tindle. Sound as an orientation cue for the pelagic larvae of reef fishes and decapod crustaceans. pp. 143–196.

Carolin E. Arndt and Kerrie M. Swadling. Crustacea in Arctic and Antarctic sea ice: Distribution, diet and life history strategies. pp. 197–315.

Volume 52, 2007.
Leys, S. P., Mackie, G. O. and Reiswig, H. M. The Biology of Glass Sponges. pp. 1–145.
Garcia E. G. The Northern Shrimp (Pandalus borealis) Offshore Fishery in the Northeast Atlantic. pp. 147–266.
Fraser K. P. P. and Rogers A. D. Protein Metabolism in Marine Animals: The Underlying Mechanism of Growth. pp. 267–362.

Volume 53, 2008.
Dustin J. Marshall and Michael J. Keough. The Evolutionary Ecology of Offspring Size in Marine Invertebrates. pp. 1–60.
Kerry A. Naish, Joseph E. Taylor III, Phillip S. Levin, Thomas P. Quinn, James R. Winton, Daniel Huppert, and Ray Hilborn. An Evaluation of the Effects of Conservation and Fishery Enhancement Hatcheries on Wild Populations of Salmon. pp. 61–194.
Shannon Gowans, Bernd Würsig, and Leszek Karczmarski. The Social Structure and Strategies of Delphinids: Predictions Based on an Ecological Framework. pp. 195–294.

Volume 54, 2008.
Bridget S. Green. Maternal Effects in Fish Populations. pp. 1–105.
Victoria J. Wearmouth and David W. Sims. Sexual Segregation in Marine Fish, Reptiles, Birds and Mammals: Behaviour Patterns, Mechanisms and Conservation Implications. pp. 107–170.
David W. Sims. Sieving a Living: A Review of the Biology, Ecology and Conservation Status of the Plankton-Feeding Basking Shark *Cetorhinus Maximus*. pp. 171–220.
Charles H. Peterson, Kenneth W. Able, Christin Frieswyk DeJong, Michael F. Piehler, Charles A. Simenstad, and Joy B. Zedler. Practical Proxies for Tidal Marsh Ecosystem Services: Application to Injury and Restoration. pp. 221–266.

Volume 55, 2008.
Annie Mercier and Jean-Francois Hamel. Introduction. pp. 1–6.
Annie Mercier and Jean-Francois Hamel. Gametogenesis. pp. 7–72.
Annie Mercier and Jean-Francois Hamel. Spawning. pp. 73–168.
Annie Mercier and Jean-Francois Hamel. Discussion. pp. 169–194.

Volume 56, 2009.

Philip C. Reid, Astrid C. Fischer, Emily Lewis-Brown, Michael P. Meredith, Mike Sparrow, Andreas J. Andersson, Avan Antia, Nicholas R. Bates, Ulrich Bathmann, Gregory Beaugrand, Holger Brix, Stephen Dye, Martin Edwards, Tore Furevik, Reidun Gangst, Hjalmar Hatun, Russell R. Hopcroft, Mike Kendall, Sabine Kasten, Ralph Keeling, Corinne Le Quere, Fred T. Mackenzie, Gill Malin, Cecilie Mauritzen, Jon Olafsson, Charlie Paull, Eric Rignot, Koji Shimada, Meike Vogt, Craig Wallace, Zhaomin Wang and Richard Washington. Impacts of the Oceans on Climate Change. pp. 1–150.

Elvira S. Poloczanska, Colin J. Limpus and Graeme C. Hays. Vulnerability of Marine Turtles to Climate Change. pp. 151–212.

Nova Mieszkowska, Martin J. Genner, Stephen J. Hawkins and David W. Sims. Effects of Climate Change and Commercial Fishing on Atlantic Cod *Gadus morhua*. pp. 213–274.

Iain C. Field, Mark G. Meekan, Rik C. Buckworth and Corey J. A. Bradshaw. Susceptibility of Sharks, Rays and Chimaeras to Global Extinction. pp. 275–364.

Milagros Penela-Arenaz, Juan Bellas and Elsa Vazquez. Effects of the *Prestige* Oil Spill on the Biota of NW Spain: 5 Years of Learning. pp. 365–396.

Volume 57, 2010.

Geraint A. Tarling, Natalie S. Ensor, Torsten Fregin, William P. Good-all-Copestake and Peter Fretwell. An Introduction to the Biology of Northern Krill (*Meganyctiphanes norvegica* Sars). pp. 1–40.

Tomaso Patarnello, Chiara Papetti and Lorenzo Zane. Genetics of Northern Krill (*Meganyctiphanes norvegica* Sars). pp. 41–58.

Geraint A. Tarling. Population Dynamics of Northern Krill (*Meganyctiphanes norvegica* Sars). pp. 59–90.

John I. Spicer and Reinhard Saborowski. Physiology and Metabolism of Northern Krill (*Meganyctiphanes norvegica* Sars). pp. 91–126.

Katrin Schmidt. Food and Feeding in Northern Krill (*Meganyctiphanes norvegica* Sars). pp. 127–172.

Friedrich Buchholz and Cornelia Buchholz. Growth and Moulting in Northern Krill (*Meganyctiphanes norvegica* Sars). pp. 173–198.

Janine Cuzin-Roudy. Reproduction in Northern Krill. pp. 199–230.

Edward Gaten, Konrad Wiese and Magnus L. Johnson. Laboratory-Based Observations of Behaviour in Northern Krill (*Meganyctiphanes norvegica* Sars). pp. 231–254.

Stein Kaartvedt. Diel Vertical Migration Behaviour of the Northern Krill (*Meganyctiphanes norvegica* Sars). pp. 255–276.
Yvan Simard and Michel Harvey. Predation on Northern Krill (*Meganyctiphanes norvegica* Sars). pp. 277–306.

Volume 58, 2010.
A. G. Glover, A. J. Gooday, D. M. Bailey, D. S. M. Billett, P. Chevaldonné, A. Colaço, J. Copley, D. Cuvelier, D. Desbruyères, V. Kalogeropoulou, M. Klages, N. Lampadariou, C. Lejeusne, N. C. Mestre, G. L. J. Paterson, T. Perez, H. Ruhl, J. Sarrazin, T. Soltwedel, E. H. Soto, S. Thatje, A. Tselepides, S. Van Gaever, and A. Vanreusel. Temporal Change in Deep-Sea Benthic Ecosystems: A Review of the Evidence From Recent Time-Series Studies. pp. 1–96.
Hilario Murua. The Biology and Fisheries of European Hake, *Merluccius merluccius*, in the North-East Atlantic. pp. 97–154.
Jacopo Aguzzi and Joan B. Company. Chronobiology of Deep-Water Decapod Crustaceans on Continental Margins. pp. 155–226.
Martin A. Collins, Paul Brickle, Judith Brown, and Mark Belchier. The Patagonian Toothfish: Biology, Ecology and Fishery. pp. 227–300.

Volume 59, 2011.
Charles W. Walker, Rebecca J. Van Beneden, Annette F. Muttray, S. Anne Böttger, Melissa L. Kelley, Abraham E. Tucker, and W. Kelley Thomas. p53 Superfamily Proteins in Marine Bivalve Cancer and Stress Biology. pp 1–36.
Martin Wahl, Veijo Jormalainen, Britas Klemens Eriksson, James A. Coyer, Markus Molis, Hendrik Schubert, Megan Dethier, Anneli Ehlers, Rolf Karez, Inken Kruse, Mark Lenz, Gareth Pearson, Sven Rohde, Sofia A. Wikström, and Jeanine L. Olsen. Stress Ecology in *Fucus*: Abiotic, Biotic and Genetic Interactions. pp. 37–106.
Steven R. Dudgeon and Janet E. Kübler. Hydrozoans and the Shape of Things to Come. pp. 107–144.
Miles Lamare, David Burritt, and Kathryn Lister. Ultraviolet Radiation and Echinoderms: Past, Present and Future Perspectives. pp. 145–187.

Volume 60, 2011.
Tatiana A. Rynearson and Brian Palenik. Learning to Read the Oceans: Genomics of Marine Phytoplankton. pp. 1–40.
Les Watling, Scott C. France, Eric Pante and Anne Simpson. Biology of Deep-Water Octocorals. pp. 41–122.

Cristián J. Monaco and Brian Helmuth. Tipping Points, Thresholds and the Keystone Role of Physiology in Marine Climate Change Research. pp. 123–160.
David A. Ritz, Alistair J. Hobday, John C. Montgomery and Ashley J.W. Ward. Social Aggregation in the Pelagic Zone with Special Reference to Fish and Invertebrates. pp. 161–228.

Volume 61, 2012.
Gert Wörheide, Martin Dohrmann, Dirk Erpenbeck, Claire Larroux, Manuel Maldonado, Oliver Voigt, Carole Borchiellini and Denis Lavrov. Deep Phylogeny and Evolution of Sponges (Phylum Porifera). pp. 1–78.
Paco Cárdenas, Thierry Pérez and Nicole Boury-Esnault. Sponge Systematics Facing New Challenges. pp. 79–210.
Klaus Rützler. The Role of Sponges in the Mesoamerican Barrier-Reef Ecosystem, Belize. pp. 211–272.
Janie Wulff. Ecological Interactions and the Distribution, Abundance, and Diversity of Sponges. pp. 273–344.
Maria J. Uriz and Xavier Turon. Sponge Ecology in the Molecular Era. pp. 345–410.

Volume 62, 2012.
Sally P. Leys and April Hill. The Physiology and Molecular Biology of Sponge Tissues. pp. 1–56.
Robert W. Thacker and Christopher J. Freeman. Sponge–Microbe Symbioses: Recent Advances and New Directions. pp. 57–112.
Manuel Maldonado, Marta Ribes and Fleur C. van Duyl. Nutrient Fluxes Through Sponges: Biology, Budgets, and Ecological Implications. pp. 113–182.
Grégory Genta-Jouve and Olivier P. Thomas. Sponge Chemical Diversity: From Biosynthetic Pathways to Ecological Roles. pp. 183–230.
Xiaohong Wang, Heinz C. Schröder, Matthias Wiens, Ute Schloßmacher and Werner E. G. Müller. Biosilica: Molecular Biology, Biochemistry and Function in Demosponges as well as its Applied Aspects for Tissue Engineering. pp. 231–272.
Klaske J. Schippers, Detmer Sipkema, Ronald Osinga, Hauke Smidt, Shirley A. Pomponi, Dirk E. Martens and René H. Wijffels. Cultivation of Sponges, Sponge Cells and Symbionts: Achievements and Future Prospects. pp. 273–338.

Volume 63, 2012.

Michael Stat, Andrew C. Baker, David G. Bourne, Adrienne M. S. Correa, Zac Forsman, Megan J. Huggett, Xavier Pochon, Derek Skillings, Robert J. Toonen, Madeleine J. H. van Oppen, and Ruth D. Gates. Molecular Delineation of Species in the Coral Holobiont. pp. 1–66.

Daniel Wagner, Daniel G. Luck, and Robert J. Toonen. The Biology and Ecology of Black Corals (Cnidaria: Anthozoa: Hexacorallia: Antipatharia). pp. 67–132.

Cathy H. Lucas, William M. Graham, and Chad Widmer. Jellyfish Life Histories: Role of Polyps in Forming and Maintaining Scyphomedusa Populations. pp. 133–196.

T. Aran Mooney, Maya Yamato, and Brian K. Branstetter. Hearing in Cetaceans: From Natural History to Experimental Biology. pp. 197–246.

Volume 64, 2013.

Dale Tshudy. Systematics and Position of *Nephrops* Among the Lobsters. pp. 1–26.

Mark P. Johnson, Colm Lordan, and Anne Marie Power. Habitat and Ecology of *Nephrops norvegicus*. pp. 27–64.

Emi Katoh, Valerio Sbragaglia, Jacopo Aguzzi, and Thomas Breithaupt. Sensory Biology and Behaviour of *Nephrops norvegicus*. pp. 65–106.

Edward Gaten, Steve Moss, and Magnus L. Johnson. The Reniform Reflecting Superposition Compound Eyes of *Nephrops norvegicus*: Optics, Susceptibility to Light-Induced Damage, Electrophysiology and a Ray Tracing Model. pp. 107–148.

Susanne P. Eriksson, Bodil Hernroth, and Susanne P. Baden. Stress Biology and Immunology in *Nephrops norvegicus*. pp. 149–200.

Adam Powell and Susanne P. Eriksson. Reproduction: Life Cycle, Larvae and Larviculture. pp. 201–246.

Anette Ungfors, Ewen Bell, Magnus L. Johnson, Daniel Cowing, Nicola C. Dobson, Ralf Bublitz, and Jane Sandell. *Nephrops* Fisheries in European Waters. pp. 247–314.

Volume 65, 2013.

Isobel S.M. Bloor, Martin J. Attrill, and Emma L. Jackson. A Review of the Factors Influencing Spawning, Early Life Stage Survival and Recruitment Variability in the Common Cuttlefish (*Sepia officinalis*). pp. 1–66.

Dianna K. Padilla and Monique M. Savedo. A Systematic Review of Phenotypic Plasticity in Marine Invertebrate and Plant Systems. pp. 67–120.

Leif K. Rasmuson. The Biology, Ecology and Fishery of the Dungeness crab, *Cancer magister*. pp. 121–174.

Volume 66, 2013.

Lisa-ann Gershwin, Anthony J. Richardson, Kenneth D. Winkel, Peter J. Fenner, John Lippmann, Russell Hore, Griselda Avila-Soria, David Brewer, Rudy J. Kloser, Andy Steven, and Scott Condie. Biology and Ecology of Irukandji Jellyfish (Cnidaria: Cubozoa). pp. 1–86.

April M. H. Blakeslee, Amy E. Fowler, and Carolyn L. Keogh. Marine Invasions and Parasite Escape: Updates and New Perspectives. pp. 87–170.

Michael P. Russell. Echinoderm Responses to Variation in Salinity. pp. 171–212.

Daniela M. Ceccarelli, A. David McKinnon, Serge Andréfouët, Valerie Allain, Jock Young, Daniel C. Gledhill, Adrian Flynn, Nicholas J. Bax, Robin Beaman, Philippe Borsa, Richard Brinkman, Rodrigo H. Bustamante, Robert Campbell, Mike Cappo, Sophie Cravatte, Stéphanie D'Agata, Catherine M. Dichmont, Piers K. Dunstan, Cécile Dupouy, Graham Edgar, Richard Farman, Miles Furnas, Claire Garrigue, Trevor Hutton, Michel Kulbicki, Yves Letourneur, Dhugal Lindsay, Christophe Menkes, David Mouillot, Valeriano Parravicini, Claude Payri, Bernard Pelletier, Bertrand Richer de Forges, Ken Ridgway, Martine Rodier, Sarah Samadi, David Schoeman, Tim Skewes, Steven Swearer, Laurent Vigliola, Laurent Wantiez, Alan Williams, Ashley Williams, and Anthony J. Richardson. The Coral Sea: Physical Environment, Ecosystem Status and Biodiversity Assets. pp. 213–290.

Volume 67, 2014.

Erica A.G. Vidal, Roger Villanueva, José P. Andrade, Ian G. Gleadall, José Iglesias, Noussithé Koueta, Carlos Rosas, Susumu Segawa, Bret Grasse, Rita M. Franco-Santos, Caroline B. Albertin, Claudia Caamal-Monsreal, Maria E. Chimal, Eric Edsinger-Gonzales, Pedro Gallardo, Charles Le Pabic, Cristina Pascual, Katina Roumbedakis, and James Wood. Cephalopod Culture: Current Status of Main Biological Models and Research Priorities. pp. 1–98.

Paul G.K. Rodhouse, Graham J. Pierce, Owen C. Nichols, Warwick H.H. Sauer, Alexander I. Arkhipkin, Vladimir V. Laptikhovsky, Marek R. Lipiński, Jorge E. Ramos, Michaël Gras, Hideaki Kidokoro, Kazuhiro Sadayasu, João Pereira, Evgenia Lefkaditou, Cristina Pita, Maria

Gasalla, Manuel Haimovici, Mitsuo Sakai, and Nicola Downey. Environmental Effects on Cephalopod Population Dynamics: Implications for Management of Fisheries. pp. 99–234.

Henk-Jan T. Hoving, José A.A. Perez, Kathrin Bolstad, Heather Braid, Aaron B. Evans, Dirk Fuchs, Heather Judkins, Jesse T. Kelly, José E.A.R. Marian, Ryuta Nakajima, Uwe Piatkowski, Amanda Reid, Michael Vecchione, and José C.C. Xavier. The Study of Deep-Sea Cephalopods. pp. 235–362.

Jean-Paul Robin, Michael Roberts, Lou Zeidberg, Isobel Bloor, Almendra Rodriguez, Felipe Briceño, Nicola Downey, Maite Mascaró, Mike Navarro, Angel Guerra, Jennifer Hofmeister, Diogo D. Barcellos, Silvia A.P. Lourenço, Clyde F.E. Roper, Natalie A. Moltschaniwskyj, Corey P. Green, and Jennifer Mather. Transitions During Cephalopod Life History: The Role of Habitat, Environment, Functional Morphology and Behaviour. pp. 363–440.

Volume 68, 2014.

Paul K.S. Shin, Siu Gin Cheung, Tsui Yun Tsang, and Ho Yin Wai. Ecology of Artificial Reefs in the Subtropics. pp. 1–64.

Hrafnkell Eiríksson. Reproductive Biology of Female Norway Lobster, *Nephrops norvegicus* (Linnaeus, 1758) Leach, in Icelandic Waters During the Period 1960–2010: Comparative Overview of Distribution Areas in the Northeast Atlantic and the Mediterranean. pp. 65–210.

Volume 69, 2014.

Ray Hilborn. Introduction to Marine Managed Areas. pp. 1–14.

Philip N. Trathan, Martin A. Collins, Susie M. Grant, Mark Belchier, David K.A. Barnes, Judith Brown, and Iain J. Staniland. The South Georgia and the South Sandwich Islands MPA: Protecting A Biodiverse Oceanic Island Chain Situated in the Flow of the Antarctic Circumpolar Current. pp. 15–78.

Richard P. Dunne, Nicholas V.C. Polunin, Peter H. Sand, and Magnus L. Johnson. The Creation of the Chagos Marine Protected Area: A Fisheries Perspective. pp. 79–128.

Michelle T. Schärer-Umpierre, Daniel Mateos-Molina, Richard Appeldoorn, Ivonne Bejarano, Edwin A. Hernández-Delgado, Richard S. Nemeth, Michael I. Nemeth, Manuel Valdés-Pizzini, and Tyler B. Smith. Marine Managed Areas and Associated Fisheries in the US Caribbean. pp. 129–152.

Alan M. Friedlander, Kostantinos A. Stamoulis, John N. Kittinger, Jeffrey C. Drazen, and Brian N. Tissot. Understanding the Scale of

Marine Protection in Hawai'i: From Community-Based Management to the Remote Northwestern Hawaiian Islands. pp. 153–204.
Louis W. Botsford, J. Wilson White, Mark H. Carr, and Jennifer E. Caselle. Marine Protected Area Networks in California, USA. pp. 205–252.
Bob Kearney and Graham Farebrother. Inadequate Evaluation and Management of Threats in Australia's Marine Parks, Including the Great Barrier Reef, Misdirect Marine Conservation. pp. 253–288.
Randi Rotjan, Regen Jamieson, Ben Carr, Les Kaufman, Sangeeta Mangubhai, David Obura, Ray Pierce, Betarim Rimon, Bud Ris, Stuart Sandin, Peter Shelley, U. Rashid Sumaila, Sue Taei, Heather Tausig, Tukabu Teroroko, Simon Thorrold, Brooke Wikgren, Teuea Toatu, and Greg Stone. Establishment, Management, and Maintenance of the Phoenix Islands Protected Area. pp. 289–324.
Alex J. Caveen, Clare Fitzsimmons, Margherita Pieraccini, Euan Dunn, Christopher J. Sweeting, Magnus L. Johnson, Helen Bloomfield, Estelle V. Jones, Paula Lightfoot, Tim S. Gray, Selina M. Stead, and Nicholas V. C. Polunin. Diverging Strategies to Planning an Ecologically Coherent Network of MPAs in the North Sea: The Roles of Advocacy, Evidence and Pragmatism in the Face of Uncertaintya. pp. 325–370.
Carlo Pipitone, Fabio Badalamenti, Tomás Vega Fernández, and Giovanni D'Anna. Spatial Management of Fisheries in the Mediterranean Sea: Problematic Issues and a Few Success Stories. pp. 371–402.

Volume 70, 2015.
Alex D. Rogers, Christopher Yesson, and Pippa Gravestock. A Biophysical and Economic Profile of South Georgia and the South Sandwich Islands as Potential Large-Scale Antarctic Protected Areas. pp. 1–286.

Volume 71, 2015.
Ricardo Calado and Miguel Costa Leal. Trophic Ecology of Benthic Marine Invertebrates with Bi-Phasic Life Cycles: What Are We Still Missing? pp. 1–70.
Jesse M.A. van der Grient and Alex D. Rogers. Body Size Versus Depth: Regional and Taxonomical Variation in Deep-Sea Meio- and Macrofaunal Organisms. pp. 71–108.
Lorena Basso, Maite Vázquez-Luis, José R. García-March, Salud Deudero, Elvira Alvarez, Nardo Vicente, Carlos M. Duarte, and Iris E. Hendriks. The Pen Shell, *Pinna nobilis*: A Review of Population Status and Recommended Research Priorities in the Mediterranean Sea. pp. 109–160.

Volume 72, 2015.

Thomas A. Jefferson and Barbara E. Curry. Humpback Dolphins: A Brief Introduction to the Genus *Sousa*. pp. 1–16.

Sarah Piwetz, David Lundquist, and Bernd Würsig. Humpback Dolphin (Genus *Sousa*) Behavioural Responses to Human Activities. pp. 17–46.

Tim Collins. Re-assessment of the Conservation Status of the Atlantic Humpback Dolphin, *Sousa teuszii* (Kükenthal, 1892) Using the IUCN Red List Criteria. pp. 47–78.

Caroline R. Weir and Tim Collins. A Review of the Geographical Distribution and Habitat of the Atlantic Humpback Dolphin (*Sousa teuszii*). pp. 79–118.

Gill T Braulik, Ken Findlay, Salvatore Cerchio, and Robert Baldwin. Assessment of the Conservation Status of the Indian Ocean Humpback Dolphin (*Sousa plumbea*) Using the IUCN Red List Criteria. pp. 119–142.

Stephanie Plön, Victor G. Cockcroft, and William P. Froneman. The Natural History and Conservation of Indian Ocean Humpback Dolphins (*Sousa plumbea*) in South African Waters. pp. 143–162.

Salvatore Cerchio, Norbert Andrianarivelo, and Boris Andrianantenaina. Ecology and Conservation Status of Indian Ocean Humpback Dolphins (*Sousa plumbea*) in Madagascar. pp. 163–200.

Muhammad Shoaib Kiani and Koen Van Waerebeek. A Review of the Status of the Indian Ocean Humpback Dolphin *Sousa plumbea* in Pakistan. pp. 201–228.

Dipani Sutaria, Divya Panicker, Ketki Jog, Mihir Sule, Rahul Muralidharan, and Isha Bopardikar. Humpback Dolphins (Genus *Sousa*) in India: An Overview of Status and Conservation Issues. pp. 229–256.

Volume 73, 2016.

Thomas A. Jefferson and Brian D. Smith. Re-assessment of the Conservation Status of the Indo-Pacific Humpback Dolphin (*Sousa chinensis*) Using the IUCN Red List Criteria. pp. 1–26.

Leszek Karczmarski, Shiang-Lin Huang, Carmen K. M. Or, Duan Gui, Stephen C. Y. Chan, Wenzhi Lin, Lindsay Porter, Wai-Ho Wong, Ruiqiang Zheng, Yuen-Wa Ho, Scott Y. S. Chui, Angelico Jose C. Tiongson, Yaqian Mo, Wei-Lun Chang, John H. W. Kwok, Ricky W. K. Tang, Andy T. L. Lee, Sze-Wing Yiu, Mark Keith, Glenn Gailey, and Yuping Wu. Humpback Dolphins in Hong Kong and the Pearl River Delta: Status, Threats and Conservation Challenges. pp. 27–64.

Bernd Würsig, E.C.M. Parsons, Sarah Piwetz, and Lindsay Porter. The Behavioural Ecology of Indo-Pacific Humpback Dolphins in Hong Kong. pp. 65–90.

John Y. Wang, Kimberly N. Riehl, Michelle N. Klein, Shiva Javdan, Jordan M. Hoffman, Sarah Z. Dungan, Lauren E. Dares, and Claryana Araújo-Wang. Biology and Conservation of the Taiwanese Humpback Dolphin, *Sousa chinensis taiwanensis*. pp. 91–118.

Bingyao Chen, Xinrong Xu, Thomas A. Jefferson, Paula A. Olson, Qiurong Qin, Hongke Zhang, Liwen He, and Guang Yang. Conservation Status of the Indo-Pacific Humpback Dolphin (*Sousa chinensis*) in the Northern Beibu Gulf, China. pp. 119–140.

Gianna Minton, Anna Norliza Zulkifli Poh, Cindy Peter, Lindsay Porter, and Danielle Kreb. Indo-Pacific Humpback Dolphins in Borneo: A Review of Current Knowledge with Emphasis on Sarawak. pp. 141–156.

Guido J. Parra and Daniele Cagnazzi. Conservation Status of the Australian Humpback Dolphin (*Sousa sahulensis*) Using the IUCN Red List Criteria. pp. 157–192.

Daniella M. Hanf, Tim Hunt, and Guido J. Parra. Humpback Dolphins of Western Australia: A Review of Current Knowledge and Recommendations for Future Management. pp. 193–218.

Isabel Beasley, Maria Jedensjö, Gede Mahendra Wijaya, Jim Anamiato, Benjamin Kahn, and Danielle Kreb. Observations on Australian Humpback Dolphins (*Sousa sahulensis*) in Waters of the Pacific Islands and New Guinea. pp. 219–272.

Alexander M. Brown, Lars Bejder, Guido J. Parra, Daniele Cagnazzi, Tim Hunt, Jennifer L. Smith, and Simon J. Allen. Sexual Dimorphism and Geographic Variation in Dorsal Fin Features of Australian Humpback Dolphins, *Sousa sahulensis*. pp. 273–314.

Volume 74, 2016.

J. Salinger, A.J. Hobday, R.J. Matear, T.J. O'Kane, J.S. Risbey, P. Dunstan, J.P. Eveson, E.A. Fulton, M. Feng, É.E. Plagányi, E.S. Poloczanska, A.G. Marshall, and P.A. Thompson. Decadal-Scale Forecasting of Climate Drivers for Marine Applications. pp. 1–68.

S.A. Foo and M. Byrne. Acclimatization and Adaptive Capacity of Marine Species in a Changing Ocean. pp. 69–116.

N.D. Gallo and L.A. Levin. Fish Ecology and Evolution in the World's Oxygen Minimum Zones and Implications of Ocean Deoxygenation. pp. 117–198.

R.J. Olson, J.W. Young, F. Ménard, M. Potier, V. Allain, N. Goñi, J.M. Logan, and F. Galván-Magaña. Bioenergetics, Trophic Ecology, and Niche Separation of Tunas. pp. 199–344.

Volume 75, 2016.
G. Notarbartolo di Sciara. Marine Mammals in the Mediterranean Sea: An Overview. pp. 1–36.
L. Rendell and A. Frantzis. Mediterranean Sperm Whales, *Physeter macrocephalus*: The Precarious State of a Lost Tribe. pp. 37–74.
G. Notarbartolo di Sciara, M. Castellote, J.-N. Druon, and S. Panigada. Fin Whales, *Balaenoptera physalus*: At Home in a Changing Mediterranean Sea?. pp. 75–102.
M. Podestà, A. Azzellino, A. Cañadas, A. Frantzis, A. Moulins, M. Rosso, P. Tepsich, and C. Lanfredi. Cuvier's Beaked Whale, *Ziphius cavirostris*, Distribution and Occurrence in the Mediterranean Sea: High-Use Areas and Conservation Threats. pp. 103–140.
R. Esteban, P. Verborgh, P. Gauffier, D. Alarcón, J.M. Salazar-Sierra, J. Giménez, A.D. Foote, and R. de Stephanis. Conservation Status of Killer Whales, *Orcinus orca*, in the Strait of Gibraltar. pp. 141–172.
P. Verborgh, P. Gauffier, R. Esteban, J. Giménez, A. Cañadas, J.M. Salazar-Sierra, and R. de Stephanis. Conservation Status of Long-Finned Pilot Whales, *Globicephala melas*, in the Mediterranean Sea. pp. 173–204.
A. Azzellino, S. Airoldi, S. Gaspari, C. Lanfredi, A. Moulins, M. Podestà, M. Rosso, and P. Tepsich. Risso's Dolphin, *Grampus griseus*, in the Western Ligurian Sea: Trends in Population Size and Habitat Use. pp. 205–232.
D. Kerem, O. Goffman, M. Elasar, N. Hadar, A. Scheinin, and T. Lewis. The Rough-Toothed Dolphin, *Steno bredanensis*, in the Eastern Mediterranean Sea: A Relict Population?. pp. 233–258.
J. Gonzalvo, G. Lauriano, P.S. Hammond, K.A. Viaud-Martinez, M.C. Fossi, A. Natoli, and L. Marsili. The Gulf of Ambracia's Common Bottlenose Dolphins, *Tursiops truncatus*: A Highly Dense and yet Threatened Population. pp. 259–296.
G. Bearzi, S. Bonizzoni, N.L. Santostasi, N.B. Furey, L. Eddy, V.D. Valavanis, and O. Gimenez. Dolphins in a Scaled-Down Mediterranean: The Gulf of Corinth's Odontocetes. pp. 297–332.
M.C. Fontaine. Harbour Porpoises, *Phocoena phocoena*, in the Mediterranean Sea and Adjacent Regions: Biogeographic Relicts of the Last Glacial Period. pp. 333–358.

G. Notarbartolo di Sciara and S. Kotomatas. Are Mediterranean Monk Seals, *Monachus monachus*, Being Left to Save Themselves from Extinction?. pp. 359–386.
T. Scovazzi. The International Legal Framework for Marine Mammal Conservation in the Mediterranean Sea. pp. 387–416.

Volume 76, 2017.
K.S. Meyer. Islands in a Sea of Mud: Insights From Terrestrial Island Theory for Community Assembly on Insular Marine Substrata. pp. 1–40.
E.M. Montgomery, J.-F. Hamel, and A. Mercier. Patterns and Drivers of Egg Pigment Intensity and Colour Diversity in the Ocean: A Meta-Analysis of Phylum Echinodermata. pp. 41–104.
F. Espinosa and G.A. Rivera-Ingraham. Biological Conservation of Giant Limpets: The Implications of Large Size. pp. 105–156.
L. Yebra, T. Kobari, A.R. Sastri, F. Gusmão, and S. Hernández-León. Advances in Biochemical Indices of Zooplankton Production. pp. 157–240.

Volume 77, 2017.
Dayv Lowry and Shawn E. Larson. Introduction to Northeast Pacific Shark Biology, Ecology, and Conservation. pp. 1–8.
David A. Ebert, Jennifer S. Bigman, and Julia M. Lawson. Biodiversity, Life History, and Conservation of Northeastern Pacific Chondrichthyans pp. 9–78.
Shawn E. Larson, Toby S. Daly-Engel, and Nicole M. Phillips. Review of Current Conservation Genetic Analyses of Northeast Pacific Sharks. pp. 79–110.
Joseph J. Bizzarro, Aaron B. Carlisle, Wade D. Smith, and Enric Cortés. Diet Composition and Trophic Ecology of Northeast Pacific Ocean Sharks. pp. 111–148.
Jonathan C.P. Reum, Gregory D. Williams, and Chris J. Harvey. Stable Isotope Applications for Understanding Shark Ecology in the Northeast Pacific Ocean. pp. 149–178.
Mary E. Matta, Cindy A. Tribuzio, David A. Ebert, Kenneth J. Goldman, and Christopher M. Gburski. Age and Growth of Elasmobranchs and Applications to Fisheries Management and Conservation in the Northeast Pacific Ocean. pp. 179–220.

Volume 78, 2017.
Shawn E. Larson and Dayv Lowry. Introduction to Northeast Pacific Shark Biology, Research, and Conservation, Part B. pp. 1–8.

Jackie King, Gordon A. McFarlane, Vladlena Gertseva, Jason Gasper, Sean Matson, and Cindy A. Tribuzio. Shark Interactions With Directed and Incidental Fisheries in the Northeast Pacific Ocean: Historic and Current Encounters, and Challenges for Shark Conservation. pp. 9–44.

Dovi Kacev, Timothy J. Sippel, Michael J. Kinney, Sebastián A. Pardo, and Christopher G. Mull. An Introduction to Modelling Abundance and Life History Parameters in Shark Populations. pp. 45–88.

Michael Grassmann, Bryan McNeil, and Jim Wharton. Sharks in Captivity: The Role of Husbandry, Breeding, Education, and Citizen Science in Shark Conservation. pp. 89–120.

Peter A. Mieras, Chris Harvey-Clark, Michael Bear, Gina Hodgin, and Boone Hodgin. The Economy of Shark Conservation in the Northeast Pacific: The Role of Ecotourism and Citizen Science. pp. 121–154.

Dayv Lowry. Conclusions: The Future of Shark Management and Conservation in the Northeast Pacific Ocean. pp. 155–164.

Volume 79, 2018.

Patricia L.M. Lee, Gail Schofield, Rebecca I. Haughey, Antonios D. Mazaris, and Graeme C. Hays. A Review of Patterns of Multiple Paternity Across Sea Turtle Rookeries. pp. 1–32.

Johanne Vad, Georgios Kazanidis, Lea-Anne Henry, Daniel O.B. Jones, Ole S. Tendal, Sabine Christiansen, Theodore B. Henry, and J. Murray Roberts. Potential Impacts of Offshore Oil and Gas Activities on Deep-Sea Sponges and the Habitats They Form. pp. 33–60.

Gianmarco Ingrosso, Marco Abbiati, Fabio Badalamenti, Giorgio Bavestrello, Genuario Belmonte, Rita Cannas, Lisandro Benedetti-Cecchi, Marco Bertolino, Stanislao Bevilacqua, Carlo Nike Bianchi, Marzia Bo, Elisa Boscari, Frine Cardone, Riccardo Cattaneo-Vietti, Alessandro Cau, Carlo Cerrano, Renato Chemello, Giovanni Chimienti, Leonardo Congiu, Giuseppe Corriero, Federica Costantini, Francesco De Leo, Luigia Donnarumma, Annalisa Falace, Simonetta Fraschetti, Adriana Giangrande, Maria Flavia Gravina, Giuseppe Guarnieri, Francesco Mastrototaro, Marco Milazzo, Carla Morri, Luigi Musco, Laura Pezzolesi, Stefano Piraino, Fiorella Prada, Massimo Ponti, Fabio Rindi, Giovanni Fulvio Russo, Roberto Sandulli, Adriana Villamor, Lorenzo Zane, and Ferdinando Boero. Mediterranean Bioconstructions Along the Italian Coast. pp. 61–136.

Alex D. Rogers. The Biology of Seamounts: 25 Years on. pp. 137–224.

Volume 80, 2018.

Michael H. Schleyer and Sean N. Porter. Drivers of Soft and Stony Coral Community Distribution on the High-Latitude Coral Reefs of South Africa. pp. 1–56.

Sara González-Delgado and José Carlos Hernández. The Importance of Natural Acidified Systems in the Study of Ocean Acidification: What Have We Learned? pp. 57–100.

Volume 81, 2018.

Melissa C. Cook, Adam May, Lucas Kohl, Geert Van Biesen, Christopher C. Parrish, and Penny L. Morrill. The Potential Impact of Hydrocarbons on Mussels in Port au Port Bay, Newfoundland pp. 1–22.

Yunwen Tao, Bing Chen, Baiyu (Helen) Zhang, Zhiwen (Joy) Zhu, and Qinhong Cai. Occurrence, Impact, Analysis and Treatment of Metformin and Guanylurea in Coastal Aquatic Environments of Canada, USA and Europe pp. 23–58.

Gong Zhang, Chun Yang, Mariam Serhan, Graeme Koivu, Zeyu Yang, Bruce Hollebone, Patrick Lambert, and Carl E. Brown. Characterization of Nitrogen-Containing Polycyclic Aromatic Heterocycles in Crude Oils and Refined Petroleum Products. pp. 59–96.

Arpana Rani Datta, Qiao Kang, Bing Chen, and Xudong Ye. Fate and Transport Modelling of Emerging Pollutants from Watersheds to Oceans: A Review. pp. 97–128.

Meijin Du, Wenwen Gu, Xixi Li, Fuqiang Fan, and Yu Li. Modification of Hexachlorobenzene to Molecules with Lower Long-Range Transport Potentials Using 3D-QSAR Models with a Full Factor Experimental Design pp. 129–166.

Weiyun Lin, Xixi Li, Min Yang, Kenneth Lee, Bing Chen, and Baiyu (Helen) Zhang. Brominated Flame Retardants, Microplastics, and Biocides in the Marine Environment: Recent Updates of Occurrence, Analysis, and Impacts. pp. 167–212.

Xixi Li, Zhenhua Chu, Jiawen Yang, Minghao Li, Meijin Du, Xiaohui Zhao, Zhiwen (Joy) Zhu, and Yu Li. Synthetic Musks: A Class of Commercial Fragrance Additives in Personal Care Products (PCPs) Causing Concern as Emerging Contaminants. pp. 213–280.

Volume 82, 2019.

Junaid S. Khan, Jennifer F. Provencher, Mark R. Forbes, Mark L. Mallory, Camille Lebarbenchon, and Karen D. McCoy. Parasites of seabirds: A survey of effects and ecological implications. pp. 1–50.

Charlotte Havermans, Holger Auel, Wilhelm Hagen, Christoph Held, Natalie S. Ensor, and Geraint A. Tarling. Predatory zooplankton on the move: *Themisto* amphipods in high-latitude marine pelagic food webs. pp. 51–92.

Angel Borja, Guillem Chust, and Iñigo Muxika. Forever young: The successful story of a marine biotic index. pp. 93–128.

Ferdinando Boero, Francesco De Leo, Simonetta Fraschetti, and Gianmarco Ingrosso. The Cells of Ecosystem Functioning: Towards a holistic vision of marine space. pp. 129–154.

Volume 83, 2019.

Dayv Lowry and Shawn Larson. Introduction: The sharks of Pacific Mexico and their conservation: Why should we care?. pp. 1–10.

Luz Erandi Saldaña-Ruiz, Emiliano García-Rodríguez, Juan Carlos Pérez-Jiménez, Javier Tovar-Ávila, and Emmanuel Rivera-Téllez. Biodiversity and conservation of sharks in Pacific Mexico. pp. 11–60.

Felipe Galván-Magaña, José Leonardo Castillo-Geniz, Mauricio Hoyos-Padilla, James Ketchum, A. Peter Klimley, Sergio Ramírez-Amaro, Yassir Eden Torres-Rojas, and Javier Tovar-Avila. Shark ecology, the role of the apex predator and current conservation status. pp. 61–114.

Jonathan Sandoval-Castillo. Conservation genetics of elasmobranchs of the Mexican Pacific Coast, trends and perspectives. pp. 115–158.

Volume 84, 2019.

Brian Morton and Fabrizio Marcondes Machado. Predatory marine bivalves: A review. pp. 1–98.

Volume 85, 2020.

James T. Ketchum, Mauricio Hoyos-Padilla, Alejandro Aldana-Moreno, Kathryn Ayres, Felipe Galván-Magaña, Alex Hearn, Frida Lara-Lizardi, Gador Muntaner-López, Miquel Grau, Abel Trejo-Ramírez, Darren A. Whitehead, and A. Peter Klimley. Shark movement patterns in the Mexican Pacific: A conservation and management perspective. pp. 1–38.

Oscar Sosa-Nishizaki, Emiliano García-Rodríguez, Christian D. Morales-Portillo, Juan C. Pérez-Jiménez, M. del Carmen Rodríguez-Medrano, Joseph J. Bizzarro, and José Leonardo Castillo-Géniz. Fisheries interactions and the challenges for target and nontargeted take on shark conservation in the Mexican Pacific. pp. 39–70.

A.M. Cisneros-Montemayor, E.E. Becerril-García, O. Berdeja-Zavala, and A. Ayala-Bocos. Shark ecotourism in Mexico: Scientific research, conservation, and contribution to a Blue Economy. pp. 71–92.

Oscar Sosa-Nishizaki, Felipe Galván-Magaña, Shawn E. Larson, and Dayv Lowry. Conclusions: Do we eat them or watch them, or both? Challenges for conservation of sharks in Mexico and the NEP. pp. 93–102.

Volume 86, 2020.

David J. Vance and Peter C. Rothlisberg. The biology and ecology of the banana prawns: *Penaeus merguiensis* de Man and *P. indicus* H. Milne Edwards. pp. 1–140.

Jeff A. Eble, Toby S. Daly-Engel, Joseph D. DiBattista, Adam Koziol, and Michelle R. Gaither. Marine environmental DNA: Approaches, applications, and opportunities. pp. 141–170.

Volume 87, 2020.

Bernhard M. Riegl and Peter W. Glynn. Population dynamics of the reef crisis: Consequences of the growing human population. pp. 1–30.

Chiara Pisapia, Peter J. Edmunds, Holly V. Moeller, Bernhard M. Riegl, Mike McWilliam, Christopher D. Wells, and Morgan S. Pratchett. Projected shifts in coral size structure in the Anthropocene. pp. 31–60.

Alex E. Mercado-Molina, Alberto M. Sabat, and Edwin A. Hernández-Delgado. Population dynamics of diseased corals: Effects of a Shut Down Reaction outbreak in Puerto Rican *Acropora cervicornis*. pp. 61–82.

Elizabeth A. Goergen, Kathleen Semon Lunz, and David S. Gilliam. Spatial and temporal differences in *Acropora cervicornis* colony size and health. pp. 83–114.

Rian Prasetia, Zi Wei Lim, Aaron Teo, Tom Shlesinger, Yossi Loya, and Peter A. Todd. Population dynamics and growth rates of free-living mushroom corals (Scleractinia: Fungiidae) in the sediment-stressed reefs of Singapore. pp. 115–140.

Joshua S. Feingold and Brandon Brulé. Population fluctuations of the fungiid coral *Cycloseris curvata*, Galápagos Islands, Ecuador. pp. 141–166.

Stuart A. Sandin, Clinton B. Edwards, Nicole E. Pedersen, Vid Petrovic, Gaia Pavoni, Esmeralda Alcantar, Kendall S. Chancellor, Michael D. Fox, Brenna Stallings, Christopher J. Sullivan, Randi D. Rotjan, Federico Ponchio, and Brian J. Zgliczynski. Considering the rates of growth in two taxa of coral across Pacific islands. pp. 167–192.

Geórgenes Cavalcante, Filipe Vieira, Jonas Mortensen, Radhouane Ben-Hamadou, Pedro Range, Elizabeth A. Goergen, Edmo Campos,

and Bernhard M. Riegl. Biophysical model of coral population connectivity in the Arabian/Persian Gulf. pp. 193–222.

S.A. Matthews, C. Mellin, and Morgan S. Pratchett. Larval connectivity and water quality explain spatial distribution of crown-of-thorns starfish outbreaks across the Great Barrier Reef. pp. 223–258.

S.A. Matthews, K. Shoemaker, Morgan S. Pratchett, and C. Mellin. COTSMod: A spatially explicit metacommunity model of outbreaks of crown-of-thorns starfish and coral recovery. pp. 259–290.

Tim R. McClanahan. Coral community life histories and population dynamics driven by seascape bathymetry and temperature variability. pp. 291–330.

William F. Precht, Richard B. Aronson, Toby A. Gardner, Jennifer A. Gill, Julie P. Hawkins, Edwin A. Hernéndez-Delgado, Walter C. Jaap, Tim R. McClanahan, Melanie D. McField, Thaddeus J.T. Murdoch, Maggy M. Nugues, Callum M. Roberts, Christiane K. Schelten, Andrew R. Watkinson, and Isabelle M. Côté. The timing and causality of ecological shifts on Caribbean reefs. pp. 331–360.

Howard R. Lasker, Lorenzo Bramanti, Georgios Tsounis, and Peter J. Edmunds. The rise of octocoral forests on Caribbean reefs. pp. 361–410.

Sascha C.C. Steiner, Priscilla Martínez, Fernando Rivera, Matthew Johnston, and Bernhard M. Riegl. Octocoral populations and connectivity in continental Ecuador and Galápagos, Eastern Pacific. pp. 411–442.

Peter W. Glynn, Renata Alitto, Joshua Dominguez, Ana B. Christensen, Phillip Gillette, Nicolas Martinez, Bernhard M. Riegl, and Kyle Dettloff. A tropical eastern Pacific invasive brittle star species (Echinodermata: Ophiuroidea) reaches southeastern Florida. pp. 443–472.

Volume 88, 2021.

Pat Hutchings. Potential loss of biodiversity and the critical importance of taxonomy—An Australian perspective. pp. 3–16.

Ferdinando Boero. Mission possible: Holistic approaches can heal marine wounds. pp. 19–38.

Iraide Artetxe-Arrate, Igaratza Fraile, Francis Marsac, Jessica H. Farley, Naiara Rodriguez-Ezpeleta, Campbell R. Davies, Naomi P. Clear, Peter Grewe, and Hilario Murua. A review of the fisheries, life history and stock structure of tropical tuna (skipjack *Katsuwonus pelamis*, yellowfin *Thunnus albacares* and bigeye *Thunnus obesus*) in the Indian Ocean. pp. 39–90.

Yan Ji and Kunshan Gao. Effects of climate change factors on marine macroalgae: A review. pp. 91–136.

Brian Morton and Fabrizio Marcondes Machado. The origins, relationships, evolution and conservation of the weirdest marine bivalves: The watering pot shells. A review. pp. 139–220.

Volume 89, 2021.

Stanislao Bevilacqua, Laura Airoldi, Enric Ballesteros, Lisandro Benedetti-Cecchi, Ferdinando Boero, Fabio Bulleri, Emma Cebrian, Carlo Cerrano, Joachim Claudet, Francesco Colloca, Martina Coppari, Antonio Di Franco, Simonetta Fraschetti, Joaquim Garrabou, Giuseppe Guarnieri, Cristiana Guerranti, Paolo Guidetti, Benjamin S. Halpern, Stelios Katsanevakis, Maria Cristina Mangano, Fiorenza Micheli, Marco Milazzo, Antonio Pusceddu, Monia Renzi, Gil Rilov, Gianluca Sarà, and Antonio Terlizzi.

Mediterranean rocky reefs in the Anthropocene: Present status and future concerns. pp. 1–52.

Johanne Vad, Kelsey Archer Barnhill, Georgios Kazanidis, and J. Murray Roberts.

Human impacts on deep-sea sponge grounds: Applying environmental omics to monitoring. pp. 53–78.

Angel Borja and Michael Elliott.

From an economic crisis to a pandemic crisis: The need for accurate marine monitoring data to take informed management decisions. pp. 79–114.

Volume 90, 2021.

Shawn Larson, Dayv Lowry, Nicholas K. Dulvy, Jim Wharton, Felipe Galván-Magaña, Abraham B. Sianipar, Christopher G. Lowe, and Erin Meyer. Current and future considerations for shark conservation in the Northeast and Eastern Central Pacific Ocean. pp. 1–50.

Ferdinando Boero. The future ocean we want. pp. 51–64.

CHAPTER ONE

Global knowledge on the commercial sea cucumber *Holothuria scabra*

Jean-François Hamel[a,*], Igor Eeckhaut[b], Chantal Conand[c], Jiamin Sun[d], Guillaume Caulier[b], and Annie Mercier[d,*]

[a]Society for the Exploration and Valuing of the Environment (SEVE), St. Philips, Newfoundland & Labrador, Canada
[b]Biology of Marine Organisms and Biomimetics unit, University of Mons, Mons, Belgium
[c]Département origines et évolution, Muséum National Histoire Naturelle, Paris, France
[d]Department of Ocean Sciences, Memorial University, St. John's, Newfoundland & Labrador, Canada
[*]Corresponding authors: e-mail address: jfhamel.seve@gmail.com; amercier@mun.ca

Contents

Advances in Marine Biology, Volume 91
ISSN 0065-2881
https://doi.org/10.1016/bs.amb.2022.04.001

Abstract

Holothuria scabra is one of the most intensively studied holothuroids, or sea cucumbers (Echinodermata: Holothuroidea), having been discussed in the literature since the early 19th century. The species is important for several reasons: (1) it is widely distributed and historically abundant in several shallow soft-bottom habitats throughout the Indo-Pacific, (2) it has a high commercial value on the Asian markets, where it is mainly sold as a dried product (beche-de-mer) and (3) it is the only tropical holothuroid species that can currently be mass-produced in hatcheries. Over 20 years have elapsed since the last comprehensive review on *H. scabra* published in 2001. Research on *H. scabra* has continued to accumulate, fuelled by intense commercial exploitation, and further declines in wild stocks over the entire distribution range. This review compiles data from over 950 publications pertaining to the biology, ecology, physiology, biochemical composition, aquaculture, fishery, processing and trade of *H. scabra*, presenting the most complete synthesis to date, including scientific papers and material published by local institutions and/or in foreign languages. The main goal of this project was to summarize and critically discuss the abundant literature on this species, making it more readily accessible to all stakeholders aiming to conduct fundamental and applied research on *H. scabra*, or wishing to develop aquaculture, stock enhancement and management programs across its geographic range.

1. Introduction

Holothuroids (Echinodermata: Holothuroidea), commonly known as sea cucumbers or holothurians, have been harvested for over 1000 years in the Indo-Pacific regions to supply markets in Asia for seafood and various biological extracts (Anonymous, 1975; Conand, 1985, 1990; Conand and Byrne, 1993; Conand and Sloan, 1989; James and James, 1994a; Purcell et al., 2010, 2012b, 2013; Toral-Granda et al., 2008; Yang et al., 2015a). The demand for sea cucumber products has been steadily growing, especially since the re-entry of China into world trade during the 1980s and the explosion of the middle class with its avid consumers.

Although between 80 and 100 holothuroid species are currently fished commercially around the world, only a few yield first-grade beche-de-mer, i.e. dried processed body wall (Conand, 1990; Conand and Byrne, 1993; Purcell et al., 2010, 2012b, 2013, 2014c; Yang et al., 2015a). *Holothuria scabra* (commonly called sandfish) is one of the most valued species; it was reported to fetch nearly 2000 US\$ kg^{-1} dry weight as beche-de-mer on the Hong Kong market in 2016 (Purcell et al., 2018a). Interestingly, *H. scabra* has not always been so popular. Before commercial harvests bloomed in the 1970s, fishermen in Sri Lanka and other countries discarded it as an unclean animal and everyone who touched one would immediately wash their hands (Anonymous, 1978b). Fisheries around *H. scabra* have since become an important source of income for local harvesters in Indonesia, Papua New Guinea, India, Madagascar, Solomon Islands, Philippines and in many other countries of the Pacific and Indian oceans (Conand, 1990, 1993; Friedman et al., 2008; Muthiga and Conand, 2014; Purcell et al., 2010, 2012b; Yang et al., 2015a).

The pressure from Asian consumers, combined with inadequate management of the sea cucumber fishery in many regions, has resulted in severe overfishing, so that natural stocks are depleted almost everywhere within the geographic distribution of *H. scabra* (Battaglene, 1999a; Battaglene and Bell, 1999; Conand, 1998a; Conand and Byrne, 1993; Preston, 1990; Purcell et al., 2013, 2014c) following the classical boom and bust pattern (Uthicke et al., 2009). In addition to being exported, *H. scabra* is also consumed locally (Adams, 1992; Conand, 1986, 1990; Conand and Byrne, 1993; Pakoa et al., 2009; Shelley, 1985b). Based on several studies, *H. scabra* is still being exploited throughout its range, even though populations are estimated to have declined by more than 90% in at least 50% of its range and are considered overexploited in at least 30% of its range, although exact declines are difficult to estimate. As a result, *H. scabra* was listed as an endangered species on the IUCN Red List (Conand et al., 2014b; Hamel et al., 2013). Hamel et al. (2013) mentioned that if fishing pressure was decreased, this species may recover due to its reproductive capacity with several annual spawning events, high fecundity and fast growth rate.

With the great demand for sea cucumber and the response from local stakeholders, the increasing pressure on natural populations of *H. scabra* has created a severe crisis. Those wishing to restore depleted populations and to develop efficient aquaculture and stock enhancement programs are quickly seeking knowledge on aspects of the biology and ecology of the species. The urgency of this situation has prompted many countries to conduct pilot studies and rearing trials on *H. scabra* over the last several decades, with the result that knowledge has accumulated rapidly but is sometimes inefficiently shared. Moreover, data were seldom assessed critically, and results were often published in internal or regional (grey) literature, if at all. The main problems encountered revolve around poorly adapted methodologies. Uncertainties and inexactitudes are widespread, and many projects were duplicated, thus constraining the overall scientific progress in the field of *H. scabra* studies.

In a previous synthesis on *H. scabra* (Hamel et al., 2001), a variety and possible subspecies that had been identified as *H. scabra versicolor* by Conand (1986) from New Caledonia was included. This variant was (and still is) known under the name golden sandfish by traders, but its taxonomic status was initially uncertain (Conand, 1989). Uthicke et al. (2005) investigated whether *H. scabra* and *H. scabra* var. *versicolor* constituted species, and whether specimens of intermediate phenotype were hybrids. Allozyme and 16S mtDNA sequence analyses showed these two sea cucumbers to be distinct. More recently, a molecular study rejected *H.* (*M.*) *scabra* var. *versicolor* (a nomen nudum), and instead identified it as a new species, *Holothuria lessoni* sp. nov. (Massin et al., 2009). It will therefore not be covered in the present synthesis.

Since the first mention of *H. scabra* in the early 1800s, the species has been mentioned or discussed in hundreds of books, theses, reports, popular and scientific articles. James (1994e) was the first to make a partial list in his annotated bibliography on sea cucumbers, and a first synthesis (Hamel et al., 2001) provided a thorough list of references published up to 2000. Over the first decades of the 21st century, research on *H. scabra* has intensified. The present document compiles and summarizes existing literature on the systematics, biology, ecology, aquaculture, fisheries and biochemical profile of *H. scabra* to make them readily accessible. This review will hopefully enable future research and commercial initiatives to focus more clearly on advancing our knowledge of this overfished species, which is both ecologically and economically valuable to many nations and communities across the Indo-Pacific.

2. Systematics

Class Holothuroidea
Order Holothuriida (formerly Aspodichirotida)
Family Holothuriidae

This family of about 170 species is typically tropical (Rowe, 1969), although a few species occur in temperate waters. The holothuriids are known as fossils at least as far back as the early Jurassic (Gilliland, 1993).

Genus *Holothuria* Linnaeus, 1767

Diagnosis: Numerous locomotory tube feet on ventral surface of body, which is often flattened; dorsal surface often arched, with more or less conspicuous conical papillae. Calcareous ring stout; interradials about half as long as wide but not so wide as to be curved. Ossicle tables in some form almost always present; in addition buttons, rods, rosettes, perforated plates present or absent (after Rowe, 1969).

Remarks: *Holothuria* comprises approximately 120 species, most of which occur in shallow tropical or subtropical waters. These are usually large and conspicuous sea cucumbers, although a few species are secretive, living under rocks or burrowing into sandy substrata. Attempts to subdivide the genus (Deichmann, 1958; Panning, 1929, 1934a,b, 1935) were only partially successful; they solved some problems and created others. In an excellent review, Rowe (1969) elaborated upon Deichmann's partial revision. The subgenera that Rowe diagnosed have been widely accepted.

***Holothuria* (*Metriatyla*) Rowe, 1969**

Diagnosis: Size small to moderate, to 200 mm [some larger to ~400 mm] in length; tentacles 20; feet arranged irregularly on flattened ventral 'sole'; dorsal surface arched; papillae conspicuous, conical, irregularly arranged dorsally; an irregular fringe of marginal papillae sometimes present; body wall usually 1–5 mm thick [may be thicker, up to 20 mm, in larger individuals] and gritty to touch; radial plates of calcareous ring up to three times as long as interradials; ossicles tables and buttons; tables with smooth disc and spire terminating in few to many small spines; buttons simple, with irregularly arranged knobs and 3–10 pairs of relatively large holes (after Rowe, 1969).

Type Species: *H. scabra* Jaeger, 1833.

Remarks: Rowe (1969) included nine species in this subgenus; all are known only from the Indo-west-Pacific.

Holothuria (Metriatyla) scabra Jaeger, 1833

- *H. scabra* Jaeger, 1833: 23; Panning, 1944: 67, figs 34-5.
- *Holothuria tigris* Selenka, 1867:333.
- *Holothuria cadelli* Bell, 1887:144.
- *Holothuria gallensis* Pearson, 1903:203.
- *Holothuria (Holothuria) scabra*. Panning, 1934a,b: 80, fig. 66a–f. (includes synonymy prior to 1934).
- *Holothuria (Halodeima) scabra*. Mortensen, 1954:6.
- *Holothuria (Metriatyla) scabra*. Rowe, 1969: 160, fig. 20; Cherbonnier, 1980:647, fig. 16A–L (includes synonymy prior to 1980); 1988: 135, fig. 55A–O; Massin, 1999:30, figs 22a–l, 23, 110f (includes synonymy prior to 1999).

Diagnosis: Tables with spires of moderate height, terminating in ca. 20 small spines. Disc of table with 8–12 holes. Buttons with 3–5 pairs of holes (after Rowe, 1969).

Remarks: The partial synonymy given above identifies publications in which complete synonymies and systematic descriptions can be found. Colour photographs of this species are provided in Féral and Cherbonnier (1986), and Massin (1999). *H. gallensis* mentioned by Pearson (1903) was shown in a later paper (Pearson, 1910) to be synonymous with *H. scabra*. The vernacular names of *H. scabra* vary markedly across its geographic range (Table 1).

Kamarudin et al. (2017) showed that identification based on ossicle shapes and non-protein-coding 12S mitochondrial rRNA gene sequences are complementary. Extensive processing of sea cucumber causes body deformation that can create difficulties in species identification. The availability of 40 new 12S mitochondrial rRNA gene sequences deposited in GenBank (including *H. scabra*) can be use by enforcement agencies to monitor and overcome the issues of species substitution and product mislabelling of processed sea cucumber products on the market (Kamarudin et al., 2018, 2019).

Table 1 Common names given to *Holothuria scabra* around the world.

Location	Common names
Everywhere	Sandfish
Burma	Pan-le-pet-kye, Pin-lehmyaw
China	Tok-sum, Tok-som, Puti-an, Thuk-su, Hai-som, Chalkyfish, Paishen, Peh-sim
Eritrea	Hedra beyda
Fiji	Dairo
Hong Kong	Sand Sea Cucumber, Hoy sum
India	Vella Attai, Kadal Attai, Kadal Vellarikka, Patos, Dalamogon, Cheena attai
Japan	Namako
Kenya	Jongoo, Jongoo jeupe
Madagascar	Zanga fotsy, Bemavo, Tricot
Malaysia	Trepang, Putih, Tepuak, Bat putih
New Caledonia	Holothurie de sable, Le gris
Palau	Rebotel, Molech

Continued

Table 1 Common names given to *Holothuria scabra* around the world.—cont'd

Location	Common names
Philippines	Rebothal, Patos, Dalamogon, Cortido, Cortido bato, Curtido, *Kagisan*, Putian
Russia	шероховатая голотурия
Samoa	Fugafuga ai
Sri Lanka	Welli attaya
Tanzania	Jongoo mchanga, Myeupe
Thailand	White sea cucumber
Viet Nam	Hải sâm cát, Hải sâm trắng, Ñoät traéng
Yemen	Chalkyfish

3. Morphology and anatomy

3.1 Size and external appearance

Table 2 summarizes some phenotypic and morphometric characteristics of *H. scabra* throughout its geographic range (see also Section 2). Adults usually measure between 150 and 400 mm in length. The reported total body weight of adults varies considerably, between 500 and 2000 g, over its geographic range (Table 2). The body wall generally accounts for over 50% of the total weight (Conand, 1989). Indriana and Kunzmann (2014) showed that the relationship between the dry and wet weights of *H. scabra* (without the intestine) can be represented by a linear equation ($y = 0.168 \times - 0.231$) with a correlation coefficient (R^2) of 0.923. However, it has been noted that the weight depends on the amount of coelomic fluid (in the hydrovascular system and perivisceral cavity), water in the respiratory tree and sediment in the alimentary tract (Baskar, 1994; Conand, 1989). Accordingly, Pitt and Duy (2004a) indicated that individuals can stretch or contract and hold variable intestinal and fluid contents, which may influence their weight (Table 3). It is not obvious which parameter changes less for a particular individual over a short period of time, and, therefore, which measurement gives the most accurate measurement. They noted that the weight measurements are generally less subjective and more accurate than lengths. Recently, Magno-Tan et al. (2016) developed a way to automatically compute the weight of *H. scabra* with a new algorithm that utilizes only the surface area

Table 2 Phenotypic and morphometric characteristics of *Holothuria scabra* across its geographic range.

Location	Body wall colour	Size (mm)	Weight (g)	Body texture and thickness	Body wall morphology	Nb of tentacles	References
General	Upper surface light grey, dull cream, olive brown, almost black also covered with fine black spots in the wrinkles, lower surface usually white	300	–	–	Short and stout flattened at the end	–	Baird and Sachithananthan (1974), Conand (1998b)
Australia	Grey or black above with dark transverse wrinkles dorsally, grey or white below	–	–	–	Somewhat flattened, firm and pliable	–	Cannon and Silver (1986)
	–	200	–	Gritty- 1 to 5 mm	Flattened ventral surface, arched dorsally	20	Rowe (1969)
	Grey white ventrally, grey or black with transverse wrinkles dorsally	–	–	–	–	–	Carter et al. (1997)
	Finely speckled grey with one large blotch on the dorsal side, ventral side white	170–300	–	Rough body surface	–	–	Clark (1931)
China	–	700	–	–	–	–	Anonymous (1991)
India	–	350	500	–	–	–	James (1973)

Continued

Table 2 Phenotypic and morphometric characteristics of *Holothuria scabra* across its geographic range.—cont'd

Location	Body wall colour	Size (mm)	Weight (g)	Body texture and thickness	Body wall morphology	Nb of tentacles	References
	Upper side grey with white or yellow horizontal bands, lower surface white with a number of fine black dots	400	500–1500	–	Body short and stout with blunt end and prominent wrinkles on dorsal surface	–	James and James (1994a)
	–	300–350	500–600	–	–	–	James (1973, 1994e), Baird and Sachithananthan (1974)
	–	300–500	500–1500	8–10 mm	–	–	Sachithananthan et al. (1986)
	–	230	100–1400	–	–	–	Baskar (1994)
	Grey to black dorsally and white ventrally	100–400	25–2000	–	Robust, elongated and cylindrical with blunt ends	20	James (1996b)
	Grey to black dorsally and white on the ventral surface	400	2000	10 mm thick	–	–	James (1989a)
	Dorsal surface grey to almost black with transverse yellow streaks	320	–	–	–	–	Gravely (1927)

	–	400	–	–	–	–	James (1991)
	–	100–400	25–2000	–	–	–	Jagadees and Sasidharan (1990)
	Black upper side with white or light yellow bands across the body, lower side white in colour with black dots	400	–	–	–	–	James (1989b), MPEDA (1989)
Indonesia	Grey dorsally with transversal greenish bands, grey-white ventrally	–	–	Rough skin, 2–3 mm thick	Body arched dorsally, more or less flat ventrally	20	Massin (1999)
Mozambique	Dorsally grey mottled with white and black above, white below	200	–	Skin very tough	–	–	Macnae and Kalk (1962)
New-Caledonia	Grey to greenish, seldom black, ventral surface pale in colour	240	485	Gritty dorsal surface, 3–13 mm thick	Pronounced wrinkles dorsally	20	Conand (1989, 1990), Féral and Cherbonnier (1986)
Papua New Guinea	From pale beige to dark brown, sometimes with dark spots dorsally, white or scattered with brown spots ventrally	–	–	–	–	–	VandenSpiegel et al. (1992)

Continued

Table 2 Phenotypic and morphometric characteristics of *Holothuria scabra* across its geographic range.—cont'd

Location	Body wall colour	Size (mm)	Weight (g)	Body texture and thickness	Body wall morphology	Nb of tentacles	References
Palau	Upper surface light grey, dull cream, olive brown or almost black, lower surface usually white	–	–	–	Short and stout and flattened at the ends	–	Baird and Sachithananthan (1974)
Philippines	Light grey to greenish dorsally sometimes with dark grey or brown bands, white ventrally	160–280	–	–	Arched dorsally, flattened ventrally	–	Leonardo and Cowan (1984)
	–	100–200	–	Gritty and thin	Cylindrical, slightly arched dorsally, flattened ventrally	20	Tan Tiu (1981b)

Table 3 Length-weight relationships in *Holothuria scabra* measured in Vietnam, using groups of sea cucumbers under various conditions.

Source	N	Taken from	Length (cm)	Weight (g)	Drying time (min)[a]
Hatchery	15	Pond; 24h in bare tank	12.1–21.3	145–400	5
Hatchery	20	Nursery tank on sand	2.0–7.9	1.16–10.93	0.5
Wild	34	Pens (>1 year); 20h in bare tank	14.5–26.0	157–574	10–20
Hatchery	8	Nursery tank on sand	5.6–8.8	17.4–46.2	3
Hatchery	32	Pond; 5h in bare tank	10.5–21.0	101–379	5

[a]Sea cucumbers were left to dry in the shade for that period.
Adapted from Pitt, R., Duy, N.D.Q., 2004a. Length–weight relationship for sandfish, *Holothuria scabra*. SPC Beche-de-mer Inf. Bull. 19, 39–40.

of an image captured by a regular camera while the individual is submerged under water. Digital images were converted to binary images and used to measure the individual length, width and surface area through pixel analysis. Results showed that the surface area and weight of *H. scabra* under normal culture condition is highly correlated ($R^2 = 0.90$). Moreover, error analysis revealed that the accuracy of the software in determining the length, width, surface area and weight was 94%, 94%, 94%, and 84%, respectively (Magno-Tan et al., 2016).

The body of *H. scabra* is robust, cylindrical, elongated, stout and flattened at the ends. The dorsal side is convex and the ventral side is flat (James, 1989a). The mouth is oval, located antero-ventrally and encircled by 20 peltate tentacles. The cloacal opening (anus) is postero-dorsal. The body wall is thick, gritty to the touch, and slimy. The dorsal surface is smooth with few thinly scattered tube feet and its colour is highly variable (Conand, 1989, 1998b, 1999b; Uthicke et al., 1998; VandenSpiegel et al., 1992), ranging from yellow to grey-brown and black, with irregular patterns of well-defined wrinkles (Fig. 1). The ventral surface is usually white or cream; it is also rough, and bears numerous locomotory tube feet arranged irregularly (Table 2) (Massin, 1999; Tan Tiu, 1981a). Each dark spot on the ventral surface represents one tube foot (James, 1989a; Massin, 1999). Overall, important variations in the morphological and phenotypic appearance of *H. scabra* are observed throughout its geographic range (Table 2), and there is a high degree of polymorphism (Conand, 1986, 1990). Two main explanations for this have been proposed: (i) identification errors and (ii) important morphological plasticity over the geographic range.

Fig. 1 Main colour types catalogued in *H. scabra*. (A) Various shades of typical zebra-type adults, dorsal view; (B) individual photographed in situ in the Seychelles, dorsal view; (C, D) typical zebra-type adult from Mauritius, in (C) dorsal view and (D) ventral view; (E) a Mauritian albino; (F, G) zebra-type adult from Madagascar, in (F) dorsal view and (G) ventral view; (H) juveniles and (I) albino from Madagascar. Scale bars: 8 cm in (A, B); 4 cm in (C–G); 3 cm in (H); 6 cm in (I). *Photos (C)–(E) from Mauritius courtesy of D. Vaïtiligon. Photo (I) courtesy of L. Gaumez.*

In the Pacific, *H. scabra* generally exhibits the characteristics of the type specimen, having a whitish to dark brown bivium, occasionally with dark or black transverse markings often referred to as zebra-type (Cherbonnier and Féral, 1984; Conand, 1998b; Féral and Cherbonnier, 1986; Massin, 1999). Mercier et al. (1999, 2000a,b) and Uthicke and Benzie (1998, 1999) also observed black body-wall phenotypes within populations of *H. scabra* in the Solomon Islands and Australia, respectively. They found that both colour morphs were present at all depths. However, the proportion of black vs grey forms varied considerably among sites (Uthicke and Benzie, 1998). Uthicke and Benzie (1998, 1999) did not find any genomic differences, and Mercier et al. (1999) did not find any behavioural differences between the colour morphs. Genetic work has confirmed that the black and grey colour morphs belong to the same species (also see Section 6.7), although the factors underlying this colour variation are not known (Uthicke and Benzie, 1999). It also remains unresolved why the colour morphs constitute varying proportions of the population, and whether the colour is environmentally or genetically determined.

The Indian Ocean morphotype differs from the above-mentioned one by the presence of yellow elongated spots (Conand, 1999a; Sachithananthan, 1994a). Moreover, Asha (2007) described *H. scabra* from the Gulf of Mannar (India), reporting a white specimen and another that was bright yellow with black dots and fine patches (Table 2). Recent observations on specimens from several sites in the Indian Ocean show that most of them present uncharacteristic colours (Chantal Conand, unpublished data), indicating that populations of *H. scabra* from this basin would benefit from molecular comparisons with those of the Pacific.

3.2 Functional anatomy

The internal anatomy of *H. scabra* is similar to that of other species in the genus Holothuria. However, while about 20 species of *Holothuria* possess Cuvierian tubules, *H. scabra* does not. The dermis includes various ossicles that form a reduced endoskeleton. A complete digestive tract opens anteriorly through a mouth and into a posterior cloaca. A respiratory system formed of two respiratory trees branches off from the posterior cloaca, representing a unique system in metazoans. Respiration is also trans-tegumentary and gas exchanges also take place through the ambulacral podia (also called tube feet). The circulatory systems consist of three elements: (i) the general cavity which is a perivisceral coelom, in which nutrients and dissolved gases are circulated, (ii) the haemal system and (iii) the hydrovascular (ambulacral) system, the latter two being synapomorphies in echinoderms. In addition to circulating fluids, these two systems are sites of immune cells synthesis. *H. scabra* is gonochoric and its genital and reproductive systems are outlined in detail below. Excretion takes place mainly in the form of ammonium ion (NH_4^+), probably through the integument as there is no dedicated excretory system. A 3D representation of the internal anatomy of *H. scabra* is available on the MetaMorphos application (Eeckhaut and Puozzo, 2018).

3.2.1 Main systems

Studies by Mary Bai (1971a,b, 1978, 1980, 1994) provided the initial outline the internal anatomy of *H. scabra*, including the digestive system, haemal system, respiratory system, water vascular system and reproductive system. Mary Bai (1980) indicates that the tentacles are covered by a thin cuticle, below which lies a dermis. Glandular cells with secretory granules are visible among the epithelial [epidermal] cells. Sensory cells are observed in the epidermis and a tentacular nerve is located beneath the nerve plexus. A lumen called the tentacular canal is bordered by a coelomic epithelium and filled with fluid and coelomocytes (see Section 3.2.2). The body wall is covered by a thin cuticle composed of dying cells from the epidermis. From the external surface, the successive layers of tissue that compose the body wall are a cuticle, an epithelium, a layer of glandular cells, a dermis containing the pigments and connective tissues, muscle strands, haemal lacunae, muscle layer and coelomic epithelium. *H. scabra* possesses five pairs of longitudinal muscles bands running along the length of its body (Fig. 2).

Hartati et al. (2016) provided a fine description and ultrastructural study of the digestive system, consisting of a pharynx (pharyngeal bulb), oesophagus, stomach, first descending loop of the intestine (foregut), ascending intestine (midgut), second descending intestine (hindgut) and cloaca. More recently, Pierrat et al. (2022) described the foregut comprising the pharyngeal bulb, oesophagus, and the first descending loop of the intestine, followed by a midgut starting beyond the first intestinal loop where the proximal tubules of the rete mirabile are attached and consisting of the ascending anterior loop of the intestine (Fig. 3). A third sphincter separates the midgut from the hindgut, which represents approximately 70% of the total length of the digestive tract and consists of the final posterior descending loop of the intestine and the cloaca from which respiratory trees emanate (Pierrat et al., 2022). Of note is the fact that descriptions of the

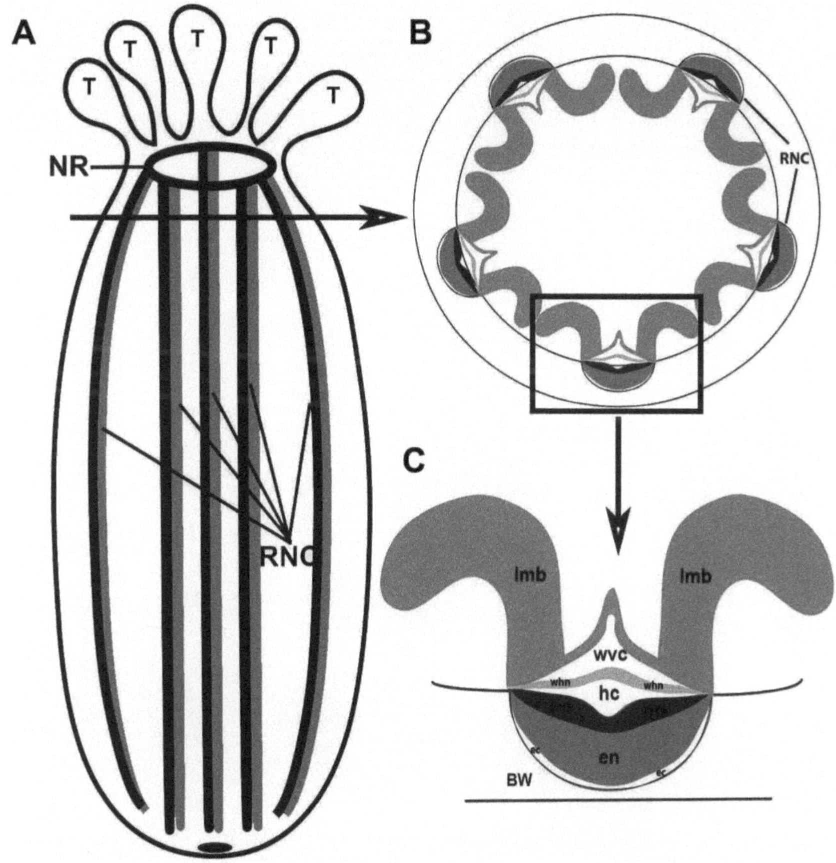

Fig. 2 Schematic illustration of the structural organization of the nervous system of *H. scabra*. (A) Nervous system comprising five radial nerve cords (RNC) running along the body wall to a nerve ring (NR) located at the anterior end below the tentacles (T). (B) Cross section of five RNCs. (C) Close up on cross section of RNC (from box in (B)); ec = epineural canal, en = ectoneural part, hc = hyponeural canal, hn = hyponeural part, lmb = longitudinal muscle band, whn = wall of hyponeural canal, wvc = water vascular canal. *Modified from Chaiyamoon, A., Tinikul, R., Chaichotranunt, S., Poomthong, T., Suphamungmee, W., Sobhon, P., Tinikul, Y., 2018. Distribution and dynamic expression of serotonin and dopamine in the nervous system and ovary of* Holothuria scabra *during ovarian maturation. J. Comp. Physiol. A 204, 391–407, with authors' permission.*

digestive system in holothuroids often include mentions of a rectum and terminal anus, which herein will preferentially be referred to as the cloaca and cloacal opening, to account for the fact that the digestive tube merges with the respiratory tree before opening to the exterior. Based on the histological analysis, the structure of the alimentary canal was found to display

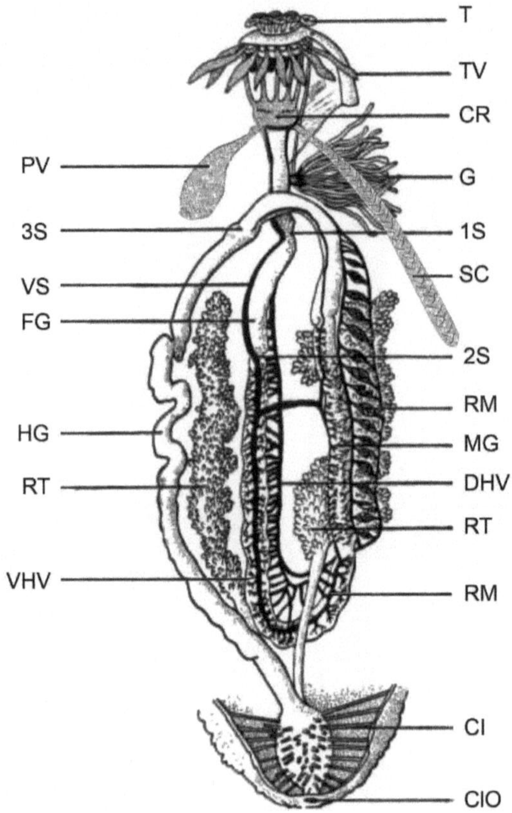

Fig. 3 Schematic drawing of the digestive system in *H. scabra*. Cl = cloaca; ClO = cloacal opening; CR = calcareous ring; DHV = dorsal haemal vessel; Fg = foregut; G = gonad; HG = hindgut; MG = midgut; RM = rete mirabile; RT = respiratory tree; PV = polian vesicle; 1S = sphincter 1; 2S = sphincter 2; 3S = sphincter 3; SC = sand canal; T = tentacles; TV = tentacle vesicles; VHV = ventral haemal vessel; VS = ventral sinus. *Modified from Pierrat, J., Bédier, A., Eeckhaut, I., Magalon, H., Frouin, P., 2022. Sophistication in a seemingly simple creature: a review of wild holothurian nutrition in marine ecosystems. Biol. Rev. 79 (1), 273–298, with authors' permission.*

three main constituents: the lumen, the intestinal villi and the connective tissue composed of muscle fibres (Hartati et al., 2016).

The haemal system in *H. scabra* consists of a ring, encircling the pharynx/oesophagus, directly behind and closely attached to the water vascular ring. Two main sinuses, one dorsal and one ventral, run along the small intestine. From the haemal ring, five radial sinuses ascend along the pharyngeal bulb, accompanying the radial water canals. They run along the body wall and lie between the hyponeural sinus and radial water canals of the body wall.

The dorsal and ventral haemal sinuses consist of an outer peritoneal epithelium, composed of columnar cells with deeply staining nuclei, a thick, uninterrupted circular muscle layer and a thick connective tissue with scattered amoebocytes. The rete mirabile consists of a thick peritoneal epithelium of elongated columnar cells, a thin circular muscle layer, and connective tissue (Mary Bai, 1980).

The main organ involved in respiration in *H. scabra* is the respiratory tree, which originates from the anterior part of the cloaca near the junction with the large intestine and is attached to the body wall by irregular strands of connective tissue. It is divided into right and left arborescent tubes that extend anteriorly in the perivisceral coelomic cavity up to the end of the pharyngeal bulb. The two main branches give rise to finer branches, which fill the entire coelomic cavity, surrounding the internal organs. The tubules are colourless, transparent, and terminate in small thin-walled vesicles. The left tree is intermingled with the lacunar network of the rete mirabile of the ascending small intestine. Apart from these two main branches, there are also two to three short, branched tubes originating from the base of the common stem of the respiratory tree. The cloaca of *H. scabra* pumps rhythmically in a motion called cloacal pumping. At this time the terminus of the digestive tract is closed. Exchange of gases also takes place through the tube feet and the integument (Mary Bai, 1980).

The perivisceral coelom contains coelomic fluid, which is circulated by the cilia of peritoneal epithelium and the movement of the animal. According to Mary Bai (1980), it is less alkaline than sea water and contains a variety of free cells known as coelomocytes (see Section 3.2.2). The water vascular system consists of a circular ring canal or water ring, a Polian vesicle, and the stone canal. The Polian vesicle is an ovoid or elongated sac 1.0–2.5 cm long, and it arises from the left ventral part of the ring canal. Mary Bai (1980) indicated that there is usually one, and very rarely two Polian vesicles, while Massin (1999) reported that *H. scabra* possessed only one Polian vesicle. Conversely, James (1996b) refers to two or three Polian vesicles. The stone canal arises from the right dorsal part of the ring canal and freely floats in the coelom. It is 1–5 cm long (sometimes to 10 cm), and the wall of the stone canal is distended with folds (Mary Bai, 1980). The calcareous ring comprises 10 plates, 5 radial and 5 interradial. There is a single very long stone canal, which is 12–15% of the body length (Massin, 1999). The intestine is highly coiled and opens into the cloaca (James, 1996b). A single bunch of gonad tubules is attached to the dorsal mesentery; the gonad opens to the exterior through a single genital pore in the mid-dorsal region near the anterior end (James, 1996b).

The nervous system consists of networks concentrated into ganglionated radial nerve cords and divided into ectoneural, deeper lying hyponeural and entoneural systems (Mary Bai, 1980). The reproductive system will be detailed in Section 8.5 dedicated to gonad morphology later in this text.

Curiously, Indap et al. (1996) indicated that they studied the effect of the Cuvierian tubules of *H. scabra* on a variety of potential predators. However, the experience of the authors, along with the work of Mary Bai (1971b, 1978, 1980), Leonardo and Cowan (1984), Soota et al. (1983) and Massin (1999) show that this species does not possess Cuvierian tubules. Indap et al. (1996) either studied a different species or misidentified an internal organ.

3.2.2 Immune cells

The coelomocytes occur in the haemal fluid and the fluid of all the coelomic compartments except the hyponeural sinuses. The coelomocytes comprise: lymphocytes, which are small spherical cells of ~4–6 μm in diameter, phagocytes ranging between 4 and 20 μm, morula cells that form aggregations, each cell measuring between 3 and 12 μm, haemocytes measuring between 2 and 6 μm, fusiform cells measuring 4 to 6 μm in length, and crystal cells that measure 6–9 μm in diameter. The fluid in the water vascular system contains all types of coelomocytes, which also occur in the body fluid, but the lymphocytes and phagocytes are more common. The haemal system also contains a large number of floating morula cells. The abundance, variety and inclusions of the coelomocytes suggest that they serve vital roles in nutrition, transport of wastes and phagocytosis (Mary Bai, 1980).

Moreover, Prompoon et al. (2015) classified coelomocytes of *H. scabra* into four types: lymphocytes (4–5 μm in diameter), phagocytes (5–10 μm), spherulocytes (10–15 μm) and giant cells (20–50 μm). Lymphocytes showed intense blue staining in the 'concentric' nucleus and apparently blocked heterochromatin at the nuclear periphery; only a small amount of cytoplasm with faint toluidine blue staining was observed. Phagocytes were mostly irregular with nuclei either eccentrically or concentrically localized; they dominated, representing ~60% of the total population. Spherulocytes had a distinct mulberry shape and intense violet-pink staining in the cytoplasm, and they made up ~12% of the total population. The giant cells were the largest coelomocytes, with a high cytoplasm-to-nucleus ratio, the cytoplasm containing a filamentous-like structure extending from perinuclear to the outer rim (Prompoon et al., 2015). An illustration of common coelomocytes is provided in Fig. 4.

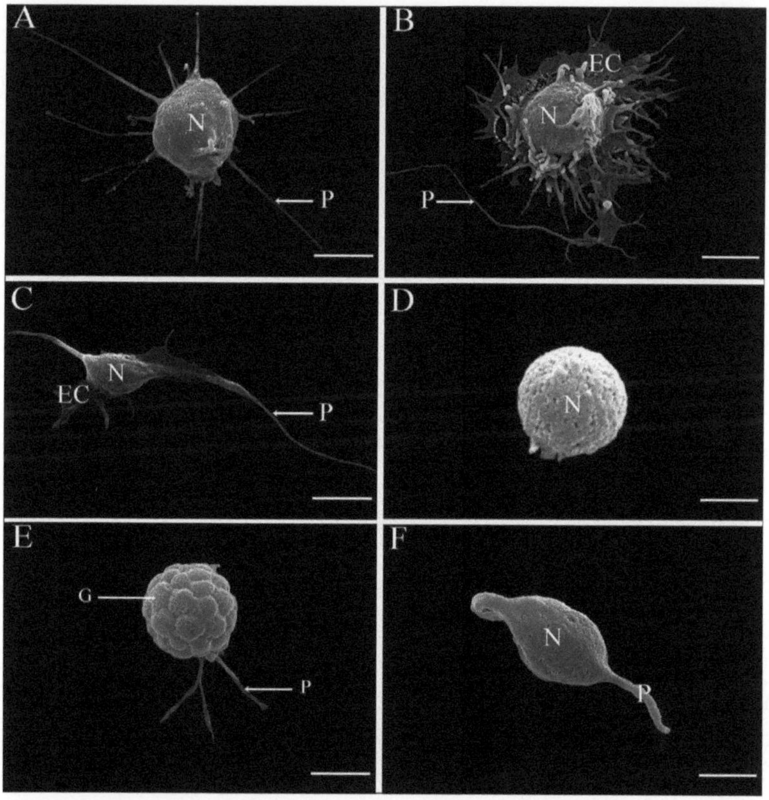

Fig. 4 Free coelomocytes in the perivisceral coelom and the hydrovascular system of *H. scabra* in scanning electron microscopy (15 kV). (A) Filipodial, (B) petaloid and (C) intermediary phagocytes; (D) progenitor cell; (E) spherulocyte; (F) fusiform cell. EC = cytoplasmic extension, G = granule; N = nucleus; P = pseudopod. Scale bars represent 3 μm in (A), 2.5 μm in (B), 3.9 μm in (C), 1.5 μm in (D), 2.9 μm in (E) and 2.3 μm in (F). *Translated from Borrello, L., 2020. Caractérisation morpho-fonctionnelle des cœlomocytes* d'Holothuria scabra *Jaeger, 1833 (Holothuroidea, Echinodermata). MSc thesis, Mons University, Belgium, with permission.*

Zhao et al. (2017) compared the phagocytic activity of coelomic cells between healthy and diseased sea cucumbers. Their results showed that the phagocytic rate and the index of healthy sea cucumbers ranged from 66 to 10% and 0.7–0.1, respectively. Phagocytic activity of diseased sea cucumbers (early stage of skin degradation) was a bit lower than that of healthy sea cucumbers, which ranged from 38 to 5% and the phagocytosis index ranged from 0.9 to 0.1, with increased tissue degradation, the phagocytic activity gradually decreased (Zhao et al., 2017).

3.2.3 Ossicles and other calcareous structures

Most studies of ossicles in *H. scabra* (Fig. 5) have been made on adults (Cannon and Silver, 1986; Cherbonnier and Féral, 1984; Conand, 1998b; Gravely, 1927; James, 1973, 1976; Mary Bai, 1980; Massin, 1999; Rowe, 1969; Tan Tiu, 1981b; VandenSpiegel et al., 1992; Wainiya, 1988) except for a note describing the ossicles of a 30-mm long specimen in India (James, 1976). Cherbonnier (1955, 1988), Cherbonnier and Féral (1984), VandenSpiegel et al. (1992) and Massin (1999) showed that the ossicles of adults varied little throughout the geographic range of *H. scabra*.

Rasolofonirina and Jangoux (2005) described various calcareous structures in larvae and early juveniles and identified the exact stage of development at which they appeared (Fig. 6). The calcareous body (five-arm star; 13 μm in diameter) is the first structural element to differentiate in the posterior part of the larva blastocoel, ~36 h post fertilization. The madreporite (a curved stalk ~40 μm long) is the second calcareous formation to differentiate at the late auricularia stage in the left dorsal-lateral part of the body, at the base of the ring canal. The peripharyngeal calcareous ring appears at the doliolaria stage around the oesophagus; it increases in size as individuals grow. Tables made up of a base plate and spire are the first type of intra-tegument ossicles to form in the doliolaria stage as small bobbin-like structures ~7 μm long (Rasolofonirina and Jangoux, 2005). Complete formation of this type of ossicle to a height of 70 μm takes about 4 days. The tables of doliolaria and pentactula larvae always appear in groups, with the first complete tables found only in juveniles (Rasolofonirina and Jangoux, 2005). All calcified structures appearing during the larval stage continue to exist in early juveniles; however, others begin to differentiate at the onset of the juvenile stage. Podia and tentacle rods, podia terminal disks and anal plates appears at the same time (~20 days after fertilization). The terminal disks of podia are pierced circular ossicles comprised of joined trabeculae that reach ~110 μm in 27-day-old juveniles. The anal plates are formed from two large, perforated ossicles appearing in 22-day-old juveniles, whereas the buttons appear in 40- to 50-day-old juveniles that measure 10–15 mm in length (Rasolofonirina and Jangoux, 2005).

Massin et al. (2000) described ossicle changes from pentactula through juveniles and adult. This precise description can be used to identify juveniles collected in the field. In general, the ossicles consist of tables and knobbed buttons (Massin et al., 2000). Ossicles of *H. scabra* vary mainly in early juveniles between 0.9 and 15 mm long. Specimens 0.9–1.5 mm long have

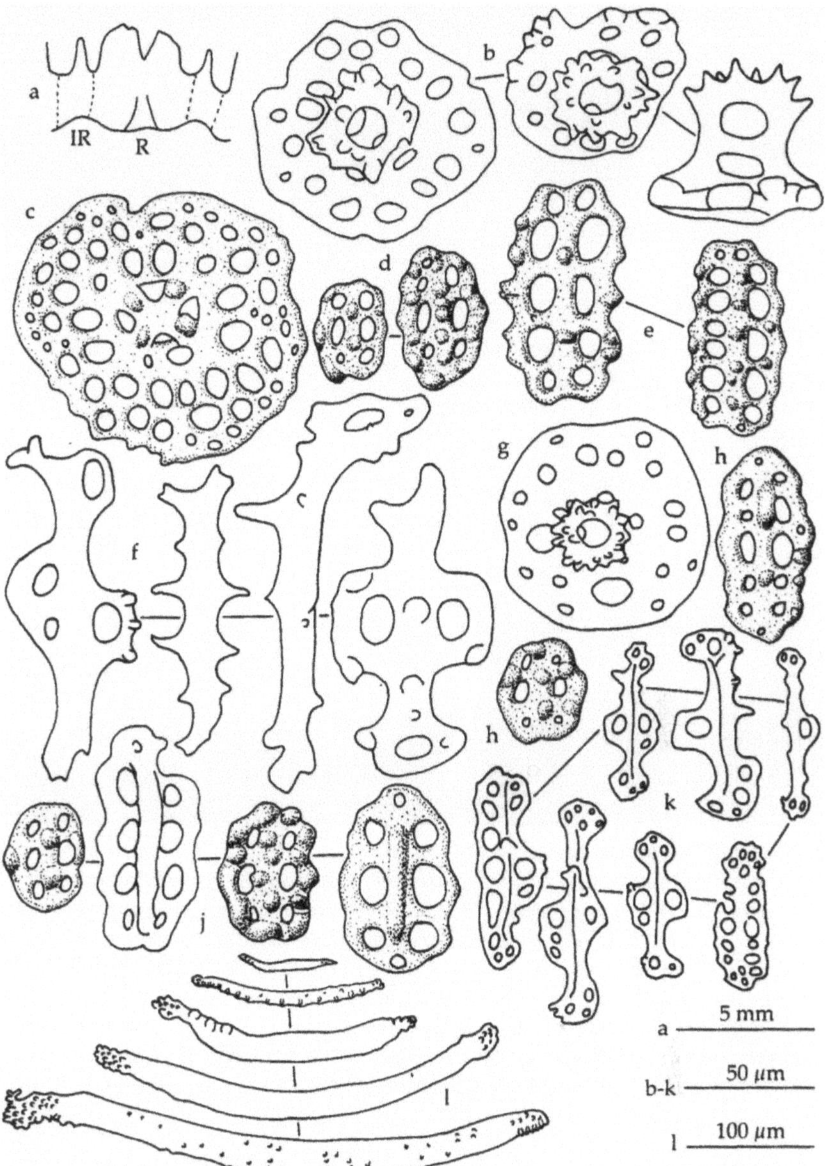

Fig. 5 Ossicles of adults of *H. scabra*: (A) calcareous ring (IR = interradial piece; R = radial piece); (B) tables from dorsal body wall; (C) large table from dorsal body wall; (D) buttons from dorsal body wall; (E) large buttons from dorsal body wall; (F) rods from dorsal body wall; (G) table from ventral body wall; (H) buttons from ventral body wall; (J) tube foot buttons; (K) perforated rods from tube feet; (L) tentacle rods. *From Massin, C., 1999. Reef-dwelling Holothuroidea (Echinodermata) of the Spermonde archipelago (South-West Sulawesi, Indonesia). Zool. Verh. 329, 1–144, with permission.*

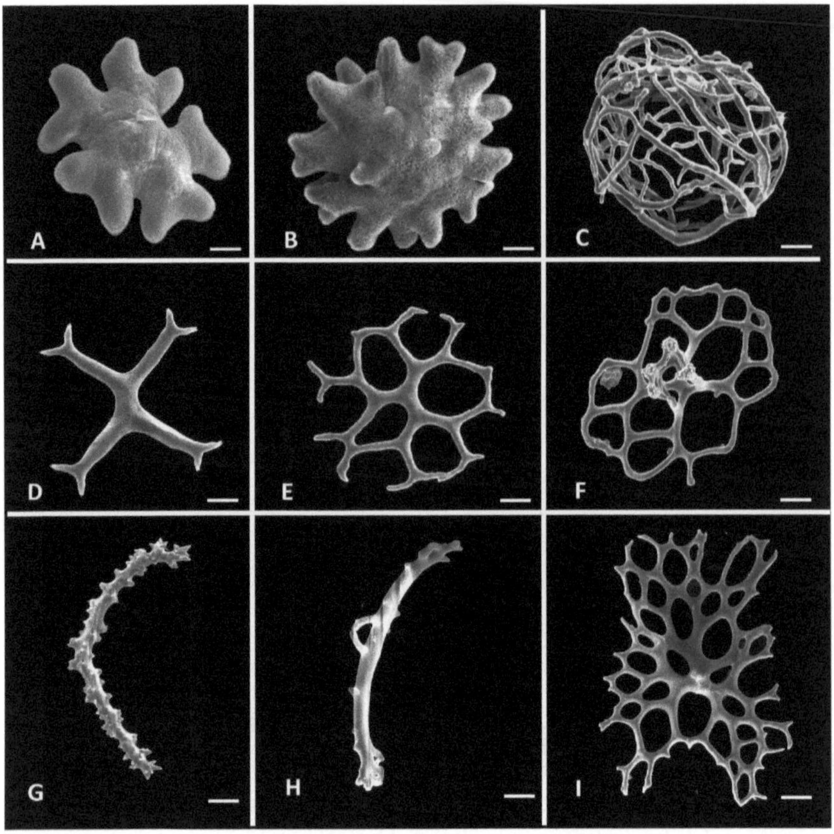

Fig. 6 Skeletal structures in larvae and juveniles of *H. scabra*. (A, B) Calcareous body from a 36-h dipleurula and a 30-day juvenile, respectively. (C) Madreporite from a 30-day juvenile. (D–F) Development of table ossicle from the peripharyngeal calcareous ring of a 17-day pentactula larva. (G–I) Rod ossicle from a tentacle and podium and perforated anal plate, respectively, from a 17-day pentactula larva. Scale bars represent 2.5 μm in (A); 3 μm in (B); 9 μm in (C); 6 μm in (D); 10 μm in (E); 8 μm in (F); 10 μm in (G–H); 16 μm in (I). *Adapted from Rasolofonirina, R., Jangoux, M., 2005. Appearance and development of skeletal structures in* Holothuria scabra *larvae and epibiont juveniles. SPC Beche-de-mer Inf. Bull. 22, 6–10, with permission.*

tables with a tall spire (4–5 cross beams), no buttons, and large irregular perforated plates. Specimens 5–6 mm long have tables with a moderate spire (2–4 cross beams) and a few smooth buttons. Specimens 9–16 mm long have tables with a low spire (1–2 cross beams) and knobbed buttons. From 30 mm, ossicles are similar to those of adults, with more buttons and fewer tables (Fig. 5). Kamarudin et al. (2017) identified five main ossicle shapes, mainly rod, plate, rosette, button and table. Several features of the ossicles of early juveniles, including their size, shape and prevalence, are unique to the species. Comparison with juveniles of other holothuroid species indicates that presence of tables with tall spire and absence of buttons are plesiomorphic characters in the Holothuriidae (Massin et al., 2000).

4. Geographic range

H. scabra is known from locations throughout the Indo-Pacific, roughly between latitudes 30°N and 30°S (Clark, 1971; Conand, 1998b; Massin, 1999) (Fig. 7). Table 4 provides a list of countries or regions where *H. scabra* was observed or where studies confirming its geographic distribution were conducted at a finer scale. Furthermore, it has been suggested that *H. scabra* is also present in several other regions of the Indian and Pacific oceans (James, 1991; Sant, 1995; Sluiter, 1901; Théel, 1886) (Fig. 7). Notably, the species has not yet been reported from Hawaii.

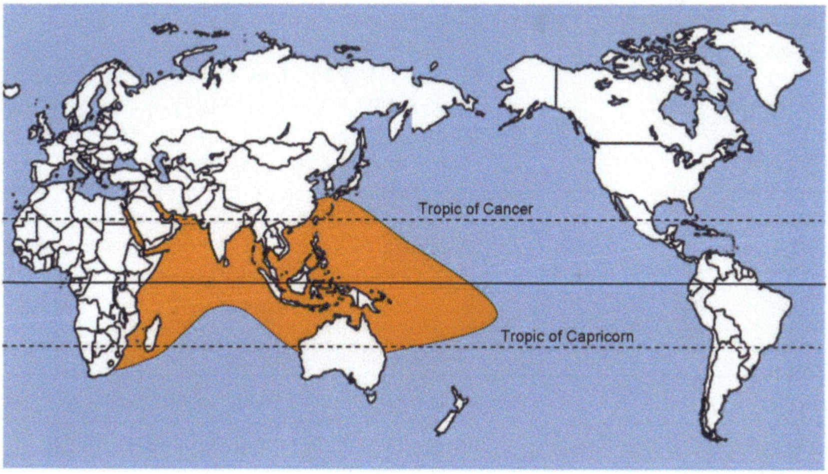

Fig. 7 Known geographic distribution of *Holothuria scabra* (in orange). *Adapted from Bell, J.D., Purcell, S.W., Nash, W.J., 2008. Restoring small-scale fisheries for tropical sea cucumbers. Ocean Coast. Manag. 51, 589–593, with permission.*

Table 4 Worldwide known geographic distribution of the sea cucumber *Holothuria scabra*.

Country name (region)	References
Australia (including Cocos, Ashmore and Cartier Islands)	Cannon and Silver (1986), Carter et al. (1997), Clark (1931, 1938, 1946), Endean (1953, 1956, 1957), Gibbs et al. (1976), Marsh et al. (1993), Marsh (1994), Massin (1999), Morgan (1996), Rowe and Gates (1995), Stephenson et al. (1958)
Chagos Archipelago	Price et al. (2010, 2013), Sheppard et al. (2012)
Cook Islands	Zoutendijk (1989)
Caroline Islands	D.L. Pawson (Personal communication)
China (southern coast)	Anonymous (1991), Clark and Liao (1995), Liao (1980, 1984, 1997)
Democratic Republic of Yemen	Amir (1985), Gentle (1985), Matthes (1983)
Egypt	Din (1986), Mortensen (1937)
Fiji	Adams (1992), Preston (1990), Stewart (1993)
Guam	D.L. Pawson (Personal communication)
Hong Kong (China)	Clark (1982)
India (including Andaman and Nicobar Islands)	Alagaraja (1994), Baskar (1994), James (1978, 1982, 1986a,b, 1987a,b, 1989b, 1991, 1994a,b, c,f, 1995a, 1999), James (1990), James and James (1993, 1994a,b), Mukhopadhyay (1988), Parulekar (1981), Satyamurti (1976), Soota et al. (1983), Tikader and Das (1985)
Indonesia	Conand and Tuwo (1996), Daud et al. (1993), Massin (1999), Moore (1998), Rachmansyah et al. (1992), Rowe and Gates (1995), Tangko et al. (1993a,b), Tuwo (1999)
Iran	Dabbagh et al. (2012)
Japan	Kobayashi et al. (1991), Wiedemeyer (1992)
Kenya	Humphreys (1981), Muthiga et al. (2009), Samyn (2000)
Kosrae	Kerr (1994)
Madagascar	Cherbonnier (1988), Conand et al. (1998)

Table 4 Worldwide known geographic distribution of the sea cucumber *Holothuria scabra*.—cont'd

Country name (region)	References
Malaysia	Baine and Choo (1999), Ridzwan et al. (1995)
Maldives	Massin (1996, 1999)
Marshall Islands	D. L. Pawson (Personal communication)
Mauritius	Massin (1996, 1999)
Mozambique	Abdula (1998), Balinski (1958), Macnae and Kalk (1962), Rowe and Richmond (1997)
New Caledonia	Cherbonnier (1952, 1980), Conand (1986, 1989, 1990, 1993, 1994), Féral and Cherbonnier (1986), Intes and Menou (1979), Northern Province Fisheries Division (1993)
Oman	Al-Rashdi et al. (2007b)
Palau	Yamanouchi (1939, 1956)
Papua New Guinea	Clark (1921), Lokani (1990, 1995, 1996), Lokani et al. (1995a,b,c, 1996), Massin (1999), Shelley (1981, 1985a,b)
Philippines	Cowan and Gomez (1982), Domantay (1936), Leonardo and Cowan (1984), Ong Che and Gomez (1985), Reyes-Leonardo (1984), Reyes-Leonardo et al. (1985), Schoppe (2000), Semper (1868), Tan Tiu (1981a,b)
Saudi Arabia	Clark (1952), Price (1981), Tortonese (1979)
Seychelles	Clark (1984)
Solomon Islands	Adams et al. (1992), Battaglene and Bell (1999), Holland (1994a,b), Mercier et al. (2000a,b), Richards et al. (1994)
Somalia	Massin (1996, 1999)
South Africa	Massin (1996, 1999)
Sri Lanka	Anonymous (1978b), Elanganayagam et al. (1980), Moiyadeen (1994), Pearson (1903)
Tanzania (including Zanzibar)	Kithakeni and Ndaro (2002), Yussuf and Yahya (2020)

Continued

Table 4 Worldwide known geographic distribution of the sea cucumber *Holothuria scabra*.—cont'd

Country name (region)	References
Thailand	Bussarawit et al. (1999), Pattarapongpan et al. (2016), Wainiya (1988)
Tonga	Anonymous (1996b), Sommerville (1993)
United Arab Emirates	Yaghmour and Whittington-Jones (2018)
Vanuatu	Chambers (1989), Preston (1993)
Vietnam	Levin and Ho (1989), Tran and Ngo (1963)

5. Habitat

Habitat characteristics of *H. scabra* are presented in Table 5. Several sources indicate that *H. scabra* occurs mainly in low-energy environments behind fringing reefs or within protected bays and shores (Conand, 1986, 1989, 1990; Domantay, 1936; Lokani et al., 1996; Mercier et al., 2000a, b; Shelley, 1981, 1985a). Only a few contradictory observations placed *H. scabra* in high-energy environments (Anonymous, 1991, 1996a), although the methods used to characterize the habitats were not provided. Numerous reports indicate that *H. scabra* is often found in areas of low salinity in India (Jagadees and Sasidharan, 1990; James, 1989a, 1994e; James and James, 1994a) and in Australia (Springhall and Dingle, 1967) but, unfortunately, no precise values of salinity were given. The work of Mercier et al. (1999) demonstrated the ability of *H. scabra* to tolerate salinity decreases down to 20 psu by burrowing into the sediment, which could explain why this species can sometimes be found near estuaries.

Tuwo et al. (2020) investigated the impact of sudden and gradual salinity changes on *H. scabra*. Results showed progressive negative impacts as salinity was reduced below 18, with a survival rate of 50% at a salinity of 14.6 psu and no survival below 14 psu. In another study, Tuwo et al. (2021) showed that this species can sustain increases in salinity up to 46 psu (47 psu being the lethal threshold) before showing signs of evisceration and mortality. Moreover, Sembiring et al. (2019) exposed juveniles measuring 4.4 cm in length (5.6 g body weight) to various salinities: 24, 29, 34, 39 and 44 psu. Over 2 months, the salinities between 24 and 34 psu yielded up to 100% survival and high growth rates, while a salinity of 44 psu resulted in low survival rate and negatively impacted growth and immunity. Accordingly, *H. scabra* is found in habitats with terrigenous inputs in New Caledonia

Table 5 Habitat characteristics of *Holothuria scabra* over its geographic range.

Location	Age	Substratum	Energy	Salinity	Depth	Comments	References
General South-Pacific	Adult	Silty sand or mud	Low	Estuarine	2–10 m	–	Baird and Sachithananthan (1974), South Pacific Commission (1994)
Australia	Adult	–	–	Estuarine	–	–	Morgan (1996)
	Adult	–	–	–	0–10 m	Inshore	Rowe and Gates (1995)
	Adult	Sand	–	–	–	Reefs and seagrass beds	Cannon and Silver (1986)
	Adult	–	–	–	–	Close to mangrove	Stephenson et al. (1958)
	Adult	–	–	–	–	In *Zostera* flats	Endean (1953)
	Adult	Mud flats	–	–	Intertidal to 18–20 m	Intertidal seagrass beds, deep bay	Uthicke and Benzie (1999, 2001)
	Juvenile	–	–	–	<2 m	In seagrass beds	Vail (1989)
China	Adult	Muddy sand	High	Coastal reef	–	Coastal reefs and where seaweeds are lush	Anonymous (1991)
India	Adult	Mud	–	–	–	–	Baskar (1994)
	Adult	Silty sand	–	Brackish waters	1–10 m	Near estuaries, on *Cymodocea* beds	James (1989a)
	Adult	Coral sand	–	–	2–10 m	Estuarine waters near rivers	Baird and Sachithananthan (1974)
	Adult	Mud	–	Low salinity	Intertidal to 20 m	–	James et al. (1994b)

Continued

Table 5 Habitat characteristics of *Holothuria scabra* over its geographic range.—cont'd

Location	Age	Substratum	Energy	Salinity	Depth	Comments	References
	Adult	Silty sand	–	Low salinity	Intertidal to 10 m	On *Cymodocea* beds	James and James (1994a)
	Juveniles	Sand and mud	–	–	–	Among algae	James (1976)
	Adult	Mud and sand	–	–	Intertidal to 20 m	Mud and sand flats	James (1994a)
	Adult	Mud and sand	–	–	10–15 m	Seagrass beds and near mangrove or estuaries	Sachithananthan et al. (1986)
	Juvenile	–	–	–	Near shore	–	Sachithananthan et al. (1986)
	Adult	Mud and sand	–	Brackish	Shallow waters	–	Jagadees and Sasidharan (1990)
Indonesia	Adult	Sandy clay	–	–	–	Seagrass *Enhalus*	Sipahutar and Soeharmoko (1989)
Japan	Adult	Mud and sand	–	–	–	–	Wiedemeyer (1992)
Kosrae	Adult	–	–	–	Shallow water	–	Kerr (1994)
Malaysia	Adult	Sandy shore	–	–	10–40 m	Reef slope and rise	Baine and Choo (1999)

Mozambique	Adult	Sand	High	–	Intertidal	Across seagrass and algae	Anonymous (1996a)
	Adult	Sand and muddy sand	Low	33–35	–	Mangrove and seagrass areas	Macnae and Kalk (1962)
New Caledonia	Adult	Muddy sand	Low	–	0–5 m	Terrigenous input (<3.5 km from the coast)	Conand (1989, 1994)
	Adult	–	–	–	–	Reef flat, terrigenous input	Conand (1990)
	Adult	–	Low	–	0–12 m	Behind fringing reef	Intes and Menou (1979)
	Juvenile	–	Low	–	Intertidal	Near estuary flat	Conand and Tuwo (1996)
	Adult	Mud and sand	Low	–	0–5 m	Inner reef flats, fringing or islet reefs, terrigenous input	Conand (1990)
	Adult	Mud and sand	–	–	Rarely >5 m	Bay, close to mangrove and sandy beach	Conand (1989, 1994)
Palau	Adult	Sand flat	Low	–	Shallow	Close to reef margin along fringing reef, seagrass	Yamanouchi (1939)
Papua New Guinea	Adult	–	Low	Brackish	Shallow	Close to mangrove and behind fringing reef	Shelley (1985a)
	Adult	–	–	–	–	Seagrass with terrigenous sediments	Long and Skewes (1997)
	Adult	Silt	–	–	6–11 m	Seagrass bed	VandenSpiegel et al. (1992)

Continued

Table 5 Habitat characteristics of *Holothuria scabra* over its geographic range.—cont'd

Location	Age	Substratum	Energy	Salinity	Depth	Comments	References
	Adult	–	–	–	–	Seagrass bed	Long et al. (1996)
	Adult	Mud, sand	–	–	LWN to ELWS	Inner reef flats	Shelley (1981)
Philippines	Adult	Sand	–	–	–	Shallow water, seagrass, algae beds, near estuary	Tan Tiu (1981a)
	Adult	Sand, coral rubbles	–	–	–	–	Leonardo and Cowan (1984)
	Adult	–	Low	–	–	Behind fringing reef	Domantay (1936)
Solomon Islands	Adult	Sand to mud	Low	Brackish water	<5 m	Protected inner-reef flats, bays and estuaries, near mangroves and terrigenous input	Battaglene and Bell (1999)
	Adult	Mud and sand	Low	30–33	Intertidal to 5 m	Close to mangrove and seagrass beds	Mercier et al. (2000a)
	Juvenile	Mud and sand	Low	30–33	<1 m	In seagrass beds	Mercier et al. (2000a)
Sri Lanka	Adult	–	–	–	6–20 m	–	Anonymous (1978b)
Yemen	Adult	Fine silty sand	Low	–	–	Shallow sheltered lagoon and turtlegrasses (*Halodule* and *Halophila*)	Gentle (1985)

(Conand, 1986, 1990, 1994), Papua New Guinea and Australia (Long and Skewes, 1997), in nutrient-rich environments in the Solomon Islands (Battaglene, 1999b; Mercier et al., 2000a,b), and close to mangrove swamps in Australia (Stephenson et al., 1958), Madagascar (Conand, 1997b), and the Solomon Islands (Mercier et al., 2000b) often in areas exposed to sustained or temporary low salinity conditions.

H. scabra is one of the rare tropical species that prefers coastal areas to coral reefs (Conand, 1989; Uthicke and Benzie, 1999). A few studies indicate that *H. scabra* is attracted to muddy-sand habitats (Mercier et al., 1999; Shelley, 1981) or mud (Baskar, 1994) (Table 5). According to laboratory experiments and field studies, juveniles show a clear preference for medium-sized grains around 0.4 mm and, to a lesser degree, for finer muddy sand, whereas coarse sand and crushed coral are avoided (Mercier et al., 1999). Associations with seagrass were observed in Papua New Guinea (Long and Skewes, 1997; Long et al., 1996) and Australia (Uthicke and Benzie, 1999; Vail, 1989). Skewes et al. (2000) observed the highest concentrations of *H. scabra* in the seagrass beds around Warrior Reef (Australia). Overall, there was a clear relationship between seagrass cover and abundance of *H. scabra*. Large populations were also found near or within beds of *Thalassia hemprichii* and *Enhalus acoroides* in the Solomon Islands (Mercier et al., 2000b), within *Zostera* beds in Australia (Endean, 1953), and within *Enhalus* beds in Indonesia (Sipahutar and Soeharmoko, 1989). Hasan (2005) showed that sandy areas were the most suitable habitat for *H. scabra* in the Red Sea. In contrast, rocky, rubble, dead and live coral habitats are considered unfavourable. The highest densities ranged from 95 ind. 100 m^{-2} in sandy areas, followed by 29 ind. 100 m^{-2} in seagrass areas, and 9.7 ind. 100 m^{-2} on mud. Densities were lower in rocky areas (1.9 ind. 100 m^{-2}), dead coral areas (1.8 ind. 100 m^{-2}), rubble (2.9 ind. 100 m^{-2}) and live coral areas (2.3 ind. 100 m^{-2}) (see Section 6.1 for more details). Mmbaga (2002) found *H. scabra* mainly in the intertidal zone among seagrasses and algae along the coast of Tanzania, with length frequency distributions showing a dominance of young individuals ranging between 7 and 16 cm in length.

Jasmadi (2018) indicated that the habitat of *H. scabra* in eastern Indonesia is optimally characterized by depth ranges between 0.5 and 1 m, a temperature around 28°C, a salinity of ~33 psu, an absence of ammonia, a pH of 8.1, water current ~0.03 m s^{-1}, underwater visibility of ±1 m, a tidal range of ~1.8 m and sandy substratum. In addition, seagrass (*T. hemprichii*) and macro algae (*Ulva lactuca*) were associated with the highest abundance of *H. scabra*.

6. Spatial distribution and population dynamics

6.1 Density and biomass

Recorded abundances of *H. scabra* are quite variable throughout the Pacific and Indian Oceans with average values ranging between 0.1 and 0.6 ind. m^{-2} (Table 6). The highest density reported is ~2 ind. m^{-2} in India (James, 1994a). Moreover, Amir (1985) indicated that *H. scabra* was the most abundant species of sea cucumber along the coast of Yemen. Other regional reports are outlined below.

In New Caledonia, *H. scabra* was found in habitats ranging from intertidal flats to depths of 10 m with muddy-sand to sandy-mud sediments, reaching densities that varied from 0.1 to 60 individuals 100 m^{-2} with a mean at 6.8, whereas the mean biomass was 2.4 kg 100 m^{-2} (Conand, 1989). *H. scabra* ranked third in density and second in biomass among the 50 holothuriid (formerly aspidochirotid) holothuroids studied by Conand

Table 6 Reports of density and biomass of *Holothuria scabra* over its geographic range.

Locations	Density (individuals m^{-2})	Biomass ($g\,m^{-2}$)	References
Egypt (Red Sea)	–	460 to 0.1 (over the years)	Hasan (2005)
India	0.4–2.0 (juveniles)	–	James (1994a)
Indonesia	0.1	–	Tuapattinaja et al. (2014)
New Caledonia	0.068[a] (maximum of 0.6)	24	Conand (1989, 1990, 1994)
Oman	0.1–04	29–39	Al-Rashdi et al. (2007a)
Papua New Guinea	0.29–1.35 0.01–0.02 0.00–0.26	– – –	Shelley (1981, 1985a,b) Lokani et al. (1995a, 1996)
Solomon Islands	0.0075[a] all substrata combined. Maximum 0.35 on muddy-sand substrata	–	Mercier et al. (2000a)
Tanzania	0.03–4.3	–	Mmbaga (2013)

[a]Average value.

(1989) in New Caledonia, already more than 40 years ago. In Papua New Guinea, Shelley (1985a) noted that *H. scabra* had an annual production of 487 kg ha^{-1} $year^{-1}$ considering the whole dry weight and 24 kg ha^{-1} $year^{-1}$ for beche-de-mer. In Australia, Skewes et al. (2000) surveyed abundances at 165 sites along Warrior Reef in January 2000, finding no more than 100 t of individuals aged 2 years or more, which was low relative to 1995 estimates (1200–1400 t) and estimates of the virgin biomass (>1600 t) mentioned by Long et al. (1996). The density of *H. scabra* in Solomon Islands was greater on mud (2.2 ind. 100 m^{-2}), muddy-sand (1.0 individual 100 m^{-2}) and sand (0.8 individual 100 m^{-2}) than on silt, coral pebbles and seagrass beds (<0.20 ind. 100 m^{-2}) (Mercier et al., 2000b). Individuals <100 mm are mainly concentrated on muddy sand (34.7 ind. 100 m^{-2}) and mud (26.8 ind. 100 m^{-2}), although data were more variable and based on rough estimates from examination of ~3% of each substratum type (Mercier et al., 2000b). The highest density measured, pooling all size-classes together, was 34 individuals on 100 m^2 of muddy substratum at a depth of ~45 cm in an area that was never exposed at low tide (Mercier et al., 2000b). Hasan (2005) indicated that at Abu Ramada Island (Red Sea) the total abundance of *H. scabra* in the study site dropped from 3362 individuals in January 2000 and 3110 individuals in July 2000, down to 1682 individuals in January 2002, with a further decrease from 802 in April 2002 to only 10 individuals in October 2003. Biomass exhibited the same patterns as abundance: high values ranging from 46 to 34 kg 100 m^{-2} from July 1999 to April 2001 decreasing during the period from July 2001 to January 2002 from 21 to 11 kg 100 m^{-2}. Finally, there was a dramatic decrease in the biomass recorded between April 2002 and October 2003 from 7 to 0.1 kg 100 m^{-2} (Hasan, 2005), suggesting that this population was disappearing due to fishing activities. In Tanzania, Mmbaga (2013) observed densities of *H. scabra* between 0.03 and 4.3 ind. m^{-2} modulated by site of collection, depth and season. In Ambon Bay (Indonesia), Tuapattinaja et al. (2014) mentioned that the density of *H. scabra* was around 0.1 ind. m^{-2}. Al-Rashdi et al. (2007a) reported that population densities varied between 1170 and 4000 ind. ha^{-1} and biomass ranged between 393 and 2903 kg ha^{-1} in Mahout Bay (Oman) before exploitation.

6.2 Distribution and size structure

6.2.1 Western Pacific

In New Caledonia, where the populations are mostly intertidal, Conand (1989, 1990, 1994) showed that size distribution was plurimodal, with poorly defined modes composed of individuals ranging from 120 to 360 mm length, with a general mean around 240 mm and no small individuals. In India, Baskar (1994) indicated that the frequency distribution of *H. scabra* had a single mode, with the smallest individuals measuring 90 mm long and the largest 370 mm, with an average of 230 mm. James (1994a) noted that specimens 300–350 mm in length were found between 5 and 10 m depth in the Gulf of Mannar, India. A similar size-frequency distribution was observed in Papua New Guinea (Lokani et al., 1995b, 1996). Lokani et al. (1996) further indicated that the size distribution was unimodal or bimodal, depending on the site studied in Papua New Guinea.

In Australia, along the Queensland coast, Uthicke and Benzie (1998, 1999) observed that the size-frequency distribution of *H. scabra* was unimodal. The average size from three shallow populations studied ranged from 98 to 178 mm, while it was much larger (269 mm) for a deeper one. However, Skewes et al. (2000) showed that the length frequency of *H. scabra* collected on Warrior Reef (Australia) in 2000 exhibited two modes, one at ~100 mm total length and the second at ~160 mm. Given the established growth rates of this species (Long et al., 1996), the smaller individuals (≤140 mm total length) were probably over 1 year old. No individuals larger than the legal minimum size of 180 mm were found during those site surveys.

In the Solomon Islands, *H. scabra* appeared to be distributed according to a depth gradient coupled with the effects of sediment granulometry and organic content (Mercier et al., 2000a,b) (see above section for more details on densities). The largest individuals (>250 mm in length) were located mainly in the deeper zone and those >120 mm on a sandy substratum. Individuals ranging from 150 to 250 mm in length occurred mainly around seagrass beds, on mud or muddy sand substrata at depths of 0.3–1.2 m. Intermediate-sized individuals ranging from 40 to 150 mm were found on mud and muddy sand in shallow water fringing the intertidal zone and in the intertidal zone itself. The individuals distributed on the exposed portion of the substratum at low tide were buried and the small depressions they created retain water. In general, *H. scabra* was not found exposed directly to the air. Similarly, James and James (1994b) observed that juveniles burrowed during low tide in Andaman Islands, although James (1994a) found that small individuals between 50 and 90 mm in length did occur on the substratum during low tide. The smallest *H. scabra* (10–40 mm) observed by Mercier et al. (2000a) outside of seagrass nursery grounds were usually found on mud and muddy sand substrata, sometimes inside the seagrass beds, in 20–120 cm of water. Most individuals occurred in areas with 5–10% organic matter, although the largest individuals appeared to prefer areas with <5%

organic matter, and some small and medium-sized individuals could be found on substrata with >10% of organic content (Mercier et al., 2000a,b). The size-frequency distribution of *H. scabra* in the Solomon Islands varied from one site to another (Mercier et al., 2000b). Multiple cohorts were detected in the population of an unfished bay called Kogu Veke with sizes varying from individuals as small as 10 to ~330 mm. Conversely, only 45 individuals between 100 and 330 mm were collected at Kogu Halingi, a harvested site, showing a population structure without recent recruitment. Thirty-five intermediate-sized individuals, 120–280 mm long, also dominated collections at another fished site (Malmaragiri Inlet), with only three specimens <100 mm found in October 1997, and seven found in February 1998.

Natan et al. (2015) found that the length of adult sea cucumbers ranged from 9.5 to 23.3 cm and their weight from 80 to 450 g in Un Bay (Indonesia). However, they showed that the population of *H. scabra* was dominated by individuals between 16.5 and 19.5 cm in length, which represented 49% of the total samples collected. Also in Indonesia (Tanjung Tiram Maluku), Ongkers et al. (2018) reported lengths ranging from 7.5 to 28.0 cm (with a mean ~ 18 cm) and weights from 38.0 to 137.6 g (mean ~ 64 g).

In Sarangani Bay (Philippines), Pitogo et al. (2016) examined the size distribution of *H. scabra* in three shallow intertidal areas over two lunar phases (full moon and new moon). Their results showed a unimodal distribution, with individuals found to be larger during the new moon than during the full moon. Population densities ranged from 0.17 ind. m^{-2} during full moon period, to 0.34 ind. m^{-2} during new moon period, suggesting that lunar phases influence the size distribution of *H. scabra*; this pattern was proposed to explain the larger individuals collected by intertidal harvesters during new moon periods (Pitogo et al., 2016).

6.2.2 Indian Ocean and Red Sea

Al-Rashdi et al. (2007a) reported that *H. scabra* in Mahout Bay (Oman) exhibited a length class around 18 cm. In the Chagos islands (British Indian Ocean Territory, BIOT), *H. scabra* was observed by Price et al. (2010) who reported them from a few sites at densities around or exceeding 10 ind. 200 m^{-2}. In Zanzibar (Tanzania) Yussuf and Yahya (2020) indicated that the overall size of all collected individuals expressed as body weight, body wall weight and total length ranged from 410 to 1957 g, 220 to 960 g, and 20 to 42 cm, respectively at 15–20 m depth (Fig. 8).

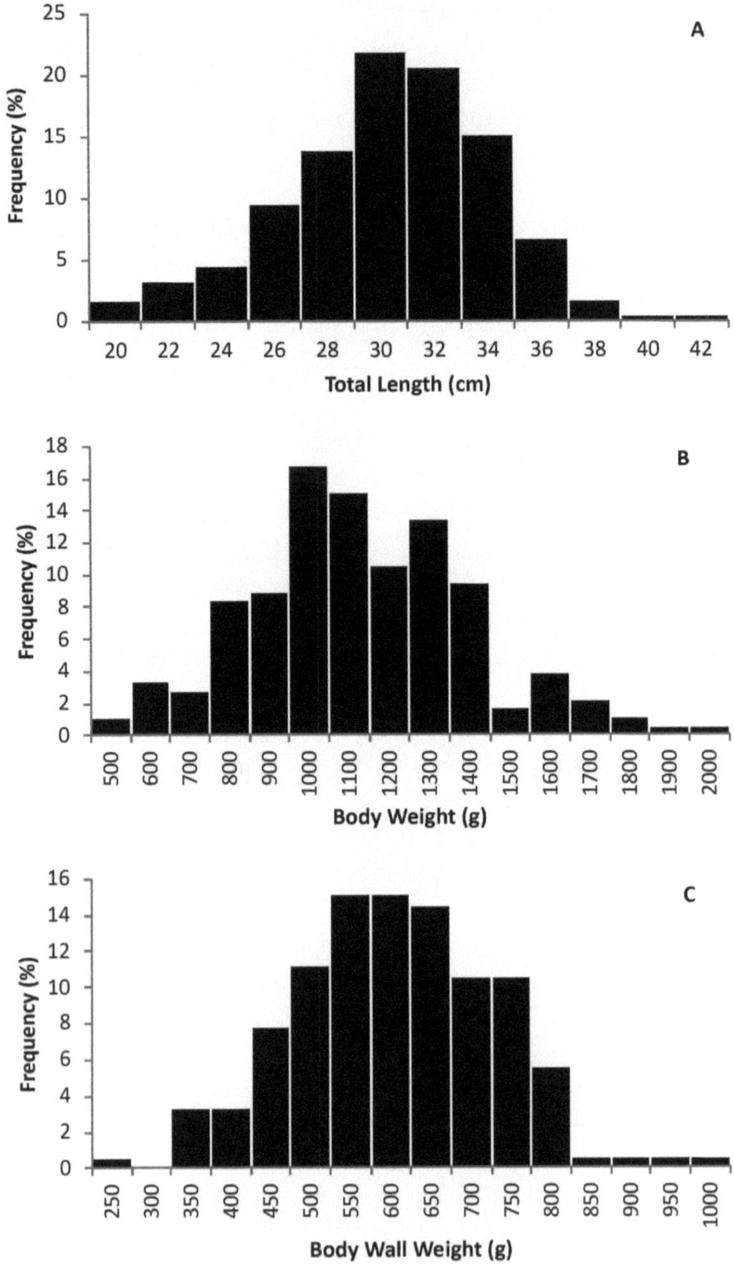

Fig. 8 Size distribution of *H. scabra* in Zanzibar, Tanzania, based on (A) total length (cm), (B) body weight (g) and (C) body-wall weigh (g). *From Yussuf, Y.S., Yahya, S.A., 2020. Size-distribution and length-weight relationship of a deep-water population of* Holothuria scabra *(Jaeger, 1833) in Zanzibar, Tanzania. West. Indian Ocean J. Mar. Sci. 19, 113–121, with permission.*

There was no difference in average size between male and female individuals. However, female individuals seemed slightly longer and heavier than males. The size distributions (total length, body weight and body wall weight) of individuals showed more than 85% of all individuals had a length between 25 and 35 cm, 9.5% were below 25 cm, 5% were above 35 cm, with no individuals smaller than 20 cm (Yussuf and Yahya, 2020). The length-frequency distribution of *H. scabra* was found to be unimodal in Dar es Salaam (Tanzania), with most individuals ranging between 8.5 and 26.5 cm (Kithakeni and Ndaro, 2002). Also in Dar es Salaam, Mmbaga (2002) indicated that the catch was generally high during the months of March, April, July, November and December. This last observation could be linked to periods during which *H. scabra* spends more time on the surface of the substratum.

A study in the Sudanese Red Sea coast mentioned an average size of 450–500 g or 16.8 cm in length (Ali et al., 2015). At Abu Rhamada Island (Egypt), also in the Red Sea, the distribution of *H. scabra* was reportedly correlated with depth, coupled with the effect of intensive exploitation of the species (Hasan, 2005). During the period between 1999 and 2001 (early at the onset of the fishery), the highest density was recorded, with 37% of all individuals found at 4 m in April 2000, and 28% of the population at 6 m in January 2001. From July 2001 to April 2002, a shift in distribution towards deeper waters was observed. The maximum recorded was at 8 m (26–28%), and low densities occurred at depths between 20 and 30 m. Between April 2002 and October 2003 (at the climax of the fishery), there was a more pronounced shift towards the deepest depths, with maximum occurrences at 20 m (5–46%), 30 m (19–66%) and 40 m (13–40%). Unusually, some individuals were also reported from depths >40 m in July (0.8%) and October (9%) 2003, which is quite deep for the species and might warrant further validation. Hasan (2005) also examined the size structure between 1999 and 2003. From July 1999 to April 2001, the size frequency of the population was trimodal with individuals varying between 10 and 390 mm in length. The first sign of recruitment appeared in July 1999, with 2.3% of individuals measuring about 10 mm; this percentage reached 11% of the population in January 2000. From July 2001 to January 2002, the cohorts became bimodal. The presence of large individuals decreased, and the whole population was dominated by medium sized individuals ranging from 151 to 270 mm and associated with the decrease of new recruits. Finally, from April 2002 to October 2003, the population became unimodal, with large sized individuals gradually disappearing.

No new recruits were observed in the population during that period. Between July and October 2003, only two size classes were found: 61–90 and 91–120 mm in length (Hasan, 2005).

6.3 Nurseries

Until the early 2000s, juveniles of *H. scabra* had rarely been documented in the wild and data on their distribution and habitat preferences were scarce. Still today, reports and investigations of juveniles in the field remain limited.

Lokani et al. (1995b) initially reported that young individuals were absent from sites supporting adults of *H. scabra* in Papua New Guinea. This turned out to be contrary to the findings of Mercier et al. (2000a,b) who found newly-settled and small individuals in the same general area as adults in Solomon Islands (see details below). A few intermediate-sized juveniles have been found on fine sand on an inner reef flat exposed at low water in Papua New Guinea (Long and Skewes, 1997; Shelley, 1985a,b).

Conand and Tuwo (1996) observed specimens of *H. scabra* about 100 mm long in the intertidal zone near an estuary flat in South Sulawesi (Indonesia). Gravely (1927) found a 50-mm long juvenile in India, and a 30-mm long juvenile *H. scabra* was discovered in India among algae by James (1976). The same investigator also found juveniles 60–160 mm long along the coast of India (James, 1983). James (1989a, 1994a,d, 1996b) noted that ~500 juveniles ranging from 65 to 165 mm were collected from the Andaman Islands, in the intertidal region during low tide. Conand (1997b) observed an artisanal fishery exploiting juveniles in the north of Madagascar. Specimens were harvested from *Syringodium* seagrass flat in front of a mangrove, and only measured from 40 to 80 mm. They were collected by women; the estimated C.P.U.E. (Catch Per Unit Effort) has been as high as 300 individuals per woman per hour, but had already decreased markedly by the time of the study.

Field studies conducted in the Solomon Islands showed that several small individuals of *H. scabra* <10 mm occurred on leaves of *E. acoroides* and *T. hemprichii* but not on other substrata (Mercier et al., 2000b). The highest densities of newly-settled and early juveniles *H. scabra*, corresponding to the smallest average size of 1.7 mm, were found typically before the full moon. By contrast, the lowest densities, comprising individuals of 5.4 mm, were usually observed 12–16 days after the full moon. This pattern was repeated each month, except between September and November in 1998. Also, samples from December 1998 and January and February 1999 revealed very low or zero recruitment (Mercier et al., 2000b). The occurrence of other fauna on the seagrass, including numerous potential predators such as polychaete worms, tectibranchs, crabs, gastropods, copepods, shrimps, flatworms and isopods, varied monthly in number and diversity. Maximum abundance of potential predators was correlated with peak recruitment of young *H. scabra* and could explain their rapid decrease in abundance (Mercier et al., 2000b).

In a review covering reports up to 2004, Shiell added some references of juveniles of *H. scabra* observed from the wild (Shiell, 2004). This included an 8-cm long individual found in Crab Island, Moreton Bay, Queensland (Australia) in a shallow bed of seagrass/algae; a 10-cm long individual found in Pouangué, Northern Province (New Caledonia) in a seagrass bed on a muddy inshore reef flat (shallow water); and individuals 1–5 cm long found in Ambandja (Madagascar) in a muddy intertidal habitat.

Several dozen juveniles measuring 2–5 cm in length were observed in tide pools on one morning in March 2018 in Koh Mook, southern Thailand (Igor Eeckhaut, Personal observation) They rested on the substrate, close to a dense seagrass bed. Similarly, numerous juveniles were found in Oman (in 2019–21) close to seagrass beds in the low intertidal zone (Michel Claereboudt, Personal communication).

6.4 Substratum selectivity

The substratum preferences of juveniles of *H. scabra* > 10–140 mm have been investigated (Mercier et al., 1999). A clear preference was shown for a sand with medium-sized grains of around 0.4 mm. The second choice was a finer muddy sand, while coarse sand and crushed coral were avoided. In all cases, preferences were firmly established within 1 or 2 h, and persisted throughout the experiments, which extended over 24 h (Mercier et al., 1999).

Although the substrata were washed prior to testing substratum preferences based on grain size, the residual organic matter contents of the different sediments did not seem to substantiate the hypothesis that some grain sizes are preferred because they provide more organic matter (Mercier et al., 1999). Medium sand was preferred even though it was not the substratum with the highest proportion of organic matter. However, organic content might not be an important variable at the low values measured in the washed sediments. Alternatively, sea cucumbers may compromise between burrowing energetics and feeding efficiency. Thus, medium sand may constitute an ideal substratum because it retains a sufficient load of organic material, and is easy to ingest and to burrow into. Coarser media may be harder to process through the gut, especially for small specimens, while organic content is probably less readily available due to the lower surface/volume ratio. Muddy sand appeared to provide a suitable substratum, but it was largely avoided when opposed to medium sand, possibly because mud may not offer the ideal conditions for burrowing or support. The fact that juveniles were burrowed deeper in muddy sand might indicates that this type of substratum has a tendency to draw them in, so that more energy may be required during surfacing (Mercier et al., 1999).

Although the influence of residual organic content was apparently not the determining factor in the grain size preference, the presence of organic matter was a strong attractant for juveniles foraging on medium sand. Most juveniles wandered to the richest sectors of tanks within an hour, suggesting a detection ability (Mercier et al., 1999). The preference for a richer substratum was firmly established after 2 h and the distribution of juveniles did not vary much after 15 h. The organic contents measured at the beginning and at the end of the experiment revealed that the poor substratum remained essentially unchanged with an initial mean organic content of 1.6% and a final mean organic content of 1.4%, while the rich substratum was depleted, according to the respective initial and final mean values of 9.4% and 3.0%. The juveniles thus removed 71% of the organic matter present in the rich substratum. Four of the tanks filled with poor substratum were enriched by 48%, whereas the other two were depleted by 58%. Control substrata holding no juveniles did not show any noticeable variation in organic matter content (Mercier et al., 1999).

Baskar (1994) measured the size of the particles present in the gut of adults of *H. scabra* collected from the wild and noted a dominance of muddy grains between 130 and 250 μm. He concluded that the species was a selective feeder but did not analyse the substratum composition around the sea cucumbers. The results of Baskar (1994) could therefore represent only preferential choice of substrata, especially since his data roughly correspond to the grain size of the preferred substratum of the juveniles in the study of Mercier et al., 1999.

Tsiresy et al. (2011) and Plotieau et al. (2014b) respectively reported the mineral and organic features of sediments in farms where *H. scabra* was reared by villagers in Madagascar (see details in Section 21). Sea cucumbers grew well on sediments with low content in total organic carbon (0.4–2.0%, measured by carbonization). Contents in particles <500 μm varied from 43% to 85% depending on the village (Plotieau et al., 2014b). In the inorganic fraction, the proportion of quartz was often high (in three out of four villages), representing >50% of the dry weight, while the proportion of carbonate was always <10%.

6.5 Movement and migration

6.5.1 Locomotion

H. scabra moves chiefly through muscular action, with the help of tube feet densely distributed on the ventral surface of the body wall (see Section 3 for details). It was additionally shown to be able to roll and float using a strategy dubbed active buoyancy adjustment (ABA) that helps it travel much faster using currents and changing tides (Hamel et al., 2019c).

Individuals are able to climb on hard surfaces such as rocks (James, 1989a). Mercier et al. (1999, 2000a,b) noted that young individuals of *H. scabra* moved between 50 and 331 cm day^{-1} in the laboratory and between 41 and 80 cm day^{-1} in the field. Lokani et al. (1995a,b, 1996) and Lokani (1996) reported that *H. scabra* moved at a mean speed of 12 cm min^{-1} and that while their movement was random, they could target a precise site. Mercier et al. (1999) observed a locomotory speed of juveniles between 3.6 and 15.4 mm min^{-1} depending on the size of the individuals. The locomotory speed of *H. scabra* was greater on non-optimal substrata such as crushed coral than on sand or muddy sand (Mercier et al., 2000b). In the laboratory, juveniles moved faster on bare surfaces or organically-poor substrata than on rich substrata, indicating that they do indeed move around in search of food if organic matter is distributed unequally (Mercier et al., 1999). Calculations of the mean speed over the active period gave approximate values of 71–331 cm day^{-1} on a poor substratum, depending on the size of the juveniles. The same estimation yielded values of 50–215 cm day^{-1} on a rich substratum. Considering that faecal pellets are emitted less frequently on the poor substratum, the rapid movements of juveniles in such environments does not appear to be associated with higher feeding rates. Instead, the animals might spend a great deal of time wandering without feeding, suggesting that the energetic cost of processing poor sediment could be higher than the cost of moving to find a more suitable feeding ground (Mercier et al., 1999).

In a study conducted in New Caledonia, the relationship between speed and body weight was determined to be linear (Fig. 9A) (Purcell and Kirby, 2006). Adult displacement overnight was ~130 cm over 16 h, although this is likely an underestimation of the speed. Purcell and Kirby (2006) also found that the movement directionality of wild individuals (12–23 cm in length; 130–690 g) was variable but non-random, yet there was no apparent movement towards a common direction. Circular patterns were rare and there

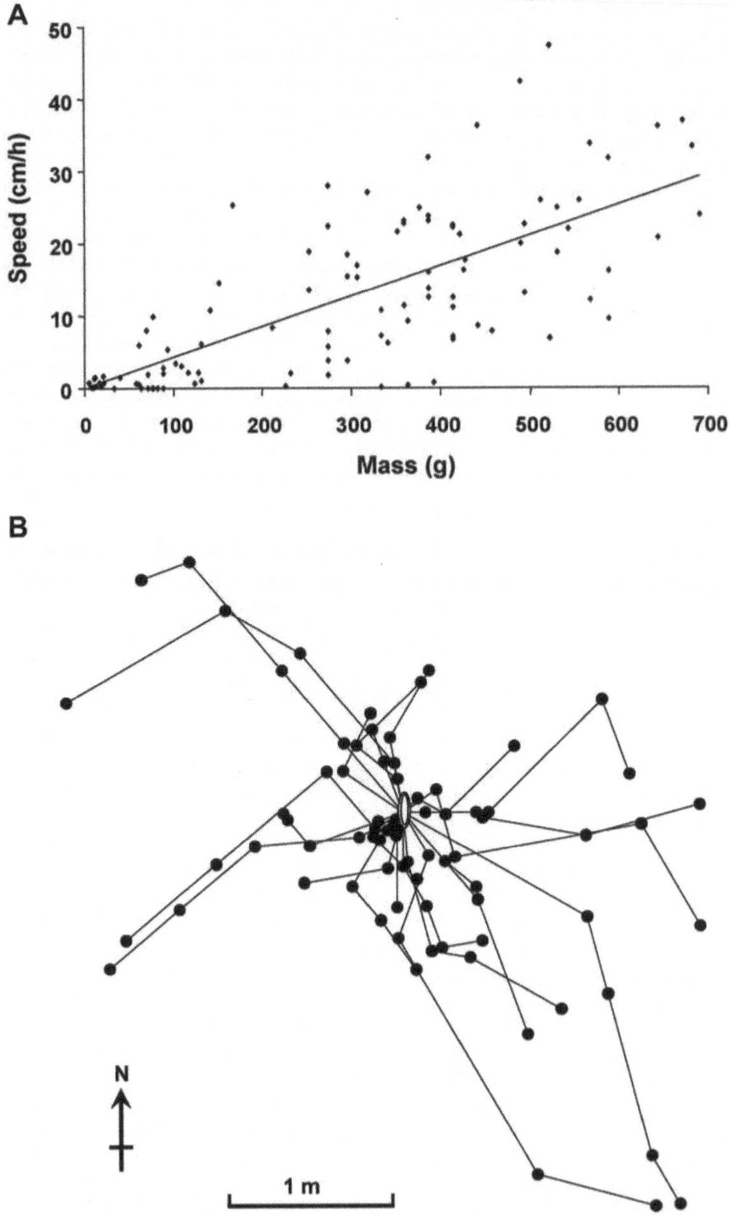

Fig. 9 (A) Average speed of individuals of *H. scabra* (over 24h) against body mass (g). Line of best fit is least-squares regression: speed (cm h^{-1}) = 0.04237 × mass (g); S.E. of slope = 0.02185, r^2 = 0.53, n = 105, P < 0.001. (B) Overhead view of paths travelled by 26 adults during the last 2 days of observation (in August 2004); each path is the track of one individual from ~8:00 to 17:00h (starting from the centre), with ~2h time steps punctuated by solid dots. *From Purcell, S.W., Kirby, D.S., 2006. Restocking the sea cucumber* Holothuria scabra: *sizing no-take zones through individual-based movement modelling. Fish. Res. 80, 53–61, with permission.*

was evidence of roaming in different directions within daylight hours (Fig. 9B). Small individuals burrowed during the day and moved during the night hours (see Section 13 for more details).

Lee et al. (2018a) indicated that *H. scabra* was active for about 10 h day^{-1} with a travel range between 2 and 8 m day^{-1}, and potentially 120 m $month^{-1}$. Although movement was measured as straight distance, movement patterns appeared to be haphazard, similar to findings by Mercier et al. (2000b). Observations of marked individuals of *H. scabra* revealed a 'home ranging' behaviour, i.e. moving inside a restricted area, roughly 100-m radius (Lee et al., 2018a), similar to that of *Bohadschia argus* and *Thelenota ananas* (Purcell et al., 2016c). In conclusion, *H. scabra* is relatively mobile, but appears to have a limited range (Lee et al., 2018a). Movement patterns of sea cucumbers may be linked to the distribution of food, as some species of sea cucumbers exhibit patch selectivity (Uthicke and Karez, 1999).

Youngs of *H. scabra* released in the field in three different habitats (sand, crushed coral, seagrass bed) exhibited apparently random movement, changing direction every day (Mercier et al., 2000b). In general, the larger the sea cucumber, the greater distance travelled, except in a seagrass bed. In all habitats, individual movement remained relatively constant over a 2-months study period, however, there was a major difference in distance travelled when the three habitats were compared both between comparable size-classes and across size-classes. The greatest displacement occurred on a mix of shells and crushed coral, with ~80 cm day^{-1}. The seagrass habitat induced the second greatest movement (~51 cm day^{-1}) and the sand substratum induced the smallest movement (~41 cm day^{-1}). Movements with respect to organic matter content of the substratum in the three habitats were less conclusive and the highest mobility was recorded on the sediment of intermediate richness (Mercier et al., 2000b).

Holothuroids are commonly assumed to disperse chiefly through pelagic larvae but the study of Hamel et al. (2019c) explored the ability of juveniles and adults to actively modify their buoyancy, leading them to tumble or float at speeds orders of magnitudes faster than through benthic crawling (see above). This active buoyancy adjustment (ABA) was documented in *H. scabra* (Hamel et al., 2019c); it is achieved through a rapid increase in water-to-flesh ratio leading to bloating, and simultaneous detachment from the substrate. In experimental trials, ABA was triggered by high conspecific density, decreasing salinity and increasing water turbidity. Based on video footage from the field, ABA-assisted movements generated speeds of up to several km per day; it was seen to allow small individuals of *H. scabra* to escape aquaculture pens and drift out at sea. These findings imply that displacement during planktonic larval stages may not supersede the locomotor capacity of benthic stages, challenging the notion of sedentarity in *H. scabra* and other holothuroids.

6.5.2 Migration patterns

Individuals of *H. scabra* found in a seagrass bed at depths <2 m were smaller than individuals found in deeper water along the coast of Australia (Uthicke and Benzie, 1998; Vail, 1989), supporting the idea that seagrass beds are nursery habitats for this species (see Section 10 for more detail). Uthicke and Benzie (1998) also observed that a greater proportion of adults were sexually mature in deeper water. Mercier et al. (2000a,b) witnessed the settlement of pentactula larvae on seagrasses of the genera *Thalassia* and *Enhalus* in the Solomon Islands, both in the laboratory and in the field. Newly settled juveniles remained on the seagrass for 4–5 weeks before migrating to sand at around 6 mm in length. Prior to this, the juveniles spent 4–5 days moving on and off the seagrass. Once on the sand, the juveniles did not show the typical burrowing behaviour of older specimens until they reached around 11 mm in length. The larvae of *H. scabra* appear to actively select seagrass, possibly through chemical detection. It was hypothesized that larvae settling on seagrass have an increased chance of growth and survival because they are provided with a suitable substratum on which to grow, and a bridge to sandy substrata (Mercier et al., 2000a,b).

In the field, juveniles remain on sand among seagrasses until they reach ~90–100 mm and then move to the open sandy area or mud flat; larger individuals are localized in deeper water >1.5 m (Mercier et al., 2000b). James et al. (1994a) noted that *H. scabra* migrated to deeper water for breeding in India. A greater number of large and sexually mature individuals were observed in deep areas by in Australia (Uthicke and Benzie, 1998, 1999), supporting the hypothesis that there is a downward migration of growing individuals. However, Harriott, 1980) and Mercier et al. (2000b) observed large mature individuals in shallow waters in Australia and the Solomon Islands, respectively, suggesting that spawning also occurs there. Uthicke and Benzie (1999) determined that the deep and shallow populations are genetically linked and demonstrated that both groups of individuals are derived from the same larval pool. In Madagascar, Conand (1997a) observed a dense bed of juveniles on an intertidal flat, near to a 20 m deep muddy bay where a prawn fishery trawls large individuals as a by-catch. Another type of migration was noted in Mozambique after a massive organ evisceration:

individuals initially found among seagrasses and algae close to the littoral moved to deeper water with the outgoing tide (Anonymous, 1996a); this was possibly achieved with the help of ABA (see above). A physical or chemical tagging method adapted for long-term studies is still needed (see next section) to confirm the downward migration hypothesis, but the recently described locomotion via buoyancy adjustments (Hamel et al., 2019c) supports that migration of individuals to deeper water could be common.

6.6 Tagging

Attempts at tracking the movements of sea cucumbers have often been thwarted by their capacity to shed foreign materials implanted inside their tissues and to quickly heal scars. Conand (1983, 1989, 1994) tested the use of floy tags, also called t-bars or anchor tags (i.e. t-shaped fasteners similar to those used in the clothing industry) for long-term capture-recapture experiments in New Caledonia. Although some individuals eviscerated, most tagged individuals were not affected by the procedure. However, this tagging method did not prove to be very effective during long-term tracking (Conand, 1989, 1991) and was found to be deleterious in other sea cucumber species (reviewed in Gianasi et al., 2015). Stewart (1993) and Mercier et al. (2000b) subsequently found that numbers scratched on the body wall had no deleterious effects and remained visible for about 2 weeks, thus providing a valuable technique for short-term monitoring of displacement and speed.

No effective method exists for tracking *H. scabra* over a period >6 months. The longevity and effect of tags were analysed by Purcell et al. (2006c, 2009a) and Taylor (2016). Immersion-staining with fluorochromes was determined to be preferable to coded-wire tags because they provide cheap and reliable batch-marks for distinguishing released and wild individuals of *H. scabra* (Purcell et al., 2006c). Dermal ossicles of sea cucumbers can be fluorescently marked with tetracycline and calcein. Sequential marking by tetracycline and calcein, in either order, showed that calcein was deposited in a higher proportion in ossicles (Purcell et al., 2009a). Under UV illumination, ossicles from the body wall of individuals immersed in solutions of tetracycline are yellow, whereas calcein-stained ossicles are green (Fig. 10). Frequency of burial in *H. scabra* (measured as an indicator of stress) increased when both fluorochromes (tetracycline and calcein) were administered at concentrations above $50\,\mathrm{mg\,L^{-1}}$ (Purcell and Blockmans, 2009). Burying behaviour was affected by concentration; at 100, 200 and $400\,\mathrm{mg\,L^{-1}}$ the sea cucumbers were

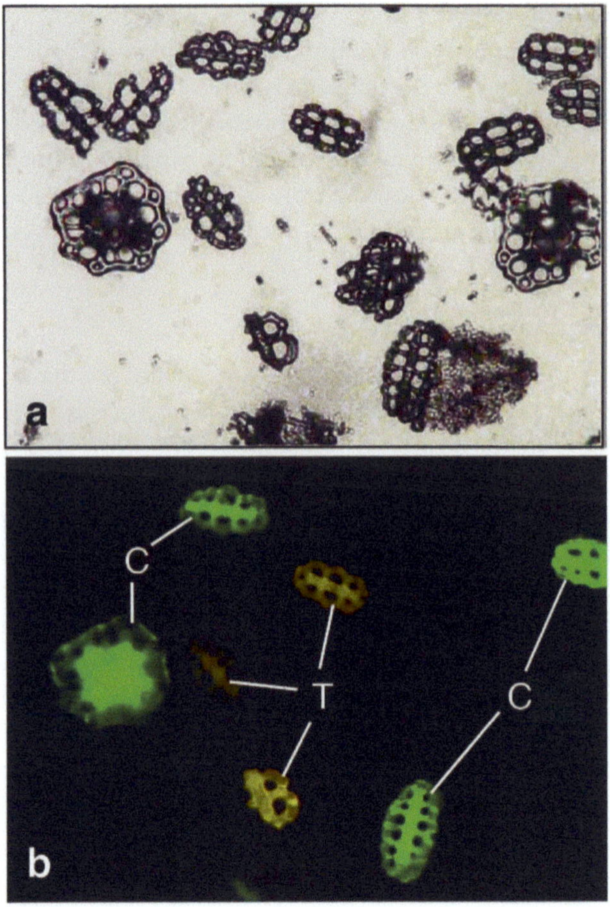

Fig. 10 Ossicles of *H. scabra* shown under (A) light microscopy and (B) UV illumination to highlight tetracycline (T; yellow) and calcein (C; green) markers. *From Purcell, S.W., Blockmans, B.F., 2009. Effective fluorochrome marking of juvenile sea cucumbers for sea ranching and restocking. Aquaculture 296, 263–270, with permission.*

buried more frequently at 14:00 h than at 50 mg L^{-1} (Purcell and Blockmans, 2009) (see Section 13 for more details). Growth rate was unaffected at concentrations of 50 and 100 mg L^{-1} but being inhibited by concentration between 200 and 400 mg L^{-1} of tetracycline or calcein. Both of them fluoresced more intensely when marked at 26 or 30°C than colder temperature.

Rakotonjanahary et al. (2015) used two types of markers, fluorescent and non-fluorescent liquid and wire elastomer implants for tagging 256 adults of *H. scabra* across different size classes: 50–100,100–150, 150–200 and 200–250 g. The results showed that all individuals were healthy after the experimentation that lasted 50 and 11 days, respectively, for liquid injection and wire elastomer implants. However, the two types of markers were rejected by all individuals (100%) within that period, whatever the size of the sea cucumber and the location of the implantation (dorsal, ventral or lateral side). It was determined that fluorescent liquid markers were the most visible (Rakotonjanahary et al., 2015).

Simoes et al. (2012) conducted an experiment using the fluorochrome oxytetracycline (OTC) to mark the dermal ossicles of juveniles of *H. scabra*; they found that marking efficiency was much lower than for tetracycline (TC) and calcein for equivalent procedures at the same water temperature (28°C). The highest marking efficiency achieved with OTC (18%) was similar to the 15% marking efficiency achieved with TC after 1 h, and made up only two thirds of the 29% marking efficiency found with calcein after 1 h. Moreover, after 24 h, TC and calcein were found to have marked 28% and 39% of dermal ossicles, respectively. Although individuals in all treatments suffered weight loss, no mortalities occurred. Preliminary results indicate that while OTC may be less toxic, it is not as effective as TC and calcein (Simoes et al., 2012).

6.7 Population genetic structure

Population genetic structure was analysed in *H. scabra* with allozyme electrophoresis, 12S rDNA haplotypes and microsatellite markers. The complete mitochondrial genome has been sequenced and many pairs of primers are now available for studying the variability inside any population.

Genetic analyses based on allozyme electrophoresis on seven polymorphic loci were performed by Uthicke and Benzie (2001) on samples of *H. scabra* collected from several locations in northeast Australia, the Torres Strait, and the Solomon Islands. They identified distinct populations from the northeast coast of Australia, representing samples from Hervey Bay, Upstart Bay and Torres Strait (Uthicke and Benzie, 2001). These were closely connected to those from the Solomon Islands. Restrictions in gene flow, observed as a high proportion of the variation in genetic distances along the east coast of Australia, was explained by isolation by distance, suggesting a low dispersal that may reduce the recovery of overfished stocks (Uthicke and Benzie, 2001). Nowland et al. (2017) evaluated population genetic structure in Papua New Guinea, and across northern Australia. They found that although microsatellite loci varied widely in the number of alleles (3–28), the allelic diversity was similar among populations sampled.

The level of genetic substructuring among populations sampled was low. Most of these differences were driven by distinctness of the Australian populations from those in PNG, and suggested that PNG populations exhibited a panmictic stock structure.

Uthicke and Purcell (2004) examined patterns of within-region genetic heterogeneity of *H. scabra* by allozyme electrophoresis to determine whether hatchery-produced juveniles collected at a single site could be released at other locations hundreds of kilometres away. They also investigated larval connectivity on a larger geographic scale by comparing New Caledonia with previously published data from other Pacific regions and other populations in the Indo-Pacific. Genetic variability within sites did not highlight differences between populations in New Caledonia, Knocker Bay (Australia), and Bali (Indonesia). However, over the geographic distances covered in Uthicke and Purcell (2004) and previous study Uthicke and Benzie (2001) (between 20 and 6000 km), Rogers' genetic distance between populations increased as a function of geographic distance. A Mantel test (Mantel, 1967) confirmed that geographic and genetic distances were correlated, supporting isolation by distance (Uthicke and Purcell, 2004).

More recently, Ninwichian and Klinbunga (2020) studied the populations of *H. scabra* from four locations (Phangnga, Satun, Trang and Krabi) in the Andaman Sea (west of peninsular Thailand). An analysis of molecular variance indicated that the gene pool of *H. scabra* in the Andaman Sea formed two genetic stocks divided between Phangnga and the remaining samples (Krabi, Satun and Trang). Phylogenetic and structure analyses highlighted separation between those populations, with possible implication for management of natural stocks of *H. scabra* in Thailand (Ninwichian and Klinbunga, 2020).

Ravago-Gotanco and Kim (2019) examined diversity and spatial patterns of genetic structure in populations of *H. scabra* across the Philippine. Analysis revealed a regional genetic structure among 15 populations across the archipelago. Genetic connectivity was strongest among populations situated in the centre of the archipelago. Peripheral locations exhibited limited gene flow (Philippine Sea, South China Sea, Sulu Sea and Celebes Sea) with relatedness suggesting relative isolation or high levels of local recruitment. Globally, there was a signal of isolation-by-distance for the rest of the Philippine populations, suggesting that distance between coastal habitats coupled with oceanographic circulation influence genetic structure (Ravago-Gotanco and Kim, 2019).

Xia et al. (2016) described the complete mitochondrial genome sequence of *H. scabra*. The mitochondrial DNA is 15,779 bp in length and contains

13 protein-coding genes (PCGs), 2 rRNA genes, 22 tRNA genes, and a 456 bp putative control region with gene order identical to the echinoderm pattern. Analyses of *H. scabra* and other holothuroids revealed that the mitochondrial genome is highly compact, exhibiting five regions with overlapping genes and fewest intergenic nucleotides among sequenced holothuroids. Mitochondrial genome in the genus *Holothuria* show high interspecies divergence at the 13 PCGs, and the genetic p-distance reaches ~26% between *H. scabra* and *H. forskali*; the incomplete stop codon T of cox2 shared with *H. forskali* may be a common feature in the genus *Holothuria* (Xia et al., 2016).

Fitch et al. (2013) isolated 18 polymorphic, microsatellite markers from *H. scabra*. Next-generation sequencing was used to identify 206 unique loci for which primers were designed. Of these loci, they trialled 65 primer pairs, and 48 amplified a product of the expected size. Eighteen loci were found to be polymorphic and reliable, and were screened for variation from a population from Croker Island (Australia). Heterozygosity ranged from 0.00 to 0.96 and the number of alleles per locus from 2 to 28. These loci could be useful for studying population structure and reproduction in *H. scabra* (Fitch et al., 2013).

Boncan et al. (2016) demonstrated that about half of microsatellites in the genome of *H. scabra* remain unidentified in perfect repeat scans. They also indicated that partial genome sequencing may be used as an efficient alternative to the traditional methods of developing microsatellite markers for *H. scabra* (Boncan et al., 2016). Moreover, combined approach of sequence similarity-based and de novo search of interspersed repeats revealed a diverse subclass/superfamily of transposable elements in the genome of *H. scabra*. The major subclass/superfamily identified include DNA/hAT-Blackjack, DNA/hAT-Tip100, DNA/Maverick, RC/Helitron, LINE/L2, LTR/Gypsy, SINE/MIR and SINE/tRNA. The interspersed repeats presented by Boncan et al. (2016) constitute a first attempt to survey the transposable elements from the genome in *H. scabra*.

Li et al. (2015) developed nine new polymorphic microsatellite loci. The number of alleles ranged from 2 to 5, and the observed and expected heterozygosities ranged from 0.1200 to 0.7391 and from 0.2408 to 0.5983, respectively. Polymorphic microsatellite loci could be used to study the genetic diversity of *H. scabra* and assist with questions related to its conservation.

Uthicke and Benzie (1999) used allozyme markers to examine whether black and grey individuals of *H. scabra* in north-eastern Australia were colour morphs of the same species, and investigate whether deep populations may be sources of recruits for shallower areas. Their results indicated no difference in allelic frequencies between the two colour morphs. Moreover, there was virtually no separation between the shallow and deeper populations, suggesting an exchange of genetic material between the two (Uthicke and Benzie, 1999). The higher number of large sized and sexually mature individuals in the deep area supports the hypothesis that there is a successive downward migration of growing individuals (Uthicke and Benzie, 1999) (see also Section 6.5.2 above).

7. Asexual propagation (fission)

To our knowledge, transversal fission has never been reported to occur naturally in *H. scabra*; however, it has been induced artificially on a few occasions. Lokani et al. (1995a) applied constriction of the body wall for a week and observed that, following fission, the two resulting sections regenerated to form fully developed sea cucumbers. Ongkers et al. (2020) induced fission using a rubber band positioned at either the anterior, middle or posterior ends in small and large individuals of *H. scabra* in Indonesia. They showed that small individuals took less time to undergo fission and to heal the wound but had lower survival compared to large individuals. The duration of fission process varied, from 12 to 36 h in small individuals to 24–48 h in large individuals (Ongkers et al., 2020). Pattinasarany and Pattiasina (2020) tied rubber bands about 60% of the distance between the anterior to the posterior ends, leading to successful fission in <2 days; however, only $\sim$14% of newly produced individuals survived. In an effort to increase the sea cucumber population, Furqon et al. (2020) also induced fission in *H. scabra* using a rubber band placed anteriorly in the first third of the sea cucumber. The results showed that 98% of individuals underwent fission to generate two new individuals; the posterior section exhibited a higher regeneration rate and a better survival rate than the anterior section.

8. Sexual reproduction

8.1 Sexual dimorphism

The sexes are separate in *H. scabra*, but it is not possible to distinguish between them externally, except during spawning when genital papilae (gonopores) are extruded and present a distinct morphology (Battaglene et al., 2002; MPEDA, 1989). Battaglene et al. (2002) mentioned that it was possible to distinguish females from males shortly before spawning from the bulge around the papilla/gonopore. Some investigators have also been able to sex *H. scabra* using a needle and syringe to sample gametes directly from the gonad through the body wall (Battaglene, 1999a; Mercier et al., 2000b; Morgan, 1999; Turner, 2016).

8.2 Sex ratio

Conand (1989, 1993) observed that males of *H. scabra* from New Caledonia were slightly more abundant than females, representing about 55% of the population. However, these percentages were not different from a 1:1 ratio (Conand, 1993). Male to female ratio did not deviate from equality in studies conducted in the Solomon Islands (Mercier et al., 2000b; Ramofafia et al., 2003a), Madagascar (Rasolofonirina et al., 2005) or Oman (Al-Rashdi et al., 2007a). In contrast, Natan et al. (2015) found 35 females and 18 males, yielding a sex ratio of 1:0.51 in Un Bay (Indonesia). The sex ratio determined for *H. scabra* at Mahout Bay (Oman) was 1.3 male for 1.0 female (Al-Rashdi and Claereboudt, 2018). In the Red Sea, the sex ratio was not different from 1:1 before fishing begun in 1999, but shifted to an increasing male bias reaching 93% males by January 2003 (Hasan, 2005). None of the small individuals remaining after January 2003 could be sexed suggesting the drastic impact of humans on the reproductive capacity of that species (Hasan, 2005).

Muthiga et al. (2009) indicated that sexable individuals were present in all monthly samples throughout the year in Kenya, and very few individuals had no gonads. The sex ratio remained balanced in the 1998–99 and 2000–01 samples; however, it became male-biased in the 2006–07 samples. The authors suggested that fishing pressure could be one of the causes of this shift in sex ratio (Muthiga et al., 2009). Fishing pressure has been shown to favour the survival of smaller individuals in fish (Ricker, 1981); accordingly individuals of *H. scabra* sampled in 2006–07 by Muthiga et al. (2009) were smaller on average and males were overall smaller than females. Fishing pressure was also suggested to promote shifts in sex ratio in favour of males in *H. scabra* studied in the Red Sea (Hasan, 2005) (see above) as well as in *Holothuria whitmaei* in the Great Barrier Reef, Australia (Shiell and Urhiche, 2006). Pitogo et al. (2016) found a sex ratio around 1:1 for *H. scabra* in Sarangani Bay (Philippines). Conversely, a study in Zanzibar (Tanzania) found a biased ratio, with 104 males (58.1%), 70 females (39.1%) and 5 sexually undifferentiated/indeterminate (2.8%) individuals (Yussuf and Yahya, 2020).

8.3 Size at sexual maturity

There is some variability in the available data on size at sexual maturity, probably reflecting the various methods used to define this metric. Some rely on formalin-preserved gonads or histological slides to identify the presence of mature gametes and others base their evaluation on gamete release in the laboratory or on field spawning observations. Since individuals can reach sexual maturity before being involved in spawning events, future studies should use standardized methods to define this metric. Interestingly, Yanti et al. (2020) recently proposed that smaller sizes at maturity in *H. scabra* could be a sign of adaptation to overfishing, through an 'r-selection' strategy driven by the progressive disappearance of large individuals.

8.3.1 Pacific Ocean

Conand (1986, 1989, 1990, 1993, 1994) indicated that the drained weight (taken after splitting the sea cucumber and draining it) at first sexual maturity was 140 g for *H. scabra* in New Caledonia and other Pacific islands, which corresponds to a total weight of 184 g, in line with the value presented by Harriott (1980) for Moreton Bay (Australia). Also in Australia (Warrior Reef), Skewes et al. (2000) suggested that *H. scabra* started breeding at about 2 years of age, and that given the minimum size limit (180 mm total length), individuals might be able to reproduce at least once before being harvested. Shelley (1981) reported that *H. scabra* in Papua New Guinea reached sexual maturity at ~136 mm length in males and at 199 mm in females. In the same region, Lokani (1995a) estimated that *H. scabra* was sexually mature at 140 mm but did not distinguish the sexes. Lee et al. (2018a) proposed that the size at sexual maturity of *H. scabra* in Fiji could be around 150 mm, based on individuals that spawned under aquaculture conditions. More recently, Tresnati et al. (2020) reported that *H. scabra* matured at a size of ~42 g gutted body weight in South Sulawesi, Indonesia. In Shioya Fishing Port, Okinawa Prefecture (Japan), Minami and Yoshikuni (2014) noted that individuals smaller than 600 g in body weight were sexually immature.

8.3.2 Indian Ocean and Red Sea

In studies conducted in India, MPEDA (1989) and James et al. (1994b) indicated that individuals of *H. scabra* reached sexual maturity after ~18 months of growth, when females were 210 mm and males 213 mm long. Baskar and James (1995) noted that size at sexual maturity was between 201 and 230 mm, based on spawning capacities. Kithakeni and Ndaro (2002) stated that the first spawning occurred at a size of 168 mm in Tanzania. The average size at first maturity was 168 mm in Dar es Salaam (Tanzania) (Kithakeni and Ndaro, 2002). In Mahout Bay (Oman) Al-Rashdi and Claereboudt (2018) found a male measuring 110 mm and a female measuring 120 mm, both with mature gonads, although most individuals seemed to reach maturity at around 150 mm, and the size at 50% population-wide maturity was determined to be 165 mm.

In the Red Sea, at Abu Rhamada Island, size at sexual maturity decreased from the prefishing period (185 mm for females and 160 mm for males) to 155 mm for females and 125 mm for males in January 2003, when the population was almost completely decimated (Hasan, 2005). Also in the Red Sea (Sudan), Ali et al. (2015) indicated that first maturity occurred at a drained weight of 450–500 g.

8.4 Fecundity

Harriott (1980) indicated that there was no relationship between the fecundity of *H. scabra* and its body size. Variability in fecundity data provided for *H. scabra* (Table 7) can probably be attributed to methodological artefacts. For instance, James and James (1993) measured fecundity using the number of spawned oocytes only, thus presenting values of ~1 million oocytes per female, whereas Conand (1989, 1990, 1993, 1994) evaluated potential fecundity by dissecting mature gonads and proposed much higher values of >2 to 18 million oocytes per female, with the higher values for the larger females. As *H. scabra* has been suggested to participate in partial spawning events, and as only some of the mature gametes are shed during artificially induced spawning (S.C. Battaglene, unpublished data), the only reliable fecundity data would be those obtained using the whole gonad and those that are effectively broadcasted. Battaglene et al. (2002) indicated that the mean number of eggs released was around 1.9 million for *H. scabra* in the Solomon Islands. Hasan (2005) showed that there was a positive relationship between fecundity and size in the Red Sea. They further mentioned that the number of oocytes per female was highest in 1999 at the onset of the exploitation in this region (0.73–1.7 million) and in 2000 (0.75–1.72 million), but decreased during 2001 (0.2–0.85 million) to reach a minimum in 2002 (0.28–0.29 million) after a few years of exploitation. The average fecundity of *H. scabra* collected in 2006–07 in Kenya by Muthiga et al. (2009) was around 1.3 million oocytes per female.

Table 7 Reports of fecundity of *Holothuria scabra* over its geographic range.

Location	Oocyte diameter (μm)	Female fecundity (× 10^6 oocytes)	References
Australia	80–125	–	Harriott (1980)
Egypt	–	0.2–1.7	Hasan (2005)
India	180–200	1	James et al. (1988), James and James (1993)
	180–200	1	James (1994h), James et al. (1989, 1994a)
	120–165	–	Baskar (1994)
	180–200 (192[a])	1	James et al. (1994b)
Kenya	–	1.3	Muthiga et al. (2009)
New Caledonia	190 (210[a])	2–9	Conand (1990)
	210		Conand (1993)
Philippines	70–80	–	Ong Che and Gomez (1985)

[a]Average value.

8.5 Gonad morphology

In the typical fashion of sea cucumbers, male and female gonads of *H. scabra* are composed of numerous thin, filamentous tubules united basally into one tuft attached to the left side of the dorsal mesentery and hanging freely in the perivisceral coelomic cavity (Conand, 1989, 1993) (see also Section 3). Ramofafia et al. (2003a) described the gonad as a single tuft of numerous elongated branched tubules found on the right side of the dorsal mesentery. They also indicated that the gonad consisted of a single cohort of tubules developing uniformly in individuals sampled in the Solomon Islands, in line with previous studies there and elsewhere (Conand, 1981, 1993, 2001; Ramofafia et al., 2000). The results presented by Ramofafia et al. (2003a) for *H. scabra* suggested that oogenesis in this species did not follow the model proposed by Smiley (1988) who mentioned distinct cohorts of tubules being progressively recruited. Somewhat differently, Rasolofonirina et al. (2005) indicated that the gonad of *H. scabra* in Madagascar consisted of a tuft of numerous tubules of unequal length, mixing both large branched and small unbranched tubules. The large tubules were mainly two- or three-branched and the proportion of three-branched morphology increased with the length of the tubules. Moreover, the number of tubules per individual clearly increased with both the weight of the gonad and of the body wall. All the tubules arose from an inconspicuous rather dense gonad base where they shared a common connective sheet (Rasolofonirina et al., 2005). Tubules merge into the gonoduct, which possesses an inner ciliated epithelium, proceeds into the mesentery and opens to the outside in the mid-dorsal region near the anterior end (Conand, 1998b; James, 1989a).

Demeuldre and Eeckhaut (2012) investigated the gonads in juveniles of *H. scabra* from Madagascar. They found the same three tissue layers seen in adults, even in very small specimens: the external mesothelium that faces the general body cavity, a layer of connective tissue with muscles and nerves, and the inner epithelium where germ cells form. The only difference between adults and juveniles <10 cm in length is that, in the latter, it is impossible to distinguish the germ cells in the inner epithelial layer, where they are probably structurally confused with somatic cells under conventional optical microscopy (Demeuldre and Eeckhaut, 2012). In specimens >10 cm in length, germ cells are found in the longer tubules (>0.6 cm long) but not in the shorter ones (<0.6 cm long). The first recognizable germ cells appear 8–9 months post fertilization in *H. scabra*. The gonad in specimens 3-cm long comprises less than five tubules, some of which are dichotomized. The gonad of specimens measuring 4 cm has 5–10 gonad tubules and there are 10–20 tubules in specimens 5–9 cm long (Demeuldre and Eeckhaut, 2012).

Several accounts describe the colour of the mature gonad in adults. Mary Bai (1980) initially noted that the female tubules were brown, uniformly thick and long and that the oocytes were visible as small white spots, while testes consisted of white, uniformly thick and long filaments. James (1989a) also indicated that ripe female gonads were brown with the oocytes visible as multiple white spots. In contrast, James and James (1993) and Conand (1986, 1989, 1990, 1993, 1994) reported that the ripe ovary was translucent. Rasolofonirina et al. (2005) simply mentioned that the tubules were often orange in females, but in Madagascar, orange to pink translucent ovaries and white opaque testicles are very characteristic of mature female and male individuals, respectively (Igor Eeckhaut, Personal observation).

Conand (1986, 1993) and Baskar (1994) defined the maturity stages of the gonad of *H. scabra*. According to the last author, immature individuals possess single and short tufts of tubules; at this stage, the sexes are not distinguishable. Maturing individuals have longer gonad tubules with a yellow tinge, and the germinal cells become visible. Young mature individuals show the presence of some spermatozoa or oocytes in the gonad. The gonad tubules in adult males are larger, yellow, and branched with round saccules. The adult female gonad tubules become yellowish red and are branched. At full maturity, males have long, pale yellow, gonad tubules with two or three ancillary branches filled with numerous spermatozoa. Baskar (1994) also noted that spent males have shortened, less abundant tubules, with only few remaining spermatozoa, while female gonad tubules become shorter and wider after spawning but they retain their colour and a few pockets of remaining oocytes. Rasolofonirina et al. (2005) showed that the diameter of the tubule changes with the gonad stage in both males and females from the south-west of Madagascar. The smallest diameters were observed in gonads at the recovery stage. The gonad length (longest tubule of the gonad) also changed with the reproductive stage, with the shortest tubules being noted during recovery. Tubules of developing gonad therefore increase both in length and width.

Maturing gonad tubules show a clear sexual dimorphism. The fecund ovarian tubules are shorter, wider and heavier than the testes. The mean ripe gonad weighs ca. 24 g in male and 31 g in female (Conand, 1993). The gonad tubules of males and females are between 80 and 82 mm in length and 10–13 mm in diameter (Conand, 1986, 1989, 1990, 1993, 1994). The gonad index (gonad weight to body weight ratio) is also higher in females than in males (Conand, 1993). The gonad tubules were also described by Shelley (1981), who reported generally shorter lengths but larger diameters.

According to Mary Bai (1980), the male spermatogonial cells are attached to the germinal epithelium. The spermatogonial cells are larger and more or less spherical in shape with a distinct nucleus in the centre. Just above the spermatogonial cells, nearer the lumen of the tubules, are the primary and secondary spermatocytes, which are much smaller with deeply staining nuclei. The mature spermatozoa are found in the centre of the lumen (Mary Bai, 1980).

The female gonoduct consists of a ciliated epithelium, cells of which are regular in size with long cilia directed towards the lumen, a broad connective tissue and externally a layer of coelomic epithelium continuous with the dorsal mesentery (Mary Bai, 1980). The innermost epithelium is the germinal epithelium, which gives rise to oogonia characterized by lightly staining nuclei with distinct deeply staining nucleoli. Primary, secondary and mature oocytes (i.e. ootids) are present in the lumen. Prior to spawning, vitellogenic oocytes are characterized by the presence of a very large germinal vesicle with a distinct nucleus (Mary Bai, 1980). Ong Che and Gomez (1985) stated that oocytes in females of *H. scabra* varied in shape from pyramidal to elongated and club-shaped and that fully grown (mature) oocytes possessed a dense, yolk-filled cytoplasm. A large nucleus with an outer basiphilic cup and an inner eosinophilic core was also noted, resting eccentrically against the nuclear membrane (Ong Che and Gomez, 1985).

More recently, Nontunha et al. (2020) showed that the ovary of *H. scabra* was composed of multiple tubules joined together at a short gonoduct. The wall of the ovarian tubule was composed of collagen fibres and smooth muscles lined internally and externally by epithelia. The outer epithelium comprised clear and granulated cells that showed numerous vesicles in the cytoplasm, evoking the appearance of hormone-producing cells. The inner wall of the tubule was lined by flat follicle cells that held early-stage oocytes, while subsequent stages of developing oocytes were freely released into the lumen.

8.6 Gametogenesis and annual reproductive cycle

Investigations of the reproductive cycle in *H. scabra* have been conducted in many regions, using various techniques. The nomenclature often differs among authors, especially when naming female germ cells or when determining sexual maturity (i.e. maturation may be applied both to descriptions of the ovary itself, and of ovarian tubules inside the ovary). Meiosis is a process of cell division that characterizes the male and female germ lines. In female sea cucumbers, the first germ cells that appear are oogonia, which divide first mitotically and then enter meiosis, dividing twice more to produce primary and secondary oocytes (the latter are also called ootids at the end of their development). In female sea cucumbers, meiosis is arrested at the prophase I stage inside the ovaries. At that time, germ cells are primary oocytes and they fill progressively with yolk to form the vitellogenic oocytes, which display a large nucleus known as the germinal vesicle (GV). By a still obscure mechanism, the arrest on meiosis is released near the spawning period and the oocytes are ovulated (transformed into ootids). Thus, it is only shortly before they get naturally expelled (spawned) that the oocytes complete GV breakdown and become fertilizable. Under hatchery conditions in Madagascar, the percentage of ootids in the ovary measured between August 1, 2017 and December 14, 2018 among 6154 females varied from 23% to 67% (of the germ cells >63 μm in diameter; i.e. excluding oogonia), the greatest proportion coinciding with the natural reproductive period (Igor Eeckhaut, Personal observation).

Reliance on the drained weight for the calculation of organ (including gonad) indices in sea cucumbers has long been a matter of debate. Some researchers remove the intestine, whereas others remove only the water content in the Polian vesicle and in the respiratory tree. It can be argued that these differences introduce important variability. The most accurate method is to remove all internal organs that could induce variability and weigh only the well-blotted body wall with the pharynx/aquapharyngeal bulb (Conand, 1986, 1989; Mercier and Hamel, 2009). The dry weight of body wall without the pharynx and muscle bands would be even more reliable, but difficult to use for this large species, especially in basic facilities or laboratories. As detailed in Table 7, the oocyte diameter seems to be very variable within the geographic range of *H. scabra*, possibly because it has been established at slightly different stages of the maturation process. Differences may also reflect the various research protocols. Although methods were not always provided, some studies were performed on histological slides, others on preserved gonads and still others on fresh samples. Methods need to be standardized, or correlations have to be calculated, to meaningfully compare the different values.

Despite the caveats, the reproductive cycle is fairly consistently described across studies, with slightly varying nomenclature (Fig. 11). It can be broadly said to comprise resting (resorption), growing, maturing, spawning and post spawning gametogenic stages (Conand, 1989, 1990, 1993, 1994; Morgan, 2000a; Ong Che and Gomez, 1985; Tuwo, 1999) or immature, mature, gravid and spent gonads (Krishnaswamy and Krishnan, 1967). Tuwo (1999) observed that mature tubules contained only vitellogenic oocytes in females and only spermatozoa in males, whereas spent tubules were filled with relicts of vitellogenic oocytes or spermatozoa (Fig. 11). Moreover, Ong Che and Gomez (1985) indicated that a rapid recovery to the growing stage occurred after spawning, once gametogenesis resumed. Ramofafia et al. (2003a) divided gametogenic stages as follows: Stage I (indeterminate) where gonads are small with white unbranched tubules (<10 mm) and sex cannot be determined; Stage II (growing), where gonad tubules are 20–70 mm long and start to bifurcate, developing gametes are present, and ovaries and testes are orange and white, respectively; Stage III (mature) where gonads possess large fecund tubules (30–200 mm long) with

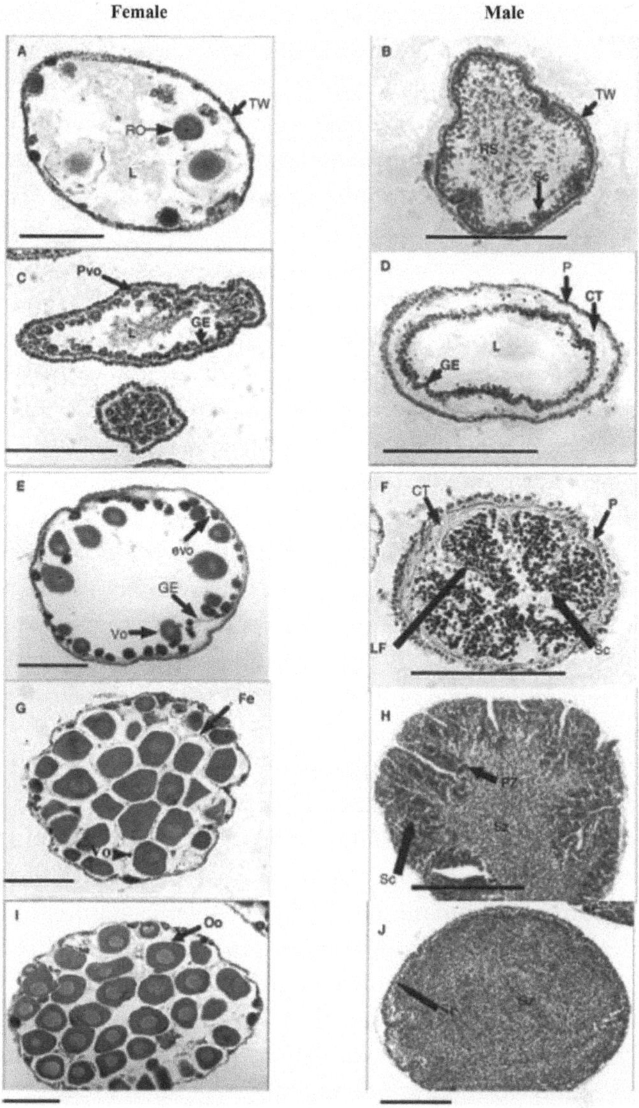

Fig. 11 Gonad stages in female (left) and male (right) of *H. scabra*. (A, B) Spent (stage 1); (C, D) Recovery (stage 2); (E, F) Growing (stage 3); (G, H) Maturation (stage 4); (I, J) Spawning (stage 5). CT: Connective tissue; Evo: early vitellogenic oocyte; Fc: follicular cells; GE: germinal epithelium; L: lumen; LF: longitudinal fold; Oo: oocyte; P: Peritoneum; Pvo: Pre-vitellogenic oocyte; PZ: proliferation zone; RO: relict oocytes; RS: residual spermatozoa; Sp: Spermatocytes; SZ: spermatozoa; TW: tubule wall; Vo: vitellogenic oocyte. Scale bar = 200 μm. *From Rasolofonirina, R., Vaïtilingon, D., Eeckhaut, I., Jangoux, M., 2005. Reproductive cycle of edible echinoderms from the south-western Indian Ocean. II The sandfish* Holothuria scabra *(Jaeger, 1833). West. Indian Ocean J. Mar. Sci. 4, 61–75, with permission.*

additional branches and a beaded appearance, ovaries are red-orange with fully grown oocytes, and testes are creamy white with active spermatozoa; Stage IV (partly spawned) where gonads display both large fecund tubules and tubules having released gametes (reduced in size and flaccid, with phagocytes in the lumen), and presence of immature oocytes in spawned and unspawned tubules indicating re-initiation of gametogenesis; Stage V (spent) where gonads are small (<1–40 g), tubules are wrinkled and shrunken in appearance, with occasional relict oocytes and unspawned spermatozoa. Conand (1993) indicated that some ovaries in the resorption stage were filled with small oocytes, some large residual oocytes and degenerating cells, while others were empty. The process of resorption could be spread over months (Conand, 1993). A finer description of the oogenic stages mentions that the early pre-vitellogenic oocyte contains partially decondensed chromatin and synaptonemal complexes; the late previtellogenic oocyte exhibits mitochondrial aggregations in cytoplasmic poles; the early vitellogenic oocyte becomes enlarged due to the presence of cytoplasmic granules; and the late vitellogenic oocyte harbours abundant cytoplasmic granules, a thick jelly coating and has undergone germinal vesicle breakdown (Nontunha et al., 2020). Both Krishnaswamy and Krishnan (1967) and Moiyadeen (1994) have indicated that the reproductive cycle was more sharply defined in females than in males.

A detailed study of the ultrastructure of the male gametes in *H. scabra* was recently conducted (Suphamungmee et al., 2018). Spermatozoa are generated through steps from germ cell proliferation and differentiation within the wall of the testicular tubule (Fig. 11). All germ cells were classified based on the heterochromatization pattern of the cell nuclei. Stage 1 exhibits a small, round nucleus with heterochromatin clumping around the nuclear envelope which, in turn, is surrounded by the mitochondrial aggregation. In Stage 2, thick chromatin fibres appear; large mitochondria fuse and the tail rudiment also starts to form at the posterior, whereas acrosome and subacrosomal structures are fully formed at the anterior region. It is speculated that a coiling of nucleosomes organized by the spermatozoa-specific histones causes the chromatin condensation in spermatids and sperm nuclei of *H. scabra*, much like what happens in sea urchin sperm (Suphamungmee et al., 2018). The authors suggested that appearance of numerous phagosomes in the Sertoli-like cells was indicative of possible roles in delivering nutrients to the proliferating germ cells and phagocytic removal of cytoplasmic debris from the maturing germ cells.

The reproductive cycle of *H. scabra* has been extensively investigated over its geographic range across the Indo-Pacific (see details in following sections). Studies of gametogenetic cycles have shown that at least a small proportion of populations in *H. scabra* seem to spawn all year round (Battaglene, 1999b; Conand, 1989, 1990; Krishnan, 1968; Ong Che and Gomez, 1985; Shelley, 1985a; Tuwo, 1999), although two major spawning peaks were noted in most regions (MPEDA, 1989). Table 8 summarizes the different studies. More precisely, it was shown to exhibit two main reproductive patterns: seasonally predictable spawning at high latitudes, and aseasonal spawning at low latitudes (Table 8). This interannual variability in reproduction may be influenced by variable local environmental conditions, particularly those associated with the coastal mangrove and seagrass habitats occupied by this species. Ultimately, spawning in *H. scabra* may be opportunistic, with some flexibility to respond to ephemeral favourable conditions (see Section 8.9). The aseasonal gametogenesis that characterizes *H. scabra* in the Solomon Islands and the Philippines may extend broadly to the central Indo-Pacific region (20°N to 20°S). At the northern and the southern limits of its range, *H. scabra* is reported to display synchronous gametogenesis culminating in an annual reproduction (Morgan, 2000a; Mortensen, 1937) potentially cued by seasonally predictable factors, such as day length and water temperature. A reproductive pattern with two peaks was observed in India (Krishnaswamy and Krishnan, 1967), Australia (Harriott, 1980), Indonesia (Turner, 2016), and Papua New Guinea (Shelley, 1981). Even when twice-yearly spawning peaks have been observed, the months differed among years (Table 8). More details on the regional patterns of gametogenesis are provided below.

8.6.1 Pacific Ocean

In New Caledonia, Conand (1986, 1989, 1990, 1993, 1994) used morphological and gonad-index variations to identify a first well marked peak from December to February, followed by a smaller more variable one between August and October. A resting period was detected around June, while the growing period overlapped July and August, and spawning occurred between August and September followed by another growing stage in mid-September and October. The maturing stage in November was followed by a long spawning period between the end of November and February, followed again by a period of growth between February and May. Females of *H. scabra* apparently do not release all mature oocytes during a single spawning event, hence the resting period is short. This supports the hypothesis that *H. scabra* can spawn throughout the year, at least in some areas.

Table 8 Estimated spawning periods of *Holothuria scabra* and proposed spawning cues.

Location	Spawning period	Spawning cue	References
Australia	September–November	–	Morgan (2000a)
	Two peaks	–	Harriott (1980)
Egypt	April–June	–	Mortensen (1937)
Fiji	October and December	Spring tides and full and new moons	Lee et al. (2018a)
India	March–May and November–December	–	MPEDA (1989)
	July and October	Change in salinity and temperature	Krishnaswamy and Krishnan (1967)
	March–May and October–December	–	James (1996a), James et al. (1994b), James and James (1993)
	March–April and September–October	–	James (1989a), Mary Bai (1980)
	July and October, but seem to spawn year round	–	Bakus (1973)
Indonesia (Sulawesi)	March–July and November–January, but seem to spawn year round	Temperature variation	Tuwo (1999)
New Caledonia	December–February and August–September	Coolest and warmest temperatures	Conand (1990, 1993, 1994)
Oman	Late spring (pre-monsoon period)	Temperature of 28°C when precipitations of 10 mm	Al-Rashdi and Claereboudt, 2018
Papua New Guinea	Austral Summer	–	Shelley (1985a)
	November to January	–	Lokani (1995)
	December to February	–	Lokani et al. (1995b)

Table 8 Estimated spawning periods of *Holothuria scabra* and proposed spawning cues.—cont'd

Location	Spawning period	Spawning cue	References
Philippines	May–June and November–January, but seem to spawn year round	When temperature and salinity change drastically	Ong Che and Gomez (1985)
	Three gonad index peaks	–	Cowan and Gomez (1982)
Red Sea	June–August	–	Boolootian (1966)
Solomon Islands	May and November	–	Battaglene et al. (1998)
	September, but spawning was evident year round	–	Mercier et al. (2000a), Ramofafia et al. (2000)
Sri Lanka	April–November (male) and March–October (female)	Salinity changes due to rainfall	Moiyadeen (1994)

Morgan (1999, 2000a) indicated that the gonad index peaked in November in Australia. Morgan (2000a) further determined that there was a continuous presence of mature oocytes that were either reabsorbed or spawned during or prior to the vitellogenic period from September to November. He concluded that it was likely that stored nutrients in the body wall were used for gametogenesis during the latter part of winter and that oocyte production was regulated by phagolytic activity.

In the Solomon Islands, reproduction is reportedly aseasonal, with mature gametes present year-round and episodic events occurring throughout the year, and a period of enhanced activity from September to December (Ramofafia et al., 2003a). Battaglene (1999b) and Ramofafia et al. (2000) found a marked spawning peak between September and October. Ramofafia et al. (2003a) proposed that the gonad indices (GI) were variable between sexes and across years in that region. The GI remained between 2 and 6, with values in female often being higher than in males. In 1997–98, the highest GI for females was recorded in October and August followed by a decrease to December in 1997 and to October in 1998. The highest GI values in males were recorded from August to October, with a different pattern in the following year. The lowest GIs for both sexes were recorded in July. The patterns of tubule length and diameter were similar between testis and ovary, and generally mirrored the trends in GI (Ramofafia et al., 2003a). The investigators concluded that gametogenesis was asynchronous and did not exhibit any seasonal pattern based on the presence of at least three different stages of oogenesis in most ovarian samples. Partly spawned and spent ovaries were the most frequently encountered stages whereas growing, mature or recovering ovaries were comparatively rare. In males, individuals with mature, growing, and partly spawned testes were common, and testis samples also exhibited three or more stages of development. Spent testes were less frequent than spent ovaries. Overall, the GI, tubule size and gonad histology suggested that *H. scabra* undergoes a continuous and asynchronous pattern of gonad development in the Solomon Islands, although maximum ovarian growth was recorded in August–October period in 2 consecutive years. This was followed by a decrease in GI due to gamete release (Ramofafia et al., 2003a).

Tuwo (1999) used similar techniques to those used by Conand (above) and found that the reproductive cycle of *H. scabra* in Indonesia showed a maturation period extending from July to March. Post-spawning stages were noted mainly at the beginning of the dry season, from March to July, and during the beginning of the rainy season, from November to January. Gonad growth occurred in two phases, first from June to October and then from February to April. A proportion of individuals in the maturation stage possessed mature tubules, so that partial spawning events could occur at any time of the year (Tuwo, 1999).

Purwati (2006b) indicated that *H. scabra* presents a uniform gonad development among various populations in Indonesia, with each tubule having equal opportunity to release gametes in the same spawning event (Fig. 12). More precisely, all sea cucumbers had mature gonads, with some showing signs of recent spawning based on the absence of visible oocytes in the tubules. In Langgur Small Kei, Southeast Maluku (also in Indonesia), Rahantoknam (2017) used gonado-somatic indices to study the reproductive cycle of *H. scabra* at the full moon and new moon in 2017. The highest index value was obtained in May and March of the full moon cycle when 90% and 70% of individuals were in the spawning stage, respectively. Overall, the results indicated that the peak spawning of *H. scabra* in that location for that year was between January and July with a peak during the full moon of May. Arsad et al. (2020) described a peak in the reproductive cycle between June and November in a population in Sabah (Malaysia).

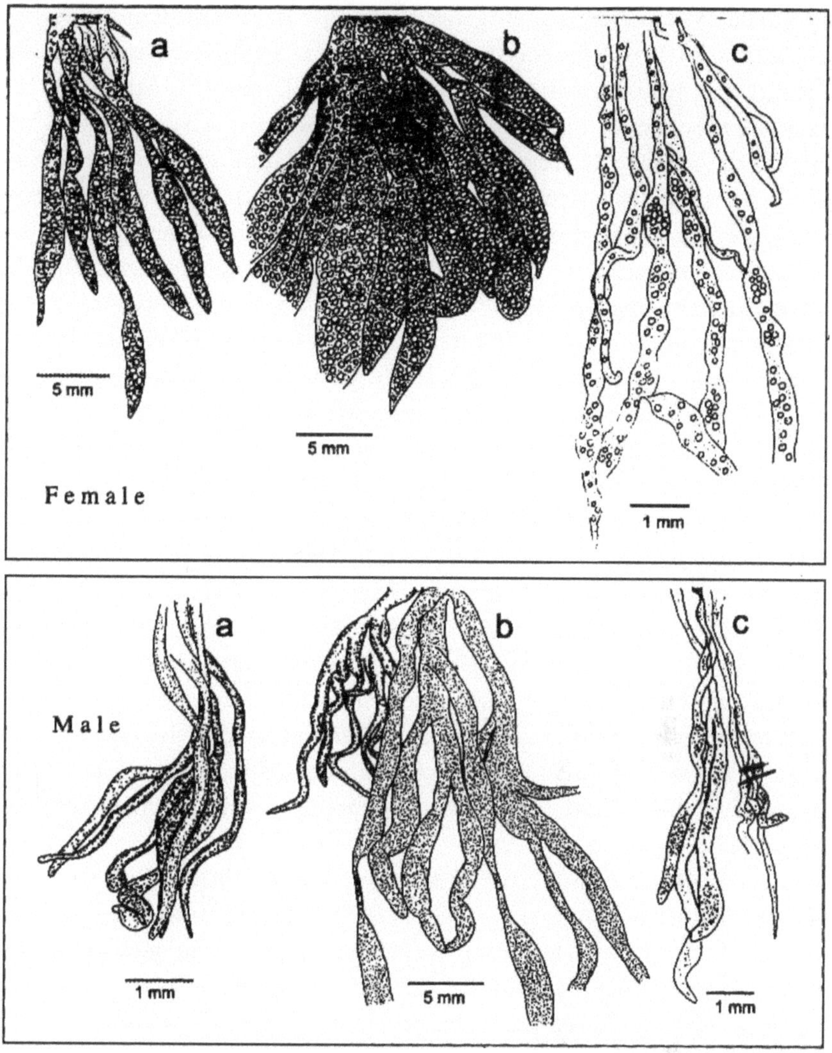

Fig. 12 Gonad tubules of females (top) and males (bottom) of *H. scabra* pstudied in Indonesia. Drawings under camera lucida; a = early stage, b = fecund/ripe stage and c = post spawning/spent stage. *From Purwati, P., 2006. Reproductive patterns of* Holothuria scabra *(Echinodermata: Holothuroidea) in Indonesian waters. Mar. Res. Indones. 47–55. under CC BY-SA 4.0.*

In the Philippines, several peaks in gonad maturity have been noted (Cowan and Gomez, 1982; Ong Che and Gomez, 1985). Two periods of enhanced spawning are reportedly superimposed on more or less perpetual reproductive maturity (Ong Che and Gomez, 1985). The latter investigators proposed that staggered gametogenesis among different individuals permitted continuous breeding for the population as a whole. However, they observed that gonad maturation occurred from January to April and from July to October, with main spawning events occurring from May to June and from October to November. Ong Che and Gomez (1985) also studied gametogenesis using seasonal fluctuations of oocyte diameter and thickness of the different layers of maturing male gametes. Both oogenesis and spermatogenesis appeared to be continuous. Analysis of the size-frequency structure of maturing oocytes showed that oocytic growth was not perfectly synchronized among females of each sample. While two or three cohorts of oocytes could be observed, unimodal distribution was noted most of the time. The size distribution of oocytes did not show any seasonal variation, with a constant peak mode around 70–80 μm in diameter. Interestingly, Battaglene (1999b) and Battaglene et al. (2002) suggested that gonad maturity of *H. scabra* (in Solomon Islands) could not be accurately gauged from oocyte diameters, but needed to be confirmed by a 5–8% change in the gonad index. Ong Che and Gomez (1985) noted that the mean thickness of the spermatozoa and spermatogenic cells layers were inversely related during the reproductive cycle. As the gonad ripened and matured, spermatozoa were recruited from the spermatogenic cell pool and the spermatogenic cell layer thinned out while the spermatozoa layer increased in thickness. During spawning and subsequent gamete growth and multiplication, the spermatozoa layer became thinner while the mean thickness of the spermatogenic cell layer increased.

The reproductive biology of *H. scabra* was studied around Shioya Fishing Port, Okinawa Prefecture (Japan) from September 2010 to 2011 (Minami and Yoshikuni, 2014). Maturity was evaluated monthly in both sexes using the GSI, oocyte size and histology of the ovaries were also examined in females. They concluded from annual transitions in these indicators that *H. scabra* spawned primarily between May and October.

The reproductive biology of *H. scabra* was studied in Hainan (China) by Yang et al. (2015b), including annual changes in gonad development and spawning pattern. Results showed that the testis and ovary could be divided into five stages: recovery, growing, mature, partly spawned and spent. Gonad development and spawning were essentially synchronous between sexes. Mature gonad stages were detected almost in every month, accounting for 35% of females and 39% of males, respectively. However, while reproduction was evidenced throughout the year, two main spawning peaks occurred from May to August and from December to February, respectively (Yang et al., 2015b).

8.6.2 Indian Ocean

Krishnaswamy and Krishnan (1967) studied the reproductive cycle of *H. scabra* in India using gonad indices and histological examination of fresh and preserved gonads. The gonad index increased drastically between June and July, followed by a sharp drop in August. In September, a second important increase was noted, and a low value was recorded in November. Although mature sea cucumbers were found almost throughout the year, greater proportions were identified before both gonad index peaks. According to Moiyadeen (1994), the gonad index showed two annual peaks in Sri Lanka, thus suggesting a bi-annual reproductive cycle that was confirmed by larval abundance in the field.

In Oman, gonad growth was very rapid and took place in March and April when maximum gonad indices were recorded (Al-Rashdi and Claereboudt, 2018). It culminated by an apparently synchronous spawning event between the end of April and middle of May. At the peak of the gonad index, the gonad represented between 9% and 12% of the total body mass of the sea cucumber (dry weight). Before and after spawning, the index was 4–5% of the total dry weight (Al-Rashdi and Claereboudt, 2018).

Muthiga et al. (2009) showed that the changes in gonad indices on the Kenyan coast exhibited a biannual pattern with two peaks, synchronized between the two sexes (Fig. 13A). One major event occurred between October and November and a minor event between March and May, during the northeast monsoon season. The pattern of gonad growth suggests that gametogenesis initiate sometime between August and October, when gonad indices start increasing, and spawning occurs between November and December, when a marked drop in gonad index occurs. Gonad growth is reinitiated between January and March as suggested by increasing gonad indices and decreases between July and September (Muthiga et al., 2009).

In nearby Tanzania, Kithakeni and Ndaro (2002) reported the absence of gonad in up to 50% of the individuals of *H. scabra* collected in summer months. The gonad indices remained relatively stable between February and August and increased markedly from October to December while spawning reportedly occurred in September and again between December and January (Kithakeni and Ndaro, 2002).

Rasolofonirina et al. (2005) reported that all tubules except the smallest ones (<1 cm in length) exhibited the same maturity stage over the reproductive cycle in a population from south-western Madagascar. It was suggested that new tubules were not recruited once a year, but rather year-round with variability among individuals (Rasolofonirina et al., 2005). New tubules that

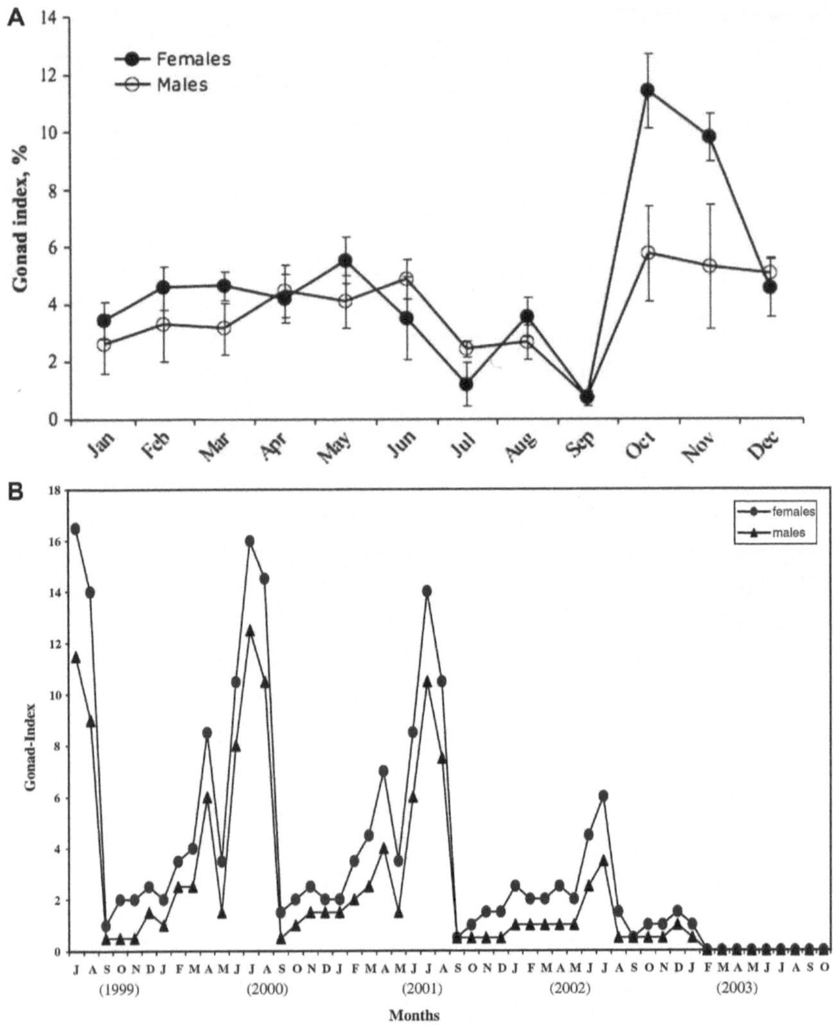

Fig. 13 (A) Monthly fluctuations in the gonad index of *H. scabra* in Kenya, shown as mean (%; ± SE, $n=3$) calculated over 3 different years (1998–99, 2000–01, 2006–07). (B) Monthly fluctuations in the gonad index (%) of *H. scabra* in Egypt between 1999 and 2003. *Panel (A) from Muthiga, N.A., Kawaka, J.A., Ndirangu, S., 2009. The timing and reproductive output of the commercial sea cucumber Holothuria scabra on the Kenyan coast. Estuar. Coast. Shelf Sci. 84, 353–360, with permission. Panel (B) from Hasan, M.H. 2005. Destruction of a* Holothuria scabra *population by overfishing at Abu Rhamada Island in the Red Sea. Mar. Environ. Res. 60, 489–511, with permission.*

did not grow or mature in the course of a gametogenic cycle may develop at the onset of the next cycle, as reported for the cold-water species *Psolus fabricii* (Hamel et al., 1993). However, large tubules clearly showed an annual gametogenic cycle and a well marked recovery (resting) period was identified, during which tubule diameter decreased from 800 to 400 μm, and tubule length from 10 to <4 cm (Rasolofonirina et al., 2005). Such regression had not been observed in the Solomon Islands population of *H. scabra* (Ramofafia et al., 2003a).

8.6.3 Red Sea

Hasan (2005) investigated the reproductive cycle of at Abu Rhamada Island over several years, from the beginning of a fishery to population collapse (Fig. 13B). Between 1999 and 2001, individuals had two clear spawning seasons, illustrated by a small peak in April and a prolonged period from June to August (with maximum activity in July). The gonad index of females was16.5 in July of 1999, 16.0 in July of 2000 and 8.5 in April of 2000, and it began to decrease in corresponding months in 2001. In 2002, the reproductive cycle changed, showing only one shorter spawning season in June and July and even lower gonad indices (4.5 and 6, respectively). Despite the peak was still visible in April 2000 and 2001 it totally disappeared in 2002. In 2003 there was no evidence of spawning and reproduction in Abu Rhamada Island collapsed. Males exhibited the same gonad index patterns as observed for females above, but with slightly lower values (Hasan, 2005).

Along the coast of Sudan, Ali et al. (2015) showed that the seasonal reproductive cycle of *H. scabra* culminated in spawning at the end of summer based on the higher gonad index in July–August and the minimum in September. This aligned with initial reports of *H. scabra* spawning between June–August in the Red Sea (Boolootian, 1966). Ali et al. (2015) also noted that the ovaries became overall heavier than the testes.

8.7 Endogenous control of gametogenesis

The gonadal levels of RNA and DNA were found to vary in different states of gonad maturity (Krishnan, 1967). In males, both RNA and DNA became more abundant as maturation progressed, while in females, maturation was associated with increasing RNA and decreasing DNA (Krishnan, 1967). Krishnan (1968, 1970) and Krishnan and Krishnaswamy (1989) observed that accumulation of organic components occurred in the gonad when they reached maturity. Proteins and lipids were found to be important storage materials. Intestinal proteins and lipids were utilized by the gonads during the reproductive cycle (Krishnan, 1968, 1970). Conversely, Morgan (1999) suggested that *H. scabra* would be able to use nutrients from gonad and gametes for somatic metabolism during food deprivation.

Chaiyamoon et al. (2018) examined the distribution and expression of serotonin and dopamine in the nervous system during various ovarian stages of *H. scabra*. The serotonin-immunoreactivity was more intense in the neurons and neuropils of the outer epineural (ectoneural) part, the inner hyponeural part, and the wall of hyponeural canal of radial nerve cord during the mature stages, whereas dopamine-immunoreactivity was detected at a higher intensity during the early stages. Both neurotransmitters were detected in the epineural part of the nerve ring. In the ovary, serotonin was more intense in the cytoplasm of late oocytes, while dopamine immunoreactivity was more concentrated in the early stages. The changes in serotonin in the radial nerve cord and oocytes were incremental towards the late stages of oocyte maturation. Dopamine in the nervous tissues and oocytes were more intense in early stages and decreased in the more advanced stages. These results suggest that serotonin and dopamine may have opposing effects on oocyte development in *H. scabra*.

Thyrotropin/gonadotropin-releasing hormone (TRH/GnRH) was described as a neurohormone playing important roles in the regulation of physiological processes in echinoderms. Chaiyamoon et al. (2020a) characterized for the first time a TRH/GnRH-like peptide in *H. scabra*. They isolated a full-length cDNA for a TRH/GnRH preprohormone gene from the radial nerve cord. The (hs)TRH/GnRH precursor comprised two types of peptides: the short TRH/GnRH-like peptides and the long TRH/GnRH-like peptides. The expression of *hsTRH/GnRH* gene was strong in radial nerve cord, nerve ring and mature ovary. By in situ hybridization the transcripts of *hsTRH/GnRH* were detected at a high level within the late gametogenic stages. Injections of the short TRH/GnRH-like peptides could accelerate gonad development as reflected by increased gonadosomatic index, and stimulated gametogenesis, as reflected by increased oocyte proliferation (Chaiyamoon et al., 2020a).

Thongbuakaew et al. (2021) indicated that sex steroids, including estrogens, androgens and progestogens, play an important role in the reproduction. They demonstrated that estradiol, progesterone, and testosterone were present in the neural and gonad of *H. scabra*. Gene expressions were similar in ovaries and testes, and they increased during gonad maturation. HscCYP17 mRNA rose during ovarian development and its expression declined at late stages. The expression of the HscCYP3A was high at the early stages of ovarian development, but not at other later stages, whereas it remained low in testes. Overall, this study showed the presence of steroids and identified genes associated with the steroidogenic pathway, tying them to the reproduction of *H. scabra* (Thongbuakaew et al., 2021).

Abayomi (2020) conducted a study that focused on GFSKLYFamide, a neuropeptide that belongs to the Echinoderm SALMFamide group, and determined that GFSKLYFamide-like immunoreactivity was expressed in the sub-epithelial fibres of the coelomic epithelium of the female gonad of *H. scabra*. It was suggested that further studies should be carried out to investigate if there is similar existence of GFSKLYFamide neuropeptide in the male gonad of *H. scabra* and also to assess the structure or activity of the peptide(s) producing GFSKLYFamide-like immunoreactivity in gonads of *H. scabra*.

Thioredoxin, as a redox-active protein, is thought to reduce target proteins within the oocyte membrane and initiate an intracellular reaction cascade that releases the block on oocyte meiosis (Léonet et al., 2019). Delroisse et al. (2021) used pharmacological agents known to have activating or inhibiting actions on oocyte maturation (forskolin, isobutylmethylxanthine, hypoxanthine, 6-dimethyaminopurine, lavendustin, genistein, roscovitine, cycloheximide) to assess additional steps in the intracellular reaction cascade induced by the action of thioredoxin on sea cucumber oocytes. The effects of these agents were analysed mainly on oocytes of *Holothuria tubulosa* but also those of *H. scabra*. Their results demonstrated that, contrary to dithiothreitol-induced oocyte maturation, thioredoxin-induced oocyte maturation is cAMP independent, but depends on the presence of calcium in seawater (Delroisse et al., 2021). Both pathways of induction require the activation of protein serine/threonine kinases.

8.8 Exogenous control of gametogenesis

Purwati (2006b) reported that the bi-annual peaks in gonad indices coincided with the end of each monsoon season in Indonesia, whereas Battaglene (1999a) and Battaglene et al. (2002) noted that the peak of the reproductive season was towards the end of the dry season in the Solomon Islands, suggesting that both salinity and temperature fluctuations could provide the proximal cue that synchronize and regulate the seasonal reproductive cycle. Similarly, the presence of a period of enhanced spawning (September to December) was consistent over 3 years of study in the same region (Ramofafia et al., 2003a) and this period coincided with the dry season preceding the late-December onset of the monsoon, and with increasing day lengths and water temperatures (Ramofafia et al., 2000, 2001). While Krishnaswamy and Krishnan (1967) suggested that salinity variations could best explain the yearly reproductive cycle of *H. scabra* in India, Conand (1989, 1993, 1994) observed a temperature effect during the warmest and coolest periods in New Caledonia, but found no clear correlation between salinity and the reproductive cycle.

From a study in Australia, Morgan (2000a) suggested that the onset of gamete maturation in *H. scabra* may be correlated with a change in photoperiod and that further development could be associated with food availability and temperature. In the Philippines, gamete synthesis increased with temperature during summer and decreased at the end of the year when the temperature dropped (Ong Che and Gomez, 1985). In Madagascar, Rasolofonirina et al. (2005) specified that temperature appeared to be involved in synchronizing gonad maturity in females, but not in males, and they suggested that other factors such as daylength or salinity may influence the annual reproduction (the warmer period in Toliara coincides with the rainy season). Although the pattern of gonad growth was not perfectly aligned with temperature and light fluctuations, monthly gonad growth was correlated with both along the coast of Kenya (Muthiga et al., 2009). Since temperature and light show strong seasonal patterns on the East African coast (McClanahan, 1988), the investigators suggested that these factors ensured reproductive synchrony in *H. scabra* in that region. The temperatures occurring in Kenya during the months of spawning was >28°C (Muthiga et al., 2009).

Recently, Arsad et al. (2020) investigated the influence of specific physicochemical parameters (temperature, salinity, chlorophyll *a*, and organic matter) on the reproductive biology of *H. scabra* at two sites in Sabah (Malaysia) using gonad indices. An annual reproductive pattern was recorded in Kudat, with the spawning period from July to November 2015, whereas continuous pattern was recorded in Kunak, with a maximum gonad index in September 2015; however, no correlation with the physicochemical parameters were detected at either site.

Finally, Mercier et al. (2000b) reported that the lunar phase played a role in the pairing behaviour and gametogenic synchrony among populations of *H. scabra* in the Solomon Islands, suggesting that gamete release is also likely to be influenced by the lunar cycle, which appears to stimulate aggregation and spawning around the new moon.

8.9 Spawning

8.9.1 Spawning behaviour

The spawning behaviour of *H. scabra* is described quite uniformly in the literature, although very few details are given to differentiate the sexes

(MPEDA, 1989). During spawning individuals lift the anterior part of their body and initiate a sweeping movement. Battaglene et al. (2002) observed that pre-spawning behaviour included rolling and twisting on the substrate; rhythmic contractions, starting at the cloaca and working along the body, were common. Males raised the anterior end of their bodies perpendicularly prior to spawning and often swayed from side to side, or in a circular motion, as they released sperm. At the same time, the genital papilla, which is located anteriorly on the dorsal surface, expands (Conand, 1993). The anterior region of the female body inflates due to accumulation of gametes in the gonoduct (James et al., 1994b; James and James, 1993). James et al. (1994a) mentioned that the male aquapharyngeal bulb (pharynx) was extended during gamete release. During spawning, sperm or oocytes are moved by contraction and ciliary action in the gonoduct and released through the genital pore (gonopore) (James, 1989a; Mary Bai, 1980). The sperm or oocytes are discharged as a continuous stream, a white cloud that is not always easily visible (Conand, 1989, 1993). However, some reports indicate that gamete broadcasting differs between males and females. For example, males have been observed to release sperm continuously, whereas females release oocytes by powerful intermittent jets or spurts over about 2–3 h (Battaglene, 1999a; Battaglene et al., 2002; James and James, 1993; James et al., 1994b; MPEDA, 1989). James (1996b) indicated that females expelled their gametes in one or two spurts. James (1994h) gave sperm release duration of 30–60 min, whereas MPEDA (1989) observed that male spawning lasted about 2 h. James et al. (1988) mentioned that males spawned during 15–20 min and Lokani (1995) indicated that females released oocytes in cycles of about 5 min. Battaglene et al. (2002) showed that males spawned before females, and spawned more often and for longer. Males spawned soon after being placed in the spawning tank and released a steady stream of sperm from a single dorsal genital papilla at the anterior end. Males remained erect and spawned continuously for up to 3 h. They usually resumed spawning when removed to smaller containers or disturbed. Females typically started spawning about an hour after the first male, by raising their bodies off the substrate, in a similar manner to males. They remained erect for 15–30 min before releasing a short powerful burst of pale yellow to pink eggs, which lasted for 10–15 s. Immediately following spawning, females returned to a horizontal position, but typically spawned two to five times over a period of 2–3 h. Shelley (1981) observed isolated spawning in *H. scabra* at 10:00 h in the field and spawning in tanks from 15:00 to 19:00 h.

The released oocytes are spherical, buoyant, white and visible to the naked eye (James and James, 1993; James et al., 1988), although James (1994d,e) also noted that oocytes are light yellow when spawned (see also Section 8.5 above).

8.9.2 Environmental cues of spawning

In studies conducted in New Caledonia (Conand, 1990, 1993, 1994), the Philippines (Ong Che and Gomez, 1985), or Indonesia (Tuwo, 1999), investigators mentioned that temperature variation was the most probable spawning cue in *H. scabra* (Table 8). In contrast, salinity variation has been proposed as the major factor triggering gamete shedding in India (Krishnaswamy and Krishnan, 1967). Moiyadeen (1994) found that a mix of salinity and temperature variations as well as food availability could trigger spawning of *H. scabra* in Sri Lanka. In Oman, spawnings in 2009 and 2010 both took place just prior to the maximum sea surface temperature recorded (28°C) at a time when precipitations of 10mm were recorded (Al-Rashdi and Claereboudt, 2018).

MPEDA (1989) mentioned that oocytes were mostly released around 15:00h, whereas Battaglene (1999a) and Battaglene et al. (2002) reported that spontaneous spawners released their gametes between 15:00 and 18:00h, with males spawning first. Both James and James (1993) in India, and Shelley (1981) in Papua New Guinea suggested that spawning of *H. scabra* in the wild occurred in the day (between 10:00 and 16:00h) and during the new moon; observations in holding tanks concurred. Battaglene (1999c) also observed that a good proportion of individuals spawned 2 days after the new moon in outdoor tanks in Solomon Islands. In the same region, Mercier et al. (2000a,b) noted that newly-settled pentactulae were present on seagrass leaves before the full moon, suggesting that spawning and fertilization had taken place around the full moon. In Indonesia Rahantoknam (2017) also reported that the most developed gonad occurred at the full moon of March and May. Lee et al. (2018a) observed *H. scabra* spawning in Fiji around 14:00 and 17:00 on 28 October 2015 and 15 December 2015, respectively; these events coincided with spring tides and the full and new moon, respectively.

Although numerous studies have concluded that spawning occurs year-round with more or less pronounced peaks, cues for synchronous spawning of males and females, which would optimize fertilization, have rarely been investigated. One indication of synchronous spawning is aggregative behaviour prior to spawning. Such behaviour was reported in the

Solomon Islands, where it is correlated with the lunar cycle (Hamel et al., 1999; Mercier et al., 2000b). The aggregative behaviour of *H. scabra* suggests that chemical communication may be occurring. In the observations from the Solomon Islands, most individuals remained away from each other in the absence of a moon, and started to form pairs, trios and larger groups progressively after the new moon. Aggregation peaked a few days before the full moon when >95% of the individuals participated, and subsequently decreased until the next new moon. Most of the time, pair formations were more common than larger aggregations, with no progressive pattern or correlation to sex. Male and female spawnings were observed in outdoor tanks, typically around the full moon. Spawning occurred both during the peak in aggregation and later on, when the individuals were more evenly distributed (Mercier et al., 2000b). No aggregation was noted in individuals <110 mm in length. Conversely, VandenSpiegel et al. (1992) mentioned that they never observed any aggregative behaviour in *H. scabra* populations from Papua New Guinea. Support for the need to reach certain densities for reproduction to occur was provided by Hasan (2005) who showed that the population of *H. scabra* at Abu Rhamada Islandin in the Red Sea spawned biannually from 1999 to 2001 (before and at the onset of fishing), and only once in 2002, following high fishing pressure, and that it had ceased completely in 2003. Reduced densities of *H. scabra* may thus inhibit the ability of this species to spawn or decrease the chances of successful spawning (fertilized oocytes), further impeding population recovery and increasing the risk of an Allee effect (Bell et al., 2008).

At the finest scale, there appears to by a slight asynchrony between the sexes, evoking the possibility that spawning itself may act as a cue. In India, James et al. (1994a,b) and MPEDA (1989) reported that males released their gametes first, followed by females minutes or hours later. James (1989a) and Lokani (1995) also observed that males of *H. scabra* spawned first, suggesting that spermatozoa could induce nearby females to spawn, although this was never demonstrated. In a contrary report, Krishnaswamy and Krishnan (1967) noted that males and females of *H. scabra* spawned simultaneously.

8.9.3 Artificial induction of spawning

Many investigators, including James (1994h, 1996b), observed that individuals of *H. scabra* collected during the breeding season may release their gametes without any external stimulus other than the stress of collection and transport. Nonetheless, thermal shock is the most widespread technique used to induce *H. scabra* to spawn. Typically, mature individuals are transferred to water that is 3–5°C warmer than that in the original holding tank, and maintained at this temperature for several hours, if necessary (Battaglene, 1999a, 2000; Battaglene et al., 1998; James, 1994e,g, 1996b; James and James, 1993; James et al., 1988, 1994a; Morgan, 1999, 2000b; MPEDA, 1989). Although results seem to vary with the protocol of thermal stimulation, the seasonal gonad maturity and lunar periodicity, the most reliable period for inducing spawning artificially was September in the Solomon Islands (Battaglene, 1999b, 2000; Battaglene et al., 2002) and October to January in Australia (Morgan, 2000b). It was reportedly easier to induce spawning in males (Battaglene, 1999b, 2000; Battaglene et al., 2002). It has also been reported that thermally shocked females could be induced to spawn even in the absence of sperm, but that males usually spawned first, suggesting that thermal stress might trigger male spawning and that females are stimulated by the presence of sperm in the water column (James, 1994e). Ramofafia et al. (2000) and Battaglene (1999b) were able to trigger spawning in mature females by adding a solution of dried algae (*Schizochytrium* sp.) to the holding tank. Spawning induction using exposure to air and a jet of water has also been used, showing that *H. scabra* can survive in dry containers and only eviscerate on average after 9 h at 30°C (Ong Che and Gomez, 1985). Stripping gonads was reportedly effective only to a certain degree (James, 1994h, 1996b; James and James, 1993), although Mercier and Hamel (unpublished data) and Battaglene (unpublished data) had no success when gametes collected by stripping gonads were used in fertilization trials.

Battaglene et al. (2002) indicated that it was easiest to induce spawning in *H. scabra* in September in the Solomon Islands, when their gonosomatic growth peaked during the dry season (see Section 8.6), and hardest during the (rainy) season commonly extending from December to April. However, they also reported that spawning of *H. scabra* in Solomon Islands can be induced artificially throughout the year. Spawning appeared to be entrained by a lunar cycle (see the role of environmental factors and timing above) and occurred more readily during the afternoon and early evening. The same study revealed that *H. scabra* spawned at each lunar phase but more individuals (18.4%) spawned during the last quarter moon than the new moon (7.0%), the first quarter moon (13.2%) and the full moon (10.5%) (Battaglene et al., 2002). Morgan (2000b,c) also reported that spawning induction was most effective around new or full moons at dusk in Australia. Further discussion related to spawning and fertilization in applied settings can be found in Section 21.

9. Development

General trends in the development of *H. scabra* are relatively uniform throughout the existing literature (Table 9) although many oddities and discrepancies appear in the fine details. Development starts immediately after the release of oocytes, fertilization occurs in the water column and larval development is planktotrophic with feeding auricularia larvae transforming into non-feeding doliolariae before initiating settlement as pentactulae. Fig. 14 schematically illustrates the life cycle of *H. scabra* from spawning to post-settlement.

A number of studies pertaining to the development of *H. scabra* have been conducted by Indian researchers. James (1996b) observed that spawned oocytes sank to the bottom, while James et al. (1988) stated that development took place at the surface of the water column. James and James (1993) noted that first cleavage occurred 15 min after fertilization and that the first polar body appeared after 20–30 min, although we find this sequence very unlikely. The gastrula (dipleurula) measured 190–256 μm (James et al., 1988) and was described as a mobile stage (James and James, 1993). After 48 h, the larvae reach the auricularia stage and they are fully formed after 5 or 6 days (James, 1994h, 1996b; James et al., 1994b). The auricularia is a slipper-shaped, transparent and pelagic larva that possesses a preoral loop anteriorly and an anal loop posteriorly; there are a number of pigment spots

Table 9 Typical embryonic development of *Holothuria scabra*.

Stage of development	Time
First cleavage	15 min
Four-cell stage	20 min
Early blastula	40 min
Fully formed blastula	3 h
Gastrula	24 h
Early auricularia	48 h
Late auricularia	5–6 days
Doliolaria	10 days
Pentactula	13 days
8 mm long juveniles	1 month

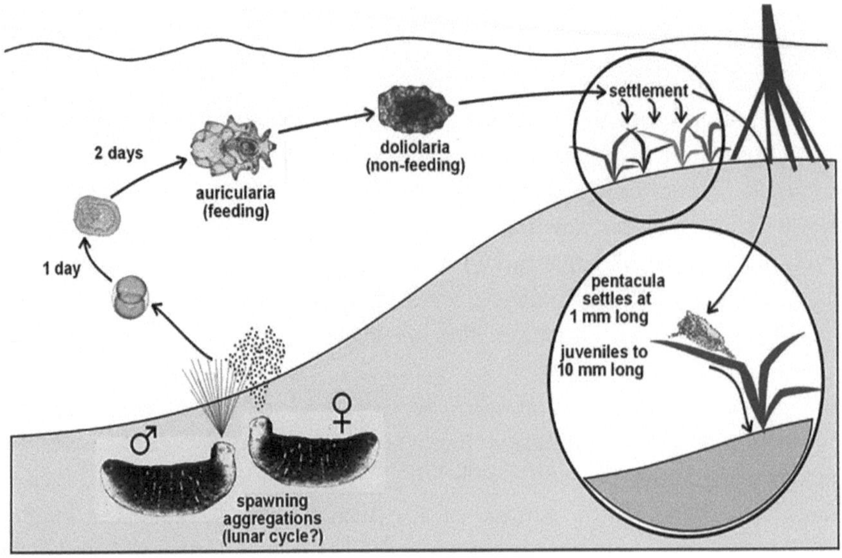

Fig. 14 Schematic drawing of the life cycle of *H. scabra*. *Courtesy of S. Battaglene; drawing by S. Pallay, ICLARM.*

on the anal section (James, 1994h, 1996b; James and James, 1993). At this stage, the digestive tract consists an elongated pharynx and sacciform stomach, enabling the larvae to feed on planktonic matter (James et al., 1988). Auricularia larvae measure about 563 μm at the early stage and 1.1 mm in length when fully developed (James, 1994h, 1996b; James and James, 1993). They are 240–690 μm wide (James et al., 1988). Duy et al. (2016a) demonstrated the importance of the development of lipid/hyaline spheres at the larval stage. In larvae of *H. scabra*, hyaline spheres are prominent in late auriculariae and early doliolariae but subsequently disappear near or during metamorphosis (Ramofafia et al., 2003b). On this basis, competent doliolariae are generally considered to have prominent hyaline spheres (Ramofafia et al., 2003b) and their presence and size is considered an indicator of larval competence (Battaglene, 1999a; Chen et al., 1991; Ramofafia et al., 2003b; Smiley et al., 1991). Despite Duy et al. (2016a) mentioned their disappearance, here we need to caution that the spheres are likely becoming compacted in the doliolaria larva and are still present in the pentactulae [as seen in *Isostichopus fuscus* (Mercier and Hamel, 2013; Figs 4D and E, 14)]. This is suggesting that they are not completely used/resorbed during that step and that they could provide energy to the early juveniles during the first days of their benthic life. Duy et al. (2016a) showed positive correlations between hyaline sphere development in late auriculariae on days 10, 11, 12 and 13 post-fertilization, and the proportion of competent doliolariae on day 15, post-settlement size (day 21) and post-settlement survival (day 25). Dietary components that influenced these relationships were carbohydrates and the n-3 polyunsaturated arachidonic acid (ARA), eicosapentaenoic acid (EPA) and docosahexaenoic acid (DHA). They confirmed the relationship between hyaline spheres formation in late auriculariae and subsequent larval competency through settlement (Duy et al., 2016a).

Continuing with data from India, the doliolaria has been described as a barrel-shaped larva with five bands around the body and two projecting tentacles (James, 1994h, 1996b; James and James, 1993). At this stage, larvae are still pelagic and measure between 460 and 620 μm in length (James, 1994h, 1996b; James and James, 1993; James et al., 1988, 1994a) and 240–390 μm in width (James et al., 1988). Larvae are generally said to metamorphose into pentactulae after 13 days, although James (1996b) indicated that metamorphosis started between the auricularia and doliolaria stages, 10 days after fertilization. The pentactula is described as tubular with five oral tentacles and a single stumpy tube foot at the posterior end (James, 1994h, 1996b; James and James, 1993). The cloacal opening is distinct, and the colour of the body is greenish brown. At this stage, the propagules become benthic. The pentactula is about 600–700 μm long (James and James, 1993; James et al., 1994a) and 250–400 μm wide (James et al., 1988). Finally, 18 days after fertilization, the tube feet and tentacles become more distinct, including two large tube feet at the posterior end. The spires of the tables (ossicles) project from the body wall (James et al., 1988).

Studies in Solomon Islands are consistent with the above reports from India, but the time frame differs. In particular, the blastula forms within 40 min and the whole planktonic cycle lasts 10–14 days at 27°C (Battaglene, 1999a,b, 2000; Battaglene et al., 2002). The pentactula forms 13 days after fertilization and marks the transition from planktonic to benthic life style (Ramofafia et al., 2000), as corroborated by most studies (MPEDA, 1989) (Table 9). Battaglene, 1999a, 2000 and Battaglene et al. (2002) indicated that the presence of distinct hyaline or lipid spheres at the late auricularia stage appeared to be a good indication of larval competency for metamorphosis (refer to the larval rearing in Section 21 for additional details).

10. Settlement

Studies conducted in the laboratory have examined settlement preferences in *H. scabra* (Mercier et al., 2000a). Independent and paired choice experiments revealed that leaves of the seagrass *T. hemprichii*, with or without their natural biofilm, yielded the highest settlement rates (4.8–10.5%). In the study, *T. hemprichii* was preferred as a settlement substratum over sand, crushed coral, several other plant species and artificial seagrass leaves with or without a biofilm. Only settlement on another seagrass, *E. acoroides*, was similar to that recorded for *T. hemprichii* (Mercier et al., 2000a). In the absence of a suitable substratum, the larvae delayed settlement for nearly 96 h and survival was <0.5%. Sand and crushed coral, either alone or together, induced settlement in <1.5% of the available larvae. The pentactula found on sand, coral and in bare containers were 10–35% smaller than those on *T. hemprichii* leaves. Introduction of soluble extracts from *T. hemprichii* and *E. acoroides* successfully induced metamorphosis and settlement onto clean plastic surfaces (Mercier et al., 2000a).

Field studies conducted in the Solomon Islands by Mercier et al. (2000a,b) showed that several individuals of *H. scabra* <10 mm were found on leaves of *E. acoroides* and *T. hemprichii* but not on other substrata. The highest density of newly-settled and early juveniles, corresponding to the smallest average size of 1.7 mm, were found typically before the full moon. By contrast, the lowest densities, comprising individuals of 5.4 mm, were usually observed 12–16 days after the full moon. This pattern was observed each month, except between September and November in 1998. Also, samples from December 1998 and January and February 1999 revealed very low or zero recruitment (Mercier et al., 2000b). The occurrence of other fauna on the seagrass, including numerous potential predators such as polychaete worms, tectibranchs, crabs, gastropods, copepods, shrimps, flatworms and isopods, varied monthly in number and diversity. Maximum abundance of potential predators was correlated with peak recruitment of young individuals of *H. scabra* and could explain their rapid decrease in abundance (Mercier et al., 2000b).

Settlement and post-settlement processes of *H. scabra* have also been studied as part of efforts to develop aquaculture methods (see Section 21 below for more details). Most studies indicate that settlement occurs at the pentactula stage (MPEDA, 1989), but some research teams indicated that

H. scabra settled at the doliolaria stage (James and James, 1993; James et al., 1994a). In the latter studies, settlement was stimulated by adding extracts of finely filtered *Sargassum* to the tank so that larvae settled on polythene sheets. James (1996b) also indicated that pentactulae settled on hard substrata with a diatom and bacterial film in the laboratory. Battaglene (1999a), Battaglene and Bell (1999) and Battaglene et al. (1999) found that *H. scabra* larvae settled onto plates conditioned with diatoms and biological films. Battaglene (1999b) noted that *H. scabra* settled on conditioned plates about 2 weeks after fertilization. The apparent discrepancy regarding the settling stage (doliolaria or pentactula) is probably related to misidentification of the stage rather than actual variability in the settlement, which occurs when the larvae develop oral tentacles and a tube foot; these features appear at the pentactula stage (see Section 21 for photos of both stages). Nevertheless, some pre-settlement behaviour may be exhibited by the doliolaria, moving up and down the water column and touching the substratum on a regular basis, pontentially adding to the confusion when there is a short transition time between the doliolaria and pentactula.

11. Growth

Numerous references to the growth rate of adults and juveniles of *H. scabra* recorded under laboratory conditions can be found in the literature (see Section 21). Average values range from 0.07 to 1.5 cm $month^{-1}$ in rearing tanks, depending on initial size and growth period (Table 10; see further information in Section 21).

In contrast, studies of growth in the field are scarce. Mercier et al. (2000b) periodically measured hatchery-reared juveniles released in the wild and found that growth rates over 2 months were dependent upon substratum type: the length of the individuals, which initially averaged 6.5 cm, increased by 385% on sand, by 327% in seagrass and by 252% on shells and crushed coral. This roughly corresponds to a growth rate of 10–15 cm $month^{-1}$ (Mercier et al., 2000b), which is 10–30 times the usual growth rate found in captivity (Table 10). Manikandan (2000) reported that hatchery-reared juveniles of *H. scabra* measuring 1.5 cm reached 10 cm after 6 months spent in an enclosed lagoon.

According to Long and Skewes (1997) *H. scabra* individuals ~18 cm long are ~2 years old and James et al. (1994b)) and MPEDA (1989) reported that *H. scabra* could live ~10 years. However, more studies are needed to establish a growth curve in the field over the entire life cycle of the species.

Table 10 Growth rate of *Holothuria scabra* according to different studies.

Conditions	Initial size	Average growth rate	Period	References
Laboratory	–	1.4 cm month^{-1}	7 months	Anonymous (1978a)
Laboratory	100–250 mm	0.5 cm month^{-1} 14 g month^{-1}	–	Shelley (1985a,b)
Laboratory	0 mm 136 mm 225 mm 284 mm 322 mm	1.13 cm month^{-1} 0.74 cm month^{-1} 0.49 cm month^{-1} 0.32 cm month^{-1} 0.22 cm month^{-1}	1 year 1 year 1 year 1 year 1 year	James et al. (1994b), MPEDA (1989)
Laboratory without food Laboratory with food	18 g 18 g	0.4 g day^{-1} 0.9 g day^{-1}	– –	Mulian (1993)
Laboratory	0 cm	0.07 cm month^{-1}	1 year	James (1994g)
Laboratory	Pentactula	0.1 cm month^{-1}	30 days	James et al. (1994b)
Laboratory	0 cm	0.7 cm month^{-1}	1 month	James et al. (1994a)
Fibreglass and concrete outdoor tanks	10 mm	0.5 mm day^{-1} 0.2 g day^{-1}	1–12 months	Battaglene (1999a,b,c, 2000), Battaglene and Bell (1999), Battaglene et al. (1999, 2002)
Field	65 mm	10–15 cm month^{-1}	2 months	Mercier et al. (2000a)
Field	15 mm	1.4 cm month^{-1}	6 months	Manikandan (2000)

Altamirano et al. (2017) examined the growth of juveniles in *H. scabra* using various substrates. During the peak hours of feeding (15:00–03:00), a greater number of juveniles preferred to feed on (21.5%) sandy mud, whereas silty mud was the least preferred substrate for feeding (13.5%). Juvenile feeding was not influenced by the presence of seagrass (*T. hemprichii*) on coarse sand. Growth rates of juveniles during the first 2 weeks were greater on coarse sand (0.59 g day^{-1} or 7.0% day^{-1}), followed by sandy mud (0.34 g day^{-1} or 4.72% day^{-1}) while organic matter content remained nearly unchanged. On silty mud, juveniles shrank, demonstrating negative growth rates (−0.02 g day^{-1} or − 0.63% day^{-1}) for 8 weeks despite organic content in sediment increased. In the field, silty mud of a mangrove pond caused mortality within 2 weeks, while sandy mud of a sand flat provided higher growth. Sandy mud to coarse sand substrates of sand flats were preferred by juveniles, while silty mud from muddy mangroves and culture ponds were deemed unsuitable (Altamirano et al., 2017).

Length-weight relationships were studied by Yussuf and Yahya (2020) in Zanzibar (Tanzania) by calculating correlation coefficient (R) values for male, female, and sexually indeterminate individuals. The calculated R was found to be larger than critical R, which indicates high correlation between length and weight for separate groups and pooled data. The coefficient of determination (R^2) value, which shows the percentage contribution of the independent variable to the dependent variable, ranged from 0.4 to 0.6 (Yussuf and Yahya, 2020). Previously calculated R^2 were generally higher, i.e. Lee et al. (2018a) (0.9), Al-Rashdi et al. (2007a) (0.8) and Natan et al. (2015) (0.4–0.7), from Fiji, Sultanate of Oman and Indonesia, respectively. The results of Yussuf and Yahya (2020) also showed that the length-weight power coefficient in male, female, indeterminate and pooled groups ranged from 1.29 to 1.69; values <3 indicate that the growth of *H. scabra* is negatively allometric. Importantly, length-weight relationships have been reported to vary with habitat, physiological condition, maturity stage and food availability (Froese, 1998; Natan et al., 2015). Moreover, variability could be attributed either to real differences existing between individuals in relation to the environmental conditions around them, or the inconsistency in procedures used to estimate length and weight between studies (Al-Rashdi et al., 2007a).

Ordoñez et al. (2021) performed de novo transcriptome assembly and characterization of fast- and slow-growing juvenile *H. scabra* from three Philippine populations. Analyses revealed 66 unigenes that were consistently differentially regulated in fast-growing sea cucumbers and found to be associated with immune response and metabolism. Further, they identified microsatellite and single nucleotide polymorphism markers potentially associated with fast growth. These findings provide insight on potential genomic determinants underlying growth regulation in early juvenile sea cucumbers which will be useful for further functional studies. The study explores the genomic basis of growth variation in juvenile by examining gene expression profiles of fast- and slow-growing early juvenile stages from three hatchery populations using RNA-seq. Sixty-six differentially regulated unigenes potentially related to growth variation are associated with several biological and molecular processes, including carbohydrate binding, extracellular matrix organization, fatty-acid metabolism, and metabolite and solute transport. A large number of potential microsatellite and growth category-associated SNP markers have been identified.

12. Evisceration and regeneration

Semper (1868) was the first to discuss the regenerative capabilities of *H. scabra*. Silver (2020) more than 150 years later described the autotomy area of the oesophagus and the mechanism of auto-evisceration. The autotomy area is characterized by a cell mass of early-stage lymphocyte-like cells in the connective tissue layer. However, Mary Bai (1971a) indicated that evisceration in *H. scabra* does not occur in nature, either spontaneously or seasonally. Nonetheless, we did observe one eviscerated *H. scabra* in the Solomon Islands during a field survey (J.-F. Hamel, unpublished data). Conand (1989) noted that 14.6% of *H. scabra* eviscerated during tagging procedures prior to field monitoring. Without giving any details, Leonardo and Cowan (1984), Sachithananthan et al. (1986), and Mary Bai (1994) observed that auto-evisceration in *H. scabra* was induced by collection and chemicals. Battaglene et al. (2002) stated that evisceration occurred in 4.5% of individuals during transport to the laboratory. Approximately 2 months are needed for *H. scabra* to regenerate its internal organs after evisceration (Cannon and Silver 1986). Mary Bai (1971a,b, 1994) reported that feeding started again 7 days after evisceration, although the alimentary canal was fully restored only after 13–35 days. The respiratory tree regenerated in about 19 days and the haemal system in 35 days (Mary Bai, 1971a,b).

13. Daily burrowing cycle

13.1 Adults

H. scabra generally moves in a sluggish manner and often remains partly or totally buried in the sediment. Battaglene et al. (1999) and Mercier et al. (1999, 2000b) showed that individuals in the Solomon Islands spent about half the day burrowed and the other half on the surface. The burrowing of adults in Palau extended from early in the morning until about 15:00 h, and individuals remained on the surface at night (Yamanouchi, 1939, 1956). A similar pattern was observed in India (Anonymous, 1978a). Yamanouchi (1939) explained this behaviour as a way to minimize daylight predation, although *H. scabra* is known to be toxic (Bakus, 1968; Halstead, 1988), and Rao et al. (1985a,b) and Bakus (1968) stated that predation on adults of *H. scabra* was low. Mercier et al. (1999) found that burrowed individuals did not show any detectable movement, and burrowing/emerging processes took about 30 min.

The burrowing cycle of *H. scabra* in the Solomon Islands varied according to environmental conditions (Mercier et al., 1999, 2000b) (Fig. 15): on a clear day, individuals <80 mm burrowed at sunrise and surfaced at sunset. Individuals >80 mm presented almost the same, but slightly dephased, burrowing pattern, with individuals beginning to surface earlier, in the middle of the afternoon. However, when it rained, 97% of individuals of both size classes remained burrowed all day (Fig. 15). On two occasions, the water temperature was found to increase to more than ~30°C (with a peak of ~37°C) and most individuals were found on the surface, remaining there until the temperature dropped below ca. 28–29°C, and they did not follow their usual burrowing cycle.

According to Yamanouchi (1939), the burrowing behaviour of *H. scabra* could be under neural control or under an internal clock. Yamanouchi (1956) determined the sensitivity of the anal region to light with a simple experiment and suggested that it played a role in the burrowing cycle. Mercier et al. (1999), in a similar experiment, modified the light pattern around juveniles of *H. scabra* and successfully triggered a new burrowing cycle.

Wolkenhauer (2008b) showed a distinct diel activity cycle under laboratory settings, with most individuals being exposed and active between the hours of 13:00 and 22:00 and most being buried and inactive between 01:00 and 09:00. When temperature decreased, fewer individuals were exposed and active, while more remained completely buried or partially buried. The trend did not change during the day, but rather the time spent buried lengthened, particularly in the morning (08:00–10:00), with only one individual out of six being buried at 24°C, compared to four out of six individuals being buried when the temperature reached 17°C (Wolkenhauer, 2008b).

The burrowing behaviour of holothuroids was hypothesized to stimulate benthic primary producers (Mercier et al., 1999; Wiedemeyer, 1992); the movement potentially aerating deeper sediment. It may release nutrients trapped in the interstitial waters, keeping sediments oxygenated and/or displace benthic microalgae (Wolkenhauer et al., 2010). However, there has been indication that burying behaviour also might have a negative impact on seagrass which may interfere with their growth and dispersion (Mosher, 1980) Differently, Wolkenhauer et al. (2010) showed that seagrass growth was affected by the exclusion of *H. scabra* in a field experiment, suggesting that seagrass systems may suffer in the absence of holothuroids

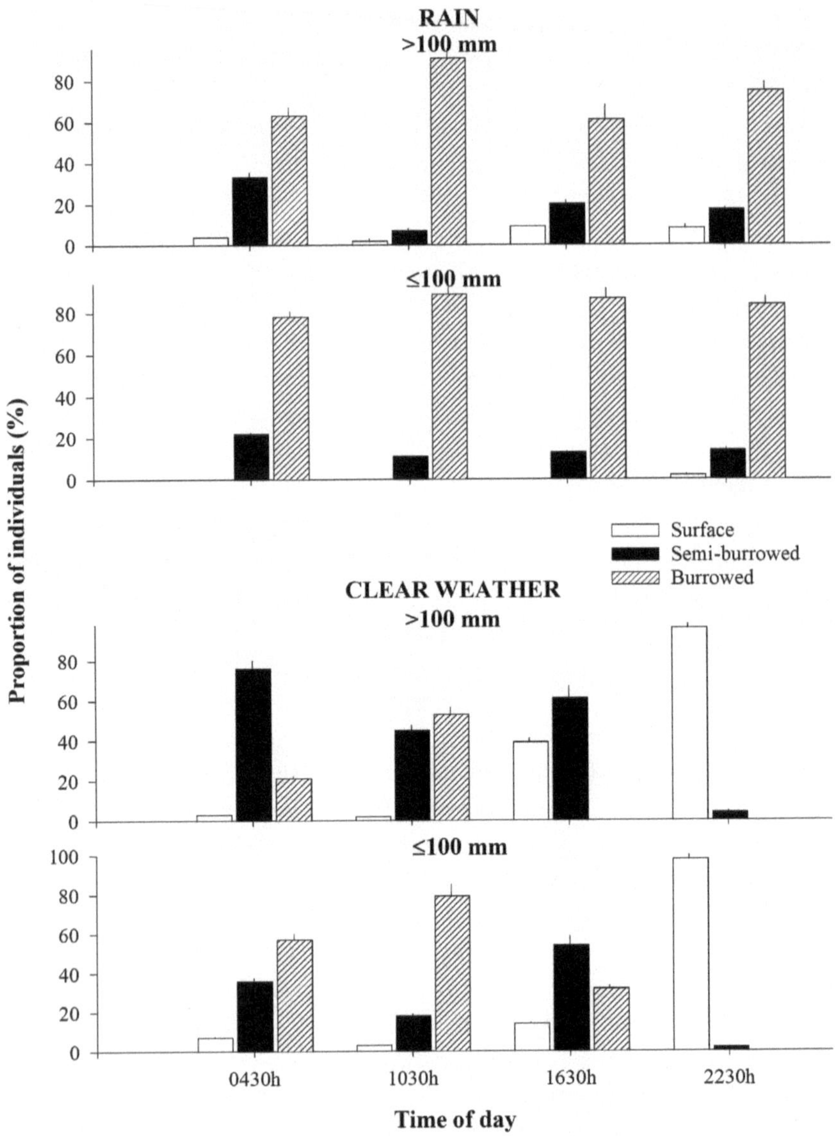

Fig. 15 Daily burrowing cycle of *H. scabra*, illustrating the behaviours in two size classes during rainy episodes and under clear weather conditions in the field. Data were recorded in October 1997 and February 1998 in Solomon Islands (mean ± SE, $n = 2$). *From Mercier, A., Battaglene, S.C., Hamel, J.-F., 2000b. Periodic movement, recruitment and size-related distribution of the sea cucumber* Holothuria scabra *in Solomon Islands. Hydrobiologia 440, 81–100, with permission.*

and also proposed that over-fishing of holothuroids could negatively impact the productivity of seagrass systems.

Skewes et al. (2000) indicated that none of the individuals of *H. scabra* held in aquaria fully burrowed during monitoring. More sea cucumbers were partially burrowed in the outside tanks (12.6%) and inside tank (5.2%); in contrast, most sea cucumbers were found burrowed in the field. This could suggest that other factors, such as tidal influence, which were not replicated in the aquaria, may be important factors influencing burrowing rates of *H. scabra*.

Yamanouchi (1956) observed the respiratory rate by counting the movements of the anus (cloaca) during different daily activities in *H. scabra*, and noted an increase when the animals were burrowed, suggesting that higher respiration rates might compensate for reduced diffusion of gases directly through the body wall. The frequency of inspiration between each spouting varied between 1 and 16 when the animals were on the surface, and it was around 3 in burrowed animals (Yamanouchi, 1956). Anal (cloacal) movements continued despite evisceration of all internal organs and loss of body fluids (Yamanouchi, 1939, 1956).

13.2 Juveniles

Mercier et al. (1999) and Battaglene et al. (1999) showed that the burrowing cycle of the smallest juveniles (10–40 mm) was linked to the light regime. These juveniles begin to burrow around sunrise and emerge close to sunset, and their burrowing behaviour is inhibited in continuous darkness (Mercier et al., 1999). Individuals 40–140 mm responded to changes in temperature, burrowing earlier around 3:00 h as temperature declined, and emerging at mid-day. Constant high temperature prevented their burrowing (Mercier et al., 1999). Burrowing and surfacing takes between 5 and 30 min.

A change in burrowing behaviour was observed when *H. scabra* individuals measuring 10–140 mm were exposed to decreasing salinity conditions (Mercier et al., 1999). In fact, when exposed to salinity ≤25 psu, juveniles of *H. scabra* burrowed rapidly in the sediment even when they would normally be at the surface. Acclimation was observed at salinities of 30, 25, and 20 psu, whereas lower values were beyond their tolerance threshold (Mercier et al., 1999). This burrowing behaviour was also observed in the field during the rainy season in the Solomon Islands (Mercier et al., 2000b) (Fig. 15). The intestinal transit time and intestinal index both decreased in all size classes of juveniles during salinity decreases (Mercier et al., 1999).

Several burrowed juveniles were observed to forcibly eject faecal pellets within the first hour. The intestinal index decreased within 30 min in small and medium juveniles but only after 10.5 h in large ones. The burrowing of juvenile *H. scabra* at low salinity may help them to better equilibrate ionic concentrations of coelomic fluid. Yamanouchi (1956) found that burrowed individuals of *H. scabra* exhibited fewer inhalations compared to wandering animals, suggesting that burrowing might reduce the entry of hypohaline water inside the respiratory tree.

Lowering of water level in the laboratory and in the intertidal also influences the burrowing behaviour (Mercier et al., 1999, 2000b). Specifically, a steadily decreasing level does not prevent the juveniles from emerging, but a constant low water level reduces emergence. This suggests that the juveniles need a few hours to determine whether the water level has become inappropriately low. However, any such mechanism does not apply to all individuals as nearly 40% of small juveniles were found on the surface during their 'normal' emergence period, even though they were barely covered with water (Mercier et al., 1999). Similarly, James (1994a) once reported that individuals living buried inside the sand came out during low tide to lie in a half-burrowed condition.

In New Caledonia Purcell (2010) examined the circadian burrowing cycle of juveniles of *H. scabra* in captivity, and both juveniles and adults in the wild. Small juveniles (1–5 g) displayed a diel cycle of epibenthic foraging in the afternoon and night, then burrowing in sediments in the morning. Burrowing was related to the combined effect of the light and temperature. Occasional or frequent handling suppressed afternoon emergence from sediments for 4 days (Purcell, 2010). In a coastal seagrass bed habitat, burrowing and emergence of *H. scabra* were monitored during three seasons. Adults rarely burrowed during the day except in the cool season. Most hatchery-produced juveniles were shown to burrow during most of the day, while larger juveniles showed little diurnal burrowing (Purcell, 2010). Combined with the results of Mercier et al. (1999) in the Solomon Islands, findings suggest that the circadian behaviours of sea cucumbers can exhibit substantial spatial variation and may be absent or different at certain sites or seasons, and can be mediated by a complexity of factors that vary over short timescales (Purcell, 2010).

Most individuals of *H. scabra* observed in the wild at low latitudes are found totally or partially buried in the morning and early afternoon (Mercier et al., 2000b; Yamanouchi, 1956). On the other hand, wild individuals in New Caledonia generally did not bury during the day, but small juveniles released into the wild did (Purcell, 2010).

Accordingly, Skewes et al. (2000) found patterns varying between sites in Torres Strait (Australia), with burial apparently circa-tidal at one site but not at another, and they speculated that habitat characteristics mediated this difference. Moreover, Wolkenhauer (2008b) found a diel cycle of afternoon and evening surfacing by adult in tanks.

Burrowing behaviour in *H. scabra* can be promoted by low seawater temperatures (Purcell, 2010; Wolkenhauer, 2008b), heavy rain or freshwater input (Mercier et al., 2000b), and potentially other environmental variables (e.g. barometric pressure or autocorrelated variables thereof). This behaviour, especially in juveniles, has been suggested to provide protection against predators (Mercier et al., 1999). Sheltering in sediments and inactivity by small individuals of *H. scabra* fit the avoidance strategy by reducing the probability of being detected by diurnal epibenthic predators (Phillips, 1977; Rochette et al., 1999). Burrowing and re-surfacing may both act to mitigate stress from environmental factors or biophysical perturbations (Purcell, 2010).

Altamirano et al. (2017) examined substrate preference for both burrowing and feeding of juveniles of *H. scabra* (3–6g wet body weight) in the Philippines and their associated daily behaviour, under laboratory and field conditions, using different substrate types such as silty mud, sandy mud, and coarse sand (Fig. 16). Peak hours of burrowing (03:00–09:00) and feeding (15:00–03:00h) were found, and a greater number of juveniles

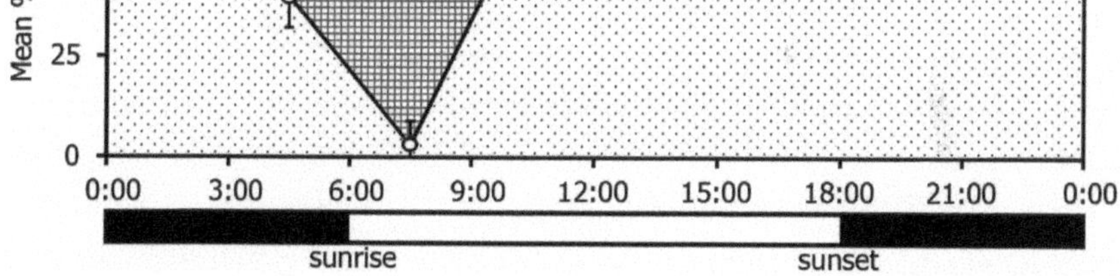

Fig. 16 Mean proportion (%, ± SE, $n=4$ sets of 15) of juveniles of *H. scabra* found at the surface of the sediment (emerged; light area) or burrowed in the sediment (buried; dark area) in the Philippines. Data consolidated into mean of 3 hourly ranges from a pool of 5-day observations. Sunrise and sunset times indicated on the bottom scale. *From Altamirano, J.P., Recente, C.P., Rodriguez Jr., J.C., 2017. Substrate preference for burying and feeding of sandfish* Holothuria scabra *juveniles. Fish. Res. 186, 514–523, with permission.*

preferred sandy mud (28.3%). Silty mud was the least preferred substrate for burying (13.8%).

The response of *H. scabra* to low water levels, combined with their ability to tolerate and acclimate to low salinity, may reflect their adaptation to shallow, tidal habitats. Data seem to demonstrate that smaller juveniles can better tolerate variable conditions than larger animals. The advantages of such adaptability may include reduced competition and predator avoidance.

14. Diet and feeding

14.1 Larvae

The planktonic larvae of *H. scabra* need to feed on pelagic microalgae to complete their development, although feeding is limited to the auricularia stage (Battaglene et al., 1999). The feeding mechanism of the auricularia larvae consists of conveying unicellular algae and suspended fragments of organic matter into the alimentary canal by ciliary movement (James, 1994h). The same observations were also made by James et al. (1988, 1994a,b), MPEDA (1989), James and James (1993) and James (1996b). Food accepted by the larvae was <40 μm, and up to 80 μm after about 2 weeks of growth (James and James, 1993; James et al., 1994b).

A number of investigators have examined the influence of microalgal diets on the growth and health of larvae in *H. scabra* in the context of aquaculture development. Full details about these studies can be found in the dedicated Section 21. For example, Makatutu et al. (1993) tested different species of algae and found that larvae grew more rapidly when fed with *Chaetoceros ceratosporum*; Morgan (1999, 2001) reported that the growth of larvae was also optimum when fed *C. cerastosporum*, followed by *Tetraselmis chuii* and *Nannochloropsis oculata*; whereas Battaglene (1999a,b) and Battaglene et al. (1999) obtained the best results with *Rhodomonas salina* followed by *Chaetoceros muelleri* and *C. calcitrans*. Using auricularia larvae of *H. scabra*, Jiang et al. (2014) studied the influence of photosynthetic bacteria on their growth, survival and digestive enzymatic activities as well as on water quality of the culture. They indicated that photosynthetic bacteria affected the activities of protease, amylase and cellulose in intestines of larvae higher than those of controls. Those results indicate that photosynthetic bacteria can not only help increase growth, survival rates and digestive enzyme activity of larvae in *H. scabra* but could also improve aquaculture water quality.

14.2 Juveniles and adults

H. scabra is generally categorized as a deposit feeder and diet descriptions are relatively uniform in the literature. When on soft bottoms, they are said to ingest large amounts of sediment from which they extract food (Baskar, 1994; Conand, 1994, 1999b). To do so, they use retractile tentacles, devoid of calcareous deposits, that can be withdrawn into the mouth (Mary Bai, 1980). *H. scabra* is also described as a detritus feeder (Morgan, 1996). Gut contents are generally composed of bacteria, copepods, diatoms and other algae, molluscans shells, foraminifera, sand and mud (Bakus, 1973; Baskar, 1994; James, 1996b; Mary Bai, 1980; Massin, 1982; Sipahutar and Soeharmoko, 1989; Wiedemeyer 1992). James (1989a, 1996a) suggested that *H. scabra* had no food preference.

When moving on compact substrata *H. scabra* are thought to feed only on deposited bacteria and algae on the surface without ingesting sediment (James 1996a). In the Solomon Islands, Battaglene and Bell (1999) observed that cultured juveniles fed on epiphytic algae and bacteria growing on the substratum. Battaglene et al. (1999) later observed that they fed on sand but also from the hard surfaces of the tanks.

Descriptions of the feeding cycle of *H. scabra* are comparatively disparate. Yamanouchi (1956) suggested that *H. scabra* fed at least one-third of the day by collecting food with their buccal tentacles while moving on the substratum. James (1989a, 1996b) specified that *H. scabra* was a continuous feeder, and James (1996b) further indicated that the peltate tentacles shovelled the sand and mud continuously into the mouth. However, Wiedemeyer (1992) reported that they stopped feeding when they were on the surface. Mercier et al. (1999) found that juveniles of *H. scabra* fed when moving, while burrowing and while emerging, but that they stopped ingesting sand when completely burrowed. These results are similar to those of Yamanouchi (1939, 1956) who observed that food transit in adults occurred mainly when they were on the surface. Similarly, Anonymous (1978a) noted that *H. scabra* generally came out of the sand for feeding in the Andaman Islands. Yamanouchi (1939, 1956) also observed that the intestinal content decreased when individuals were in the burrowing stage. More precisely 52% of the intestine was full of sediment in recently burrowed individuals, the value dropped to 32% when animals were burrowed, and increase to 87% when individuals were on the surface.

Mercier et al. (1999) found that the marked daily cycle and the intermittent digestive transit render intestinal indices unsuitable markers for in situ investigations of feeding status. This inadequacy may explain why

Wiedemeyer (1992) obtained variable intestinal indices in individuals of different drained body weight collected from the field in Japan, leading him to conclude that *H. scabra* fed during the night when burrowed and spent 8 h during the day without ingesting any sediment. He specified that small individuals had two short feeding periods, corresponding to the highest intestinal index values found after sunset and before sunrise (Wiedemeyer, 1992). The earlier studies of Yamanouchi (1939, 1956), which were mostly based on field observations of adults, more closely resemble the behaviour observed by Mercier et al. (1999): individuals of *H. scabra* began to burrow between 0200 and 0400 h, at which time they stopped eating and were not very active, emerging again between 1200 and 1800 h. Yamanouchi (1939, 1956) also pointed out that the cycle was not perfectly correlated with light and that the factors regulating it would be complex. In fact, the cycle of adults is likely to be similar to the one observed in larger juveniles and thus dependent on temperature variations (Mercier et al., 1999).

The highest feeding rate, 19.8–79.1 mg dw min^{-1}, was measured during the emerging behaviour of juveniles, and the maximum intestinal transit of 2.0–4.5 mm min^{-1} occurred just a few moments later. Complete passage of ingested material through the gut took between 30 and 60 min (Mercier et al., 1999). Yamanouchi (1939) estimated the time needed for the gut to fill was around 2 h for an adult. However, Wiedemeyer (1992) estimated that the transit was considerably lower at 2.6 cm h^{-1}. Following digestive transit, faeces are ejected from the cloacal opening in characteristic balloon-shaped strings. The egestion rate of material is irregular (Yamanouchi, 1939).

It is generally believed that *H. scabra* plays an important role in bioturbation through its feeding activities (Hammond, 1982a, b; Massin, 1982; Wiedemeyer, 1992). Maximum feeding rates found in juveniles of *H. scabra* measuring 40–140 mm were around 60–80 mg dw min^{-1} (Mercier et al., 1999). This would yield a daily ingestion of approximately 115 g dw sediment, or an annual intake of over 41 kg dw for a single individual. However, if the average feeding rate recorded during the active hours is used, we obtain a value of 22.1 ± 3.3 mg dw min^{-1}, a daily ingestion of about 32 g dw sediment or <12 kg dw processed annually. Other studies on adult *H. scabra* reported in situ values of 196 g dw ingested by an individual every day (Yamanouchi, 1939).

Battaglene et al. (1999) indicated that ~10% of the total weight of juveniles could be attributed to sand content in the intestine. The total amount of material found in the digestive tract could reach 5 g in adults and the pH level in a stomach full of material was around 6 (Yamanouchi, 1939).

Wiedemeyer (1992) stipulated that *H. scabra* ingested between 23% and 31% of its drained body weight every day. The assimilation of organic matter varied between 0.13% and 0.18% of drained body weight. Considering that the material ingested by different sizes of individuals varied and that feeding behaviour was not a constant process, Mercier et al. (1999) recommended that calculation of the assimilation of organic material should be performed using the period when the sea cucumber is on the surface. The proportion of organic matter (relative to total gut contents) was estimated by Baskar (1994) to be 2.1% in the oesophagus, 2.3% in the stomach and 1.7% in the intestine, suggesting that most absorption occurred in the stomach. Wiedemeyer (1992) and Mercier et al. (1999) both indicated that *H. scabra* ingested its own faecal pellets. Wiedemeyer (1992) also noted that there was 21% more organic matter in the pellets than in the surrounding sediment, whereas Mercier et al. (1999) found that the faecal material of *H. scabra* was alternatively enriched and depleted compared to surrounding sediment over their daily cycle, depending on whether or not active feeding was taking place. In fact, all components of feeding, including foraging activities, locomotive speed, feeding rate, intestinal transit, intestinal index and organic matter absorption efficiency were discovered to be dependent upon the time of the day (Mercier et al., 1999). Specifically, organic content was usually highest in the last third of the intestinal tract towards the end of the burrowing cycle, e.g. after 6 h spent burrowed or while emerging and settling on the surface. Organic content was typically higher in the first third of the intestine than in the last third during the feeding hours (Mercier et al., 1999). Wiedemeyer (1992) found that the organic matter content of the posterior intestine in adult *H. scabra* was enriched by 7% compared to surrounding sediment. However, his data were obtained from the combined results of 476 individuals of various sizes collected on different dates at different times during the day, without consideration for the marked daily cycle.

As discussed earlier in Section 6.4, Mercier et al. (1999) demonstrated that *H. scabra* was able to choose between different types of substrata: when given a choice, individuals preferred sand and muddy sand. Moreover, *H. scabra* was able to identify sediments containing higher organic content. Also, Plotieau (2012) and Plotieau et al. (2013) showed that the number of bacteria in the food of *H. scabra* increased by a factor of 1.5 in the foregut as compared to the external sediments, also suggesting that individuals are able to select bacteria-rich sediment. Some believe that *H. scabra* is also able to discriminate between sediments on the basis of grain size

(Wiedemeyer, 1992). For instance, Baskar (1994) found that gut contents of adults of *H. scabra* collected from the wild were dominated by muddy grains between 125 and 250 μm in diameter. However, his conclusions that *H. scabra* feeds selectively can be debated since gut contents were not compared with surrounding sediment; findings may reflect a substratum preference (for certain grain sizes) but not necessarily selectivity within a given substratum. Wiedemeyer (1992) indicated that the feeding mode of the different size classes of *H. scabra* varied mainly with the time of digestive transit, feeding cycles, particle size and organic matter assimilation. Claereboudt (2021) analysed Y-tube experiment choices of juveniles of *H. scabra*, demonstrating that they were attracted to the odour of their food, which suggests that the selective feeding of *H. scabra* is, at least partly, mediated by olfactory cues.

Wolkenhauer (2008b) showed a correlation between feeding activity and temperature under laboratory settings. Daily periods of feeding decreased by 9 h from 9.8 h at 24°C to about 0.8 h at 17°C. Modelling the daily feeding and burying cycle showed that the probability of feeding increased directly with increasing temperature, irrespective of time of day (Wolkenhauer, 2008b).

Plotieau et al. (2013) reported a reduction of the bacterial population, a depletion of fine grains (<250 μm in diameter), and reduced carbonate proportion (aragonite, calcite, magnesian calcite) in the sediment where *H. scabra* was found. They inferred that the adults consume mainly bacteria and, to a lesser extent, organic matter. Collard et al. (2014) showed that the dissolution of $CaCO_3$ into seawater was facilitate by the sea cucumbers which secrete digestive acids to dissolve minerals present in their digestive tract.

15. Digestion and metabolic activity

Yao et al. (2006) studied the main digestive enzymes (protease and amylase) in the gut of *H. scabra*, reporting that activity of amylase was higher than that of protease, consistent with the diet of micro-algae and detritus. Protease activity was highest at a pH of 3 in the foregut and midgut, but highest at a pH of 10 in the hindgut, whereas the activity of amylase peaked at a pH of 3 in the foregut and at a pH of 7 in the midgut and hindgut. The study showed that protein and starch can be digested in the whole gut, but the latter is mainly digested in the midgut and hindgut (Yao et al., 2006).

Zarate et al. (2012) also studied the digestive enzymes present in *H. scabra* and performed basic characterization to correlate their activity with nutritional status. The following enzymes were detected: alpha amylase, protease (as well as chymotrypsin), cellulase, mannanase, agarase, and xylanase. The presence of trypsin was inconclusive, and no alginate lyase or laminarinase were detected. The optimum pH of protease was reported to be 5 and that of alpha amylase to be 7. A 2-week fasting experiment showed changes in alpha amylase and protease levels, but no differences in cellulase, mannanase, agarose or xylanase activity (Zarate et al., 2012). Another experiment wherein intestines were sampled every 3 days for 15 days showed that only alpha amylase levels changed with fasting; all samples had lower alpha amylase activity after fasting began compared to day 0. These results suggest that alpha amylase activity could be used as an indicator of nutritional status, particularly food deprivation, in *H. scabra*.

Using a differential gas-volumeter method, Mukai et al. (1990) established that *H. scabra* had oxygen to nitrogen ratios ranging from 19.3 to 44.2, and that the total energy demand for respiration was between 6.56 and 12.29 kcal day^{-1}. Kodama et al. (2015) examined the oxygen consumption rate and ammonium excretion rate in *H. scabra* (in co-culture with shrimp). Both rates exhibited a negative correlation to body weight. The oxygen consumption rate of shrimp tank sediment was reduced considerably (4.5–1.0 mg O_2 g^{-1} dry h^{-1}) when measured before and after ingestion and excretion by sea cucumber. Acid volatile sulphide concentration was also decreased (0.7–0.3 mg S mg^{-1} dry), despite the reduction rates of organic carbon and nitrogen contents (0.19–0.14 mg C mg^{-1} dry and 0.022–0.019 mg N mg^{-1} dry, respectively). The authors suggested that components of sediment that had high oxygen consumption potential were removed by *H. scabra* (Kodama et al., 2015).

Kühnhold (2017) and Kühnhold et al. (2017) studied juveniles to determine the upper temperature limit at which growth and feeding activity would collapse. Controls held at 27°C were compared to individuals exposed to sub-lethal cold (21°C) and warm (33°C) temperatures over 30 days; the physiological responses to thermal acclimation, cellular energy allocation in combination with oxygen consumption rate and metabolic enzyme activity were monitored. Cellular energy allocation was shown to be a sensitive marker to detect changes in metabolic status (Kühnhold et al., 2017). Juveniles were affected by the initial temperature changes of ±6°C (1°C per day) prior to the start of the experiment and quickly recovered their energy balance (days 15–30) at both temperatures tested. Differently, oxygen consumption indicated an enhanced metabolism at warmer temperatures, supported by a shift from a mainly anaerobic to a more aerobic energy metabolism, combined with increasing foraging activity in warm acclimatized juveniles (Kühnhold et al., 2017). The authors concluded that the increased metabolic activity in juveniles of *H. scabra*, as a response to warmer temperature, was well within the homeostatic range, suggesting that a rearing temperature of 33°C could be presumably favourable.

Kühnhold et al. (2019a) later indicated that the aerobic scope (AS), which reflects the functional capacity for biological fitness, is a relevant proxy to determine thermal tolerance in *H. scabra*. Temperature dependent peak and bottom O_2 consumption rates, defined as temperature-induced maximal (TMMR) and standard metabolic rates (TSMR) were induced through acute temperature change (2°C per hour; range of 17–41°C) until critical warm (WTcrit) and cold (CTcrit) temperatures were reached. Hsp70 gene expression linked to respiration rates served as markers to confirm critical threshold temperatures. Oxygen (O_2) consumption peaked distinctly at WTcrit of 38°C (TMMR $= \sim 33.2\,\mu gO_2\,g^{-1}\,h^{-1}$). A metabolic bottom line was reached at CTcrit of 22°C (TSMR $= \sim 2.2\,\mu gO_2\,g^{-1}\,h^{-1}$). Within the thermal window of 22–38°C, the species sustained positive aerobic capacity, with assumed optimal performance range between 29 and 31.5°C (14–19 $\mu gO_2\,g^{-1}\,h^{-1}$). Between 39 and 41°C respiration decreased progressively, while gene expression levels of Hsp70 increased at 41°C, suggesting prioritization of heat shock response and homeostatic disruption. At the cold end (17–22°C) homeostatic disruption was detected through increasing energetic expenses to fuel basal maintenance costs, but no Hsp70 overexpression occurred (Kühnhold et al., 2019a). The investigators suggested that the study could help define lethal temperatures for aquaculture and predict the effects of environmental stress, such as ocean warming, in *H. scabra* (Kühnhold et al., 2019a).

16. Biochemistry and biotoxicity

16.1 Composition and nutritional properties

As described in Table 11, the gross biochemical composition of *H. scabra* is fairly well known (Sachithananthan, 1972; Springhall and Dingle, 1967). Krishnan and Krishnaswamy (1970) and Krishnan (1971) established the sugar pathway transport in different tissue compartments, showing the key role of the haemal system, amoebocytes and coelomocytes. *H. scabra* possesses a variety of carbohydrates, 1-2-glycol groups, acid mucopolysaccharides and muco-glycoprotein in their tissues (Krishnan, 1968). Elanganayagam et al. (1980) studied the nutritional composition of tissues

Table 11 Composition (%) of *Holothuria scabra* according to studies conducted in different countries.

Composition	Australia[a]	Egypt[b]	India[c]	India[d]
Moisture	1–4.9	27	27.1	15
Crude protein	31.5–58	43	43.1	65
Fat	1.2–2	2	2.2	1
Ash	43.7–79.9	–	27.6	11
Calcium $100\,g^{-1}$ tissue	6–12	–	1.2	–
Phosphorus $100\,g^{-1}$ tissue	0.1	–	0.7	–
Sodium $100\,g^{-1}$ tissue	5.2	–	2.5	–
Potassium $100\,g^{-1}$ tissue	0.6	–	–	–
Iron $100\,g^{-1}$ tissue	–	–	0.16	–
Insoluble ash	–	7	7	<0.5
Energy Cal kg^{-1}	2050	–	–	–
Minerals	–	21	–	–

[a]Springhall and Dingle (1967).
[b]Din (1986).
[c]Sachithananthan (1972).
[d]Durairaj (1982).

in *H. scabra*. Springhall and Dingle (1967) experimented with a diet based on boiled *H. scabra* and studied its pathology on the growth of chickens. Elanganayagam et al. (1980, 1982) and Mahendran et al. (1983) discussed the saponin (holothurin) content, while Sarma et al. (1987) isolated and described two sapogenin compounds and two glycosides from Indian individuals. Stonik et al. (1998) outlined the free sterol composition of *H. scabra* and Anjaneyulu et al. (1995) described a new steroid glycoside. Ridzwan et al. (1995) screened for antibacterial agents in individuals from Sabah and Kobayashi et al. (1991) reported several antifungal oligoglycosides. More recently, Assawasuparerk et al. (2016) showed that the body wall extract of *H. scabra* contains scabraside D, a sulphated triterpene glycoside, found to inhibit tumour growth in mice.

16.1.1 Proximate composition

When processed as beche-de-mer, *H. scabra* was determined to have a low water content of 13.8% (weight based) (Andriamanamisata and Telesphore, 2019). Comparatively, the moisture content of fresh sea cucumber is approximately 76–79% (weight based) of edible material (Oedjoe, 2017), which means that approximately 63 g of water is removed during treatment. This low water content corresponds to the standard requirement for their quality and conservation, 30% reportedly being the maximum tolerable level (Baird and Sachithananthan, 1974).

From a nutritional point of view, sea cucumbers are considered as ideal nutraceutical food sources, with higher protein and lower fat contents than most other seafood (Conand, 2004). Because of its commercial interest, *H. scabra* has been the subject of many studies aiming to ascertain its nutritional value. Generally, *H. scabra* has high percentage of protein and moisture, while low lipid content in their body wall. For the fresh meat (body wall), the proximate content is 83–88% water, 5–35% ash, 5–7% protein, 0.5–1.2% fat, and 0.3–2% carbohydrate (Ardiansyah et al., 2020; Nishanthan et al., 2018). The dry body wall usually consists of 7–9% water, 3–32% ash, 60–71% protein, 1–1.6% fat, and 13–15% carbohydrate (Ding and Jian, 2010; Ibrahim et al., 2015; Karnila et al., 2011; Rahael et al., 2019).

Andriamanamisata and Telesphore (2019) indicated that the raw ash content (~29%) provides a rough estimate of the total inorganic matter in the samples of *H. scabra*; proteins make up the bulk of the tissues (48%). Comparatively, Ibrahim et al. (2015) found a protein content of about 60% and Sroyraya et al. (2017) found a value of 22.5% in the whole body and 55% in the body wall. Low lipid content of 1–2% are consistently reported (Andriamanamisata and Telesphore, 2019; Oedjoe, 2017; Sroyraya et al., 2017). Carbohydrates represent ~8% of the flesh (Andriamanamisata and Telesphore, 2019; Sroyraya et al., 2017). In a study conducted in Madagascar on 63 naïve consumers, hedonic values (organoleptic properties) were >5 on a scale of 9, meaning that the product was moderately appreciated and accepted; with men valuing the product more than women, and no difference across age groups (Andriamanamisata and Telesphore, 2019).

Rasyid et al. (2020) also examined the proximate composition of *H. scabra* (fresh) collected from an aquaculture in Bali (Indonesia). The moisture content was ~87% based on the wet weight, similar to reports (84–87%) from wild stocks by Omran (2013), Özer et al. (2004) and Al Azad et al. (2017). The ash content measured by Rasyid et al. (2020) was ~7%, i.e. higher than in *Holothuria mammata* and *H. tubulosa* (5.1%) (Aydın et al., 2011), *A. japonicus* (3–3.3%) (Gao et al., 2011) and *Isostichopus* sp. (3.2–3.8%) (Vergara and Rodríguez, 2016), but lower than in *H. polii* (8%) (Aydın et al., 2011). The protein content was 5.1% (Rasyid et al., 2020). Özer et al. (2004) also reported that the protein content between

5.5% and 5.8%. Rasyid et al. (2020) indicated that no fat content was detected in *H. scabra* was detected, whereas Özer et al. (2004) reported it to be 0.2–0.37%. Carbohydrate content was between 0.45% (Oedjoe, 2017) and 0.96% (Rasyid et al., 2020).

Hair et al. (2018a) compared similar-sized ocean-cultured (hatchery bred) and wild-harvested *H. scabra* from New Ireland Province (Papua New Guinea); they found recovery rate from fresh whole weight to dried weight (beche-de-mer) of 5.0% and 5.6%, respectively. Other ratios were comparable, including the recovery rate from fresh gutted weight to dried weight (8.8% vs 9.6%); and ratio of dried to fresh length (52% vs 50%). The thickness of the fresh body wall was different (5.6 vs 6.6 mm), but the thickness of the dried body wall was not different (2.7 vs 2.9 mm) nor was collagen content of the dried body wall (95 vs 104 $mg\,g^{-1}$) or ash content of the dried body wall (15% vs 14%). Overall, while differences between wild and cultured *H. scabra* were found, several key determinants of beche-de-mer quality were not different between the two (Hair et al., 2018a).

Ram (2017) indicated processing with dry salting in 15–25% brine solutions produced reduced protein contents in beche-de-mer compared to non-salted products (~846 $mg\,g^{-1}$). Beche-de-mer produced after 48 h of dry salting (~718 $mg\,g^{-1}$) and 15% brine salting (~714 $mg\,g^{-1}$) had the highest protein contents. The lipid contents did not differ from the non-salted treatment, but ash contents of salted products were generally greater. Carbohydrate in beche-de-mer increased with increasing duration of dry salting. Inclusion of a salting step resulted in reduced levels of eicosapentaenoic acid in some treatments but did not reduce polyunsaturated fatty acid (Ram et al., 2017a). As brine concentration increased, there was a trend of decreasing collagen content and increasing effort required to shear reconstituted beche-de-mer. Beche-de-mer obtained from most treatments was less firm than cooked, partially processed tissue. The most abundant protein-bound amino acids in beche-de-mer were glycine, glutamic acid, proline, arginine, aspartic acid, alanine and hydroxyproline, but their levels did not vary across treatments. The crude collagen fibrils of the body wall were extracted with acid and digested using the porcine pepsin to extract the pepsin solubilized collagen (Ram et al., 2017b) A decrease in collagen solubility was found in between 2% and 4% NaCl concentration (Ram, 2017).

Sari (2020) indicated that the high collagen protein content of sea cucumber suggest a potential to be used in cosmetics and health. Sea cucumber collagen with the 6% HCl extraction process gave optimal collagen yield of ~21%, white degree ~744%, moisture level~ 11%, protein ~98%, ash ~0.9%, fat ~0.6%, and collagen pH of 3.92. The nano-collagen obtained was in the form of a creamy white powder (~79%), ~9% moisture, ~0.7% ash, ~0.5% fat, ~99% protein, pH of 3.98, and particle size of 285 nm (Sari, 2020).

16.1.2 Amino acids and fatty acids

The analyses carried out on *H. scabra* show that 16–18 amino acids (AA) are detected in the body wall, half of which are essential and the other half non-essential. Dominant amino acids vary among analyses. Likewise, analyses of the fatty acid composition of the body wall are highly variable. They often indicate the presence of saturated fatty acids (SFA), polyunsaturated fatty acids (PUFA) and monounsaturated fatty acids (MUFA). However, the importance of one in relation to the other is not consistent from one study to another, perhaps reflecting spatial and seasonal modulations in food supply, or the variable age of the individuals analysed.

Amino acids are the structure units that make up proteins and play an essential role in the metabolism of living cells of sea cucumbers (Rahael et al., 2019). The critical flavours of sea cucumber products, such as sweetness, bitterness, sourness and umami taste, were derived from the amino acid components (Sicuro et al., 2012).

Analysis of total amino acids in *H. scabra* has shown that 18 amino acids were detected, including 9 essential amino acids and 8–9 nonessential amino acids (Rahael et al., 2019; Sroyraya et al., 2017). Glycine, proline and glutamic acid were the most abundant constituent in both the whole body and body wall (Ardiansyah et al., 2020; Karnila et al., 2011; Ram et al., 2017a; Sroyraya et al., 2017) and higher glutamic acid levels are considered responsible for the distinctive flavour (Sroyraya et al., 2017). Other studies indicated that *H. scabra* was rich in various of fatty acids (Al Azad et al., 2017; Ardiansyah et al., 2020; Nishanthan et al., 2018; Sroyraya et al., 2017; Yahyavi et al., 2012). A recent study by Al Azad et al. (2017) revealed that primary fatty acids in the body wall of *H. scabra* were saturated fatty acids (SFA) and polyunsaturated fatty acids (PUFA) was absent. However, other studies revealed that *H. scabra* contains higher values of PUFA (37–53%) and SFA (28–42%) followed by monounsaturated fatty acids (MUFA, 19–25%) (Nishanthan et al., 2018; Yahyavi et al., 2012).

Another study found 16 amino acids in fresh tissues of *H. scabra*; threonine was identified to be the major essential amino acid, followed by leucine, phenylalanine, lysine, valine, isoleucine, histidine and tryptophan respectively (Rasyid et al., 2020). Histidine was the major essential amino acid reported in an investigation of *H. scabra* and *Holothuria leucospilota* (Omran, 2013). Glycine was the major non-essential amino acid, followed by glutamic acid, arginine, proline, alanine, aspartic acid, serine and tyrosine respectively. Omran (2013) also reported that glycine was the major non-essential amino acid in *H. scabra*, *H. leucospilota*, *B. marmorata* and *Actinopyga mauritiana*. In Bali, Indonesia, the same nutritional contents were found in cultured and wild individuals of *H. scabra* (Rasyid et al., 2020).

In a similar study, Ridwanudin et al. (2018) compared the nutritional profiles of farmed and wild juveniles of *H. scabra*, as well as amino acid and fatty acid contents. The protein, lipid and carbohydrate contents of the body wall were similar. Amino acids of wild and farmed juveniles consisted of glycine, glutamic acid and alanine. Polyunsaturated fatty acids (PUFAs) were the major fatty acids in the body wall of the juveniles, irrespective of their provenance; with a dominance of arachidonic acid (C20:4n6). The total contents of PUFAs, omega-3 and omega-6 were slightly higher in cultured than in wild individuals (Ridwanudin et al., 2018).

Differences in fatty acid composition of sea cucumbers are influenced by food sources (Taboada et al., 2003), which may translate in variations across ontogenetic stages (age classes). In a study conducted on the fresh body wall of cultured (presumably adult) individuals of *H. scabra* in Bali (Indonesia), monounsaturated fatty acids (MUFAs) were found to dominate, followed by polyunsaturated PUFAs and saturated fatty acids (SFAs) (Rasyid et al., 2020). Analysis of MUFAs showed that oleic acid was the major component, followed by heptadecanoic acid, nervonic acid, eicosenoic acid, palmitoleic acid, erucic acid and miristoleic acid. Özer et al. (2004) also identified oleic acid as the major MUFA in *H. scabra*, *H. leucospilota*, *B. marmorata* and *A. mauritiana*, whereas Ridzwan et al. (2014) and Al Azad et al. (2017) reported that palmitoleic acid was dominant. Linoleic acid was the major PUFA observed by Rasyid et al. (2020), followed by arachidonic acid, eicosapentanoic acid, eicosatrienoic acid and docosahexaenoic acid. In contrast, the latter was the major PUFA in *H. scabra* and *H. leucospilota* according to Yahyavi et al. (2012). Ridzwan et al. (2014) showed that palmitic acid was the dominant SFA, followed by stearic acid, behenic acid, arachidic acid, lignoceric acid, myristic acid, tridecanoic acid, lauric acid, heneicosanoic acid, butyric acid and pentadecanoic acid; in line with a later study where palmitic acid was the major SFA in *H. scabra* (Al Azad et al., 2017).

16.1.3 Minerals and vitamins

It has also been reported that *H. scabra* is the functional food with a source of vitamins and minerals (Ardiansyah et al., 2020; Mageswaran and Balakrishnan, 1985; Sroyraya et al., 2017). The principal mineral content in *H. scabra* were calcium, sodium, magnesium, iron, potassium, and phosphorus (Ardiansyah et al., 2020; Mageswaran and Balakrishnan, 1985), and low levels of copper, manganous, zinc, lead, nickel and cadmium are also present. A study conducted by Ardiansyah et al. (2020) indicated that the contents of Vitamin B2 and E in the fresh body wall of *H. scabra* were 0.04–0.06 mg and 0.18–0.35 mg per 100 g, respectively. Another study based on the dry weight showed that the contents of vitamin C and E were 3.19 and 2.82 mg per 100 g in the whole body, and 1.32 and 4.94 mg per 100 g in the body wall (Sroyraya et al., 2017).

Evaluation of mineral composition of *H. scabra* showed that calcium (1812 mg $100\,g^{-1}$) was the major component, followed by sodium, magnesium, phosphor, potassium and iron with 666 mg, 305, 88, 61 and 2.8 mg $100\,g^{-1}$, respectively (Rasyid et al., 2020).

Sea cucumbers have an impressive profile of high-value nutrients such as vitamin A, vitamin B1 (thiamin), vitamin B2 (riboflavin) and vitamin B3 (niacin) (Bordbar et al., 2011). Rasyid et al. (2020) showed that *H. scabra* contained vitamin E (0.48 $mg\,kg^{-1}$ wet weight), while vitamins A, B1, B2 and B12 were not detected. However, Sroyraya et al. (2017) reported that *H. scabra* from wild stocks contained vitamin E and vitamin C.

16.1.4 Contaminants

The concentrations of heavy metals like cadmium, copper, zinc and lead were investigated in the sediments and in individuals of *H. scabra* in northern Qeshm Island, Persian Gulf by Mohammadizadeh et al. (2016). The highest levels of zinc and lead was measured in respiratory tree organ. Copper and cadmium were the most abundant elements in the gonad and body wall. While the copper and zinc concentrations were below permissible limits for human consumption, the cadmium and lead were above permissible limits for human consumption (Mohammadizadeh et al., 2016).

16.1.5 Bioactive components

16.1.5.1 Saponins (triterpene glycosides)

Among bioactive substances, saponins are the typical predominant secondary metabolites in many holothuroid species. It is structurally known as triterpene glycosides composed of an aglycone (sapogenin) and a glycosidic part (Caulier et al., 2013b). Saponins can be used to deter predation, and regenerate and repair tissues by sea cucumbers (Maier, 2008; Van Dyck et al., 2011). To date, more than 100 of saponins have been detected in the body wall, internal organs and Cuvierian tubules of sea cucumbers (Caulier et al., 2011). The saponins obtained from sea cucumbers showed a wide spectrum of biological activities, such as antifungal, cytotoxic, hemolytic, cytostatic and immunomodulatory functions (Dang et al., 2007; Han et al., 2008, 2009b). Saponins can also modulate the intracellular signalling pathways of the essential physiological process such as proliferation, migration/invasion, and angiogenesis of cancer cells (Yurasakpong et al., 2021). Therefore, saponins from sea cucumbers have been identified as potential drugs in the pharmaceutical industry (Kim and Himaya, 2012).

More than 10 types of saponins were found in various tissues of *H. scabra* (Table 12). Specifically, scabraside A and B, and holothurin A3 extracted from the body wall exhibited strong cytotoxic effect to A172 and U87MG cells (Sangpairoj et al., 2016). Holothurin A3 and A4 were found to be strongly cytotoxic to both cancer cell lines, KB and Hep-G2 (Dang et al., 2007). Scabraside C, holothurin A1, echinoside A, echinoside B, 24-dehydroechinoside A, holothurin A and holothurin B had remarkable in vitro cytotoxicity against human tumour cell lines (Han et al., 2010; Yan et al., 2005). Nine saponins obtained from *H. scabra* have detected the cytotoxicity against four human cell lines, and one of them (echinoside A) also inhibited cell viability and induced apoptosis in HepG2 cells (Wang et al., 2014). A recent study revealed that the compounds of holothurin A (Fig. 17A) and holothurin B (Fig. 17B) have target proteins that play an important role in the process of apoptosis, cell cycle, and suppressing tumour (Wargasetia et al., 2020).

Some studies have also reported that saponins from *H. scabra* showed the antifungal, antioxidant and antiaging activities. Scabraside A, scabraside B, echinoside A and holothurin A1 showed antifungal activity against the fungi strains tested (Han et al., 2008, 2009b). Saponins obtained from the whole body, body wall, and viscera promoted stress resistance and lifespan extension in the *Caenorhabditis elegans* model, exhibiting the antioxidant and antiaging properties (Jattujan et al., 2018). Moreover, the presence of saponins in the extract of *H. scabra* body wall was found to be effective against mosquito larvae, which will be helpful for the discovery of novel insecticidal molecules (Thakur et al., 2004).

Table 12 Summary of compounds extracted from *Holothuria scabra* and their potential usage.

Extract chemical classification	Extract name	Tissue source	Usage	Extraction methodology	References
Alkaloids	Not specific	Body wall	Antibacterial capabilities	Liquid extract	Pranoto et al. (2012)
		Body wall	Antioxidant capabilities	Liquid extract	Murniasih et al. (2015)
		Not specific	Antioxidant capabilities	Liquid extract	Setiyanto (2012)
Anthraquinones	Not specific	Body wall	Antioxidant capabilities	Liquid extract	Murniasih et al. (2015)
Carotenoids	Not specific	Coelomic fluid	Antibacterial capabilities	Not specific	Ibrahim (2012)
		Body wall	Antibacterial capabilities	Liquid extract	Ibrahim (2012)
Elements/ions	Cadmium	Whole animal	Not specified	Not specified	Mageswaran and Balakrishnan (1985)
	Calcium	Whole animal	Not specified	Not specified	Ardiansyah et al. (2020), Mageswaran and Balakrishnan (1985)
	Copper	Whole animal	Not specified	Not specified	Mageswaran and Balakrishnan (1985)
	Iron	Whole animal	Not specified	Not specified	Ardiansyah et al. (2020), Mageswaran and Balakrishnan (1985)
	Lead	Whole animal	Not specified	Not specified	Mageswaran and Balakrishnan (1985)

	Magnesium	Whole animal	Not specified	Not specified	Ardiansyah et al. (2020), Mageswaran and Balakrishnan (1985)
	Manganese	Whole animal	Not specified	Not specified	Mageswaran and Balakrishnan (1985)
	Nickel	Whole animal	Not specified	Not specified	Mageswaran and Balakrishnan (1985)
	Phosphorus	Whole animal	Not specified	Not specified	Ardiansyah et al. (2020), Mageswaran and Balakrishnan (1985)
	Potassium	Whole animal	Not specified	Not specified	Ardiansyah et al. (2020), Mageswaran and Balakrishnan (1985)
	Sodium	Whole animal	Not specified	Not specified	Ardiansyah et al. (2020), Mageswaran and Balakrishnan (1985)
	Zinc	Whole animal	Not specified	Not specified	Mageswaran and Balakrishnan (1985)
Fatty acids	General	Body wall	As a supplementary food	Liquid extract	Al Azad et al. (2017), Ardiansyah et al. (2020), Nishanthan et al. (2018), Sroyraya et al. (2017), Yahyavi et al. (2012)
		Whole animal	As a supplementary food	Liquid extract	Sroyraya et al. (2017)

Continued

Table 12 Summary of compounds extracted from *Holothuria scabra* and their potential usage.—cont'd

Extract chemical classification	Extract name	Tissue source	Usage	Extraction methodology	References
Flavonoids	Not specified	Body wall	Antioxidant capabilities	Liquid extract	Murniasih et al. (2015)
		Body wall	Suppressed the viability of breast cancer cells	Liquid extract	Yurasakpong et al. (2021)
		Not specific	Antioxidant capabilities	Liquid extract	Setiyanto (2012)
Gamma aminobutyric acid (GABA)	Not specified	Body wall	Functional food	Liquid extract	Sroyraya et al. (2017)
		Whole animal	Functional food	Liquid extract	Sroyraya et al. (2017)
Glycosaminoglycans	Fucosylated chondroitin sulphate (HsG)	Body wall	Anticoagulant activities	Liquid extract	Yang et al. (2018)
Glycosides	Phenolic glycosides	Whole animal	Inhibited pro-inflammatory cytokines synthesis	Liquid extract	Pranweerapaiboon et al. (2019)
	Saponins (holothurinoside C, desholothurine A, pervicoside C, scabraside A, scabraside B, holothurinoside G)	Body wall	Not specified	Liquid extract	Caulier et al. (2013a)
	Saponins (scabraside A and B)	Body wall	Antifungal and antitumor capacities	Liquid extract	Han et al. (2008)

	Saponins (scabraside C, holothurin A1, echinoside A, 24-dehydroechioside A, holothurin A)	Body wall	Showed inhibition activities on HL-60, MOULT-4 and A-549 cells	Liquid extract	Han et al. (2009a)
	Saponins (holothurin A, echinoside A, 24-dehydroechinoside)	Body wall	Not specified	Liquid extract	Kobayashi et al. (1991)
	Saponins (holothurinoside C, desholothurin A, and pervicoside C)	Body wall	Not specified	Liquid extract	Mitu et al. (2017)
	Saponins (scabraside A, scabraside B, and holothurin A_3)	Body wall	Exhibited anticancer activity	Liquid extract	Sangpairoj et al. (2016)
	Saponins (holothurin A3 and A4)	Whole animal	Cytotoxicity against cancer cell lines	Liquid extract	Dang et al. (2007)
	Saponins (scabraside A, echinoside A, holothurin A_1)	Whole animal	Antifungal capabilities	Liquid extract	Airoldi et al. (2009)
	Saponins (scabraside A and B)	Whole animal	Cytotoxicity against four human tumour cell lines	Lipid extract	Han et al. (2010)

Continued

Table 12 Summary of compounds extracted from *Holothuria scabra* and their potential usage.—cont'd

Extract chemical classification	Extract name	Tissue source	Usage	Extraction methodology	References
	Saponins (holothurin A, B and B3)	Whole animal	Cytotoxicity against four human tumour cell lines	Lipid extract	Wargasetia et al. (2020)
	Saponins (holothurinogenin B, holothurin B, holothurin A, and holothurinA_2)	Not specified	Cytotoxicity against cancer cell lines	Lipid extract	Van Thanh et al. (2006)
	Saponins (echinoside A, 24-dehydro echinoside A, holothurin B, holothurin B4, HS-1, holothurin A, holothurin A1, scabraside D, scabraside B)	Not specified	Cytotoxicity against four human tumour cell lines	Lipid extract	Wang et al. (2014)
	Saponins (24-dehydro echinoside A, echinoside B, holothurin B, holothurin A)	Not specified	Cytotoxicity against four human tumour cell lines	Lipid extract	Yan et al. (2005)
	Saponins (not specified)	Cuvierian tubules	Mosquito larvicidal activity	Liquid extract	Thakur et al. (2004)
		Body wall	Anti-oxidant and anti-ageing capabilities	Liquid extract	Jattujan et al. (2018), Murniasih et al. (2015)

		Body wall	Antibacterial capabilities	Liquid extract	Pranoto et al. (2012)
		Body wall	Mosquito larvicidal activity	Liquid extract	Thakur et al. (2004)
		Body wall	Suppressed the viability of breast cancer cells	Liquid extract	Yurasakpong et al. (2021)
		Viscera	Anti-oxidant and anti-ageing capabilities	Liquid extract	Jattujan et al. (2018)
		Whole animal	Anti-oxidant and anti-ageing capabilities	Liquid extract	Jattujan et al. (2018)
		Whole animal	Inhibited pro-inflammatory cytokines synthesis	Liquid extract	Pranweerapaiboon et al. (2019)
	Not specified	Body wall	Anti-Parkinson	Liquid extract	Chalorak et al. (2018)
		Viscera	Anti-Parkinson	Liquid extract	Chalorak et al. (2018)
		Whole animal	Anti-Parkinson	Liquid extract	Chalorak et al. (2018)
Mucopolysaccharide	Acidic mucopolysaccharide	Body wall	Not specified	Liquid extract	Chen et al. (2006)
Phenols	Friedelin	Whole animal	Antioxidant capabilities	Liquid extract	Nobsathian et al. (2017)
	3-hydroxybenzaldehyde	Whole animal	Antioxidant capabilities	Liquid extract	Nobsathian et al. (2017)
	4-hydroxybenzaldehyde	Whole animal	Antioxidant capabilities	Liquid extract	Nobsathian et al. (2017)

Continued

Table 12 Summary of compounds extracted from *Holothuria scabra* and their potential usage.—cont'd

Extract chemical classification	Extract name	Tissue source	Usage	Extraction methodology	References
	Not specified	Body wall	Antioxidant and anticancer	Liquid extract	Althunibat et al. (2009), Murniasih et al. (2015)
		Body wall	Anti-Parkinson	Liquid extract	Chalorak et al. (2018)
		Viscera	Anti-Parkinson	Liquid extract	Chalorak et al. (2018)
		Whole animal	Anti-Parkinson	Liquid extract	Chalorak et al. (2018)
Protein	General	Body wall	As a supplementary food with high content of protein	Not specified	Al Azad et al. (2017), Andriamanamisata and Telesphore (2019), Ardiansyah et al. (2020), Ibrahim et al. (2015), Karnila et al. (2011), Ram et al. (2017a)
		Body wall	Increased the proliferative rate of mesenchymal stem cells	Liquid extract	Saengsuwan (2017)
		Radial nerve	Increased the proliferative rate of mesenchymal stem cells	Liquid extract	Saengsuwan (2017)
		Viscera	Increased the proliferative rate of mesenchymal stem cells	Liquid extract	Saengsuwan (2017)

	Amino acids	Body wall	Functional food	Liquid extract	Ardiansyah et al. (2020), Ding and Jian (2010), Sroyraya et al. (2017)
		Body wall	Not specified	Not specified	Rahael et al. (2019), Ram et al. (2017b)
		Whole animal	Functional food	Liquid extract	Sroyraya et al. (2017)
	Collagen	Body wall	They can be used in cosmetics and pharmaceutical industries	Not specified	Desmelati et al. (2020)
		Body wall	Functional food	Liquid extract	Sroyraya et al. (2017)
		Body wall	Not specified	Not specified	Ram et al. (2017b)
		Whole animal	Functional food	Liquid extract	Sroyraya et al. (2017)
	Lectins	Coelomic fluid	The hemagglutination activity of lectins does not require any divalent metal ions	Not specified	Gowda et al. (2008a)
		Coelomic fluid	Antibacterial capabilities	Not specified	Gowda et al. (2008b)
		Coelomic fluid	Useful tool for cancer diagnostic purpose	Not specified	Gowda et al. (2013)

Continued

Table 12 Summary of compounds extracted from *Holothuria scabra* and their potential usage.—cont'd

Extract chemical classification	Extract name	Tissue source	Usage	Extraction methodology	References
		Viscera	Have hemolytic and mitogenic activities and inhibit the germination of radish seeds	Liquid extract	Mojica and Merca (2005a)
		Not specified	Modify the carbon paste electrode	Not specified	Mojica and Merca (2005b)
	Neuropeptides (GFSKLYFamide)	Body wall	Not specified	Not specified	Ajayi et al. (2020)
		Tentacles	Not specified	Not specified	Ajayi et al. (2020)
		Tube foot	Not specified	Not specified	Ajayi et al. (2020)
	Peptides	Coelomic fluid	Inhibit the biofilm formed by the ESBL-producing *E. coli*	Not specified	Bais et al. (2018)
Sterols	Testosterone	Body wall	Not specified	Liquid extract	Dewi et al. (2011)
		Whole animal	Antibacterial capabilities	Liquid extract	Dewi et al. (2010)
	Sterols (60)	Not specified	Not specified	Liquid extract	Avilov et al. (1998)
	Not specified	Body wall	Antibacterial capabilities	Liquid extract	Pranoto et al. (2012)
		Body wall	Increase the estradiol hormone in blood serum of female rats	Not specified	Riani et al. (2016)

Sulphated polysaccharides	Glycosaminoglycans (GAGs)	Body wall	Reduce the risk of atherosclerosis and hyperlipoproteinemia	Liquid extract	Liu et al. (2002)
Terpenoids	Not specified	Body wall	Antioxidant capabilities	Liquid extract	Murniasih et al. (2015)
Vitamins	B2	Body wall	Functional food	Liquid extract	Ardiansyah et al. (2020)
	C	Body wall	Functional food	Liquid extract	Sroyraya et al. (2017)
		Whole animal	Functional food	Liquid extract	Sroyraya et al. (2017)
	E	Body wall	Functional food	Liquid extract	Ardiansyah et al. (2020), Sroyraya et al. (2017)
		Whole animal	Functional food	Liquid extract	Sroyraya et al. (2017)
Not specified	Not specified	Body wall	Antioxidant capabilities	Not specified	Ansharullah et al. (2020)
		Body wall	Antifungal and antibacterial capabilities	Liquid extract	Mashjoor and Yousefzadi (2017), Mashjoor et al. (2018), Mohammadizadeh et al. (2013)
		Body wall	Cytotoxicity against cell lines	Liquid extract	Mashjoor and Yousefzadi (2019), Mashjoor et al. (2019), Mohammadizadeh et al. (2013)

Continued

Table 12 Summary of compounds extracted from *Holothuria scabra* and their potential usage.—cont'd

Extract chemical classification	Extract name	Tissue source	Usage	Extraction methodology	References
		Body wall	Did not show antibacterial activity	Liquid extract	Ridzwan et al. (1995)
		Body wall	Showed antifouling activity	Liquid extract	Selvin and Lipton (2004)
		Body wall	Enhanced the innate immunity of fish	Liquid extract	Andriawan et al. (2019)
		Coelomic fluid	Cytotoxicity against cell lines	Liquid extract	Mashjoor and Yousefzadi (2019), Mashjoor et al. (2019)
		Cuvierian tubules	Cytotoxicity against cell lines	Liquid extract	Mashjoor and Yousefzadi (2019), Mashjoor et al. (2019)
		Gonads	Antifungal and antibacterial capabilities	Liquid extract	Mashjoor and Yousefzadi (2017), Mashjoor et al. (2018), Mohammadizadeh et al. (2013)
		Gonads	Cytotoxicity against cell lines	Liquid extract	Mashjoor and Yousefzadi (2019), Mashjoor et al. (2019), Mohammadizadeh et al. (2013)

	Intestine tract	Antifungal and antibacterial capabilities	Liquid extract	Mashjoor and Yousefzadi (2017), Mashjoor et al. (2018)
	Intestine tract	Cytotoxicity against cell lines	Liquid extract	Mashjoor and Yousefzadi (2019), Mashjoor et al. (2019)
	Respiratory tree	Antifungal and antibacterial capabilities	Liquid extract	Mashjoor and Yousefzadi (2017), Mashjoor et al. (2018), Mohammadizadeh et al. (2013)
	Respiratory tree	Cytotoxicity against cell lines	Liquid extract	Mashjoor and Yousefzadi (2019), Mashjoor et al. (2019), Mohammadizadeh et al. (2013)
	Not specified	Cytotoxicity against cell lines	Liquid extract	Wargasetia and Widodo (2018)

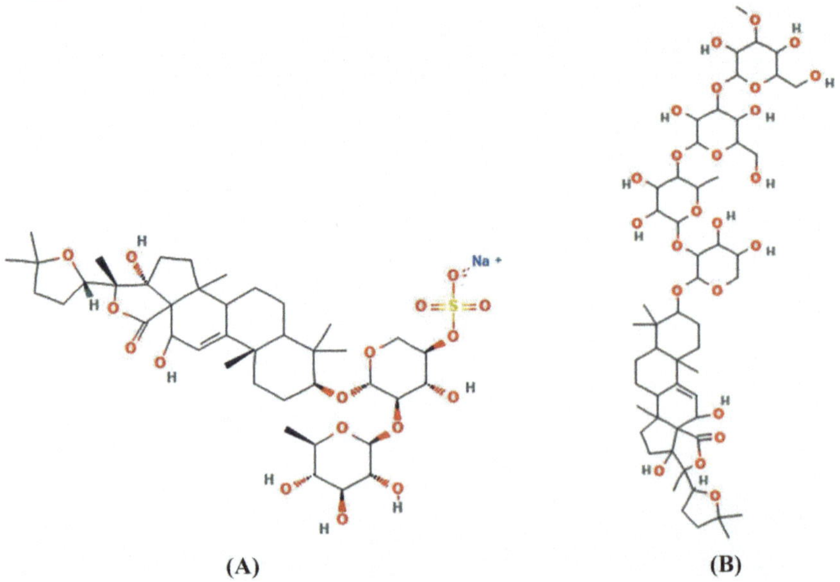

Fig. 17 The structure and characteristics of two sulphated saponins, (A) holothurin A and (B) holothurin B, extracted from *H. scabra*. *From Wargasetia, T.L., Ratnawati, H. Widodo, N., 2020. Anticancer potential of holothurin A, holothurin B, and holothurin B3 from the sea cucumber* Holothuria scabra. *In: AIP Conference Proceedings. AIP Publishing LLC. p. 040084, with permission.*

16.1.5.2 Collagens and lectins

Collagen is a protein where amino acids bind together to form a triple helix of elongated fibrils and forms the primary structural component of the connective tissues found in various parts of sea cucumber (Desmelati et al., 2020). Much of the protein content of the body wall of sea cucumbers, is composed of collagen, that provides them with body shape and form, and assists during feeding, respiration, burrowing and in defence (Ram et al., 2017b; Yamada et al., 2010). Previous studies have shown that *H. scabra* had a high content of collagen and the extraction of collagen is used in the pharmaceuticals and cosmetics (Desmelati et al., 2020; Ram et al., 2017b; Sroyraya et al., 2017).

Lectins are carbohydrate-binding proteins that are highly specific for sugar groups that are part of other molecules. A novel lectin (HSL) purified from coelomic fluid of *H. scabra* showed strong antibacterial activity against Gram-positive and Gram-negative bacteria under in vitro conditions (Gowda et al., 2013). Also, the higher affinity to T-antigen makes HSL a potential tool for cancer diagnostic purpose (Gowda et al., 2013). Another lectin isolated from the internal organs of *H. scabra* was found to have hemolytic and mitogenic activities and is a good inhibitor of seed germination (Mojica and Merca, 2005a). In addition to the biological properties, the carbon paste electrode (CPE) modified by a purified lectin isolated from *H. scabra* was found to have a high affinity towards mercury (Mojica and Merca, 2005b).

16.1.5.3 Others

Apart from those discussed above, more bioactive compounds have been isolated and identified in *H. scabra*, as shown in Table 12. Alkaloids and flavonoids isolated from the body wall of *H. scabra* had antioxidant activities and suppressed the viability of cancer cells (Murniasih et al., 2015; Pranoto et al., 2012; Setiyanto, 2012; Yurasakpong et al., 2021). Phenols obtained from the whole body, body wall and viscera of *H. scabra* can be a possible preventive and therapeutic intervention against Parkinson's disease (Chalorak et al., 2018). These compounds are potential sources of natural antioxidant and anticancer agents (Althunibat et al., 2009; Murniasih et al., 2015; Nobsathian et al., 2017). Sulphated polysaccharides obtained from *H. scabra* had the potential of being used for reducing the risk of atherosclerosis and hyperlipoproteinemia (Liu et al., 2002).

Nimah et al. (2012) showed that extracts from *H. scabra* acted against the bacteria *Pseudomonas aeruginosa* and *Bacillus cereus*. The results showed that ethyl acetate extract had the highest inhibition against *P. aeruginosa* and *B. cereus* at a concentration of $450\,g\,mL^{-1}$. The results also showed the presence of saponins, alkaloids, steroids and triterpenoids in the extract (Nimah et al., 2012).

Abraham et al. (2002) showed the antibacterial and antifungal activities of alcohol extracts of *H. scabra*. They found that except for *Bacillus* sp., other strains, namely *E. coli*, *Aeromonas hydrophila*, *Enterococcus* sp., *P. aeruginosa*, *Klebsiella pneumoniae*, *S. aureus*, *Salmonella typhi* and *Vibrio harveyi*, as well as fish- generated *Aspergillus* sp., were sensitive to the tested sea cucumber extracts. Mohammadizadeh et al. (2013) showed that extracts from *H. scabra* were not effective against any of the bacterial species they tested [*Escherichia coli* (ATCC 1763), *Staphylococcus aureus* (ATCC 25923), *P. aeruginosa* (ATCC 25853), *Candida albicans*], even at a concentration of $18\,\mu g\,mL^{-1}$. A higher cytotoxic activity was found for methanolic extracts, with LC_{50} values of about $50\,\mu g\,mL^{-1}$ with the gonad and about $69.96\,\mu g\,mL^{-1}$ with the respiratory tree.

The study of Ajayi and Withyachumnarnkul (2015) provided evidence for the presence of GFSKLYFamide neuropeptide in the digestive tract of *H. scabra*.

Docoxahexaenoic acid (DHA) is an omega-3 fatty acid that plays a key role in the development of human eye and nerve tissues; it might also reduce the risk circulatory disease by decreasing the thickness of the blood and lowering blood levels of triglycerides. Wahyuni et al. (2020) examined the ingestion of capsules made from *H. scabra* on the level of DHA of breast milk with preterm child birth. Results showed difference in DHA level between treatment and control group. After the intervention, the average increase was 187.02 $\mu g\,mL^{-1}$ for the treatment group and 7.05 $\mu g\,mL^{-1}$ for control group (Wahyuni et al., 2020).

Bais et al. (2018) described the urinary tract infection as one of the most common bacterial infections in humans, with a recurrence and infectivity that can be attributed to formation of biofilms. They evaluated the antibiofilm activity of coelomic fluid from *H. scabra* against extended spectrum beta-lactamase producing bacteria (ESBL) *E. coli*. Biofilm inhibition may be attributed to the presence of low molecular weight antimicrobial peptides in the coelomic fluid ranging from 2 to 4 kDa (Bais et al., 2018).

Kharisma (2012) aimed to determine if water extract of *H. scabra* could have an effect on the total cholesterol in hyperlipidemic rats but found it ineffective in reducing total cholesterol levels.

16.2 Biotoxicity

Although very few studies have examined tissue toxicity, Halstead (1988) established that *H. scabra* was a toxic sea cucumber. This toxicity is mainly due to the permanent secretion of saponins which are cytotoxic molecules (e.g. Eeckhaut et al., 2015) whose mode of action on cell membranes was analysed by Claereboudt et al. (2018). In this work, Claereboudt et al. (2018) also showed that sea cucumbers are tolerant to their own saponins. To elucidate the molecular mechanisms behind this tolerance, a suite of complementary biophysical tools was used, firstly using liposomes for in vitro techniques then using in silico approaches for a molecular-level insight. The holothuroid saponin Frondoside A caused less permeabilization in liposomes containing a Δ7 holothuroid sterol than those containing cholesterol. Lipid phases simulations revealed that Frondoside A has an agglomerating effect on cholesterol domains, however, induced small irregular Δ7 sterol clusters. Findings suggest that the structural peculiarities of holothuroid sterols (wider proportion of Δ7 sterol compared to other metazoans that have mostly cholesterol) provide sea cucumbers with a mechanism to mitigate the sterol-agglomerating effect of saponins, and therefore to protect their cells from the cytotoxicity of the saponins they produce.

Rao et al. (1985a,b) noted that extracts from the body wall and viscera of *H. scabra* inflicted distress, loss of equilibrium and death in fish. In addition, tests on mice induced immediate reversible paralysis. The death of fish from exposure to saponins has been analysed in other sea cucumbers (Eeckhaut et al., 2015). It results from the fact that saponins mainly act at the level of the secondary lamellae of the gills where they induce the detachment of the epithelia from the inner tissues, create oedema at the level of the epithelia and induce pores in the epithelial cells leading to their destruction and in the invasion of inner cells (pillar cells and red blood cells).

Aras (2013) assessed the toxicity of various extracts of *H. scabra* using brine shrimp (*Artemia salina*) lethality tests and found an LC_{50} value of 0.5886 ppm for the methanol and ethyl acetate extracts and a value of 0.8556 ppm for the n-hexane extract, indicative of highly toxic properties and potential as an anti-cancer drug. The aqueous fraction extract had an LC_{50} value of 1380.22 ppm indicating that it was non-toxic (Aras, 2013).

17. Ecosystem services

Lee (2016) and Lee et al. (2018b) studied *H. scabra* in Fiji by manipulating densities within enclosures on a reef flat, i.e. high density (350 $g\,m^{-2}$ wet weight), natural density (60 $g\,m^{-2}$), and exclusion (0 $g\,m^{-2}$), thereby highlighting its important role in the cycling and remineralization of organic matter in reef sands through feeding, defecation, and bioturbation processes. Findings revealed that under natural density, individuals rework ~10,590 kg dry sediment 1000 m^{-2} $year^{-1}$. In the absence of *H. scabra*, the O_2 penetration depth decreased by 63%, and the sedimentary oxygen consumption rates increased about two-fold within the first month, being consistently higher than in the high-density treatment. These results suggest that the removal of sea cucumbers can reduce the capacity of sediments to buffer organic matter pulses, impeding the function and productivity of shallow coastal ecosystems (Lee et al., 2018b).

18. Predators in nature

James et al. (1994b) and MPEDA (1989) indicated that the main predators of the larval forms of *H. scabra* were copepods and ciliates that attacked the larvae, causing injury and death. These organisms also harmed juveniles, especially the newly-settled ones, by indirectly competing for food (Battaglene 1999a,b; Battaglene et al., 1999). To control copepods and ciliates, James et al. (1994b) added chemicals containing organo-phosphorus to rearing tanks. Copepods were killed with 2 ppm dipteryx in 2 h with no harmful effect to sea cucumber larvae. Mercier et al. (2000a) observed that tectibranchs and some species of amphipods were feeding on pentactula larvae and newly-settled juveniles in the laboratory. Many predators are known from experiments carried out to optimize the aquaculture of *H. scabra* (see Section 21); on the other hand, very few predators have been observed in the wild. Juveniles of *H. scabra* are reportedly preyed upon by fishes, crabs and sea stars (Dance et al., 2003; Purcell and Simutoga, 2008). In the Solomon Islands, 30–100% of juveniles of *H. scabra* (20–70 mm) released on sand near coral reefs were eaten within 24 h by a variety of fish (including Lethrinidae, Balistidae and Nymipteridae), whereas there was no predation of juveniles of the same size released on subtidal sandy substrata near mangrove forests and sea grass beds away from coral reefs (Dance et al., 2003). Although *H. scabra* is known to be toxic (see Section 16.2 above; Bakus, 1968; Rao et al., 1985a,b; Stott et al., 1974), the fact that small juveniles forage only at night seems to suggest a possible predator-avoidance measure. Adults were never observed to be preyed upon in the field (Mercier et al., 2000b).

19. Symbiotic community

Because of their diverse external morphologies, echinoderms represent an important resource providing refuge to smaller organisms that may obtain benefits (e.g. protection, food) from the association. Echinoderms are known to interact with many symbiotic organisms belonging to various phyla (Barel and Kramers, 1977; Britayev et al., 2019), including other echinoderms (Fourgon et al., 2007; Purcell and Eriksson, 2015) and sea cucumbers are recognized for the diversity and number of their associates. Symbionts can be found externally (on the ventral or dorsal body wall or among the tentacles) or internally (in the mouth or cloaca, the digestive tract, the gonad, the respiratory tree and the perivisceral coelom) (Eeckhaut et al., 2004; Jangoux, 1987; Purcell et al., 2016b). Hundreds of publications, mainly on commercial holothuroids, describe more than 200 symbiotic species belonging to a minimum of nine phyla (Purcell et al., 2016b). Symbioses are here considered as close and long-term inter-specific associations including mutualism, commensalism and parasitism as originally defined by De Bary (1879). In order to avoid any confusion, the largest associate is considered the host and the smallest the symbiont. The type of symbiotic association is determined considering the general impact on the host fitness (i.e. lifetime reproductive success) and not only trophic interactions, as a symbiont feeding from few external host tissues would not necessarily negatively impact its hosts. Therefore, a symbiont generating a negative impact on its host's fitness is considered a parasite, associates that have a positive impact on each other's fitness are classified as mutualistic, and a symbiont that has no impact on the host is considered a commensal (De Bary, 1879; Eeckhaut, 2003; Parmentier and Michel, 2013).

This section only describes the metazoan symbionts that are hosted by *H. scabra*. Bacteria, protozoa and other parasites impacting wild and cultured individuals are treated in Section 21.

19.1 Taxonomic groups

Since 1900, around 30 different studies have described the metazoan symbiotic community associated to *H. scabra*. Adults of *H. scabra* provide shelter to at least 17 species belonging to 4 metazoan taxa, i.e. annelids, crustaceans, gastropods and fishes, which live on their external surface or inside their coeloms and respiratory trees (Table 13). Most of the species associated with *H. scabra* were not found to adversely affect their hosts, although minor injuries to the respiratory trees and gonad atrophy have been recorded (Hamel et al., 1999; Jones and Mahadevan, 1965; VandenSpiegel et al., 1992).

Two species of symbiotic annelid polychaetes (Table 13; Fig. 18A–C) were found to occur externally on the surface of the body wall. They move around their host with their parapods and can enter the pharyngeal or cloacal cavities to hide when threatened. Symbiotic polychaetes prefer to shelter on the ventral side (in contact with the sediment) of large sea cucumbers (Martin and Britayev, 1998). *Gastrolepidia clavigera* (Fig. 18A) belongs to the Polynoidae family. It is the commonest Indo-Pacific ectocommensal associated to holothuroids of the families Holothuriidae and Stichopodidae (Britaev and Lyskin, 2002; Eeckhaut et al., 2004).

Table 13 Description of symbiotic interactions involving *Holothuria scabra* as a host.

Phylum	Species name	Geographic location	Observations	References
Annelida	*Gastrolepidia clavigera*	Indo-Pacific	On the body wall and in the buccal/cloacal cavities	Britaev and Lyskin (2002), Martin and Britayev (1998), Simon et al. (2014)
	Syllis unzima	South Africa	On the body wall and in the buccal/cloacal cavities	Simon et al. (2014)
Crustacea (Brachyura)	*Hapalonotus pinnotheroides*[a]	Papua New Guinea, Solomon Islands	In the respiratory tree	Castro (2015), Chia and Ng (1999), Hamel et al. (1999), VandenSpiegel et al. (1992)
	Lissocarcinus orbicularis	Papua New Guinea, Madagascar, Solomon Islands	On the body wall and in the cloaca	Ayotte (2005), Caulier et al. (2012, 2013a, 2014), Crosnier (1962), VandenSpiegel et al. (1992)
	Buergeres deccanensis[a]	India	In the respiratory tree and cloaca, rarely in the coelomic cavity	Adithiya (1969), Chopra (1931), James (1989a, 1995b), Jangoux (1987), Jones and Mahadevan (1965), Pinn et al. (2014)
	Holotheres halingi[a]	Solomon Islands, Indonesia	In the respiratory tree	Hamel et al. (1999, 2019b), Hsueh (2001), Martin et al. (2016), Tresnati et al. (2021)
	Holotheres semperi[a]	Singapore, Indonesia	In the cloacal cavity	Hamel et al. (2019b), Lanchester (1900), Martin et al. (2016), Ng and Manning (2003)
Crustacea (Natantia)	*Zenopontonia rex*[a]	Papua New Guinea	On the body wall	VandenSpiegel et al. (1992)

Crustacea (Copepoda)	*Lichothuria mandibularis*	Madagascar	On the body wall	Humes (1980), Humes and Ho (1969), Jangoux (1987)
	Scambicornus idoneus	Madagascar	On the body wall	Humes (1967), Humes (1980)
Mollusca	*Mucronalia* sp.[a]	Mozambique	Gall around the anus	Bakus (1973), Macnae and Kalk (1962)
	Odostomia sp.	Mozambique	Attached to the buccal membrane or to the body wall	Bakus (1973), Macnae and Kalk (1962)
	Prostilifer sp.	India	On the body wall or imbedded in a gall	James (1998)
Chordata (Pisces)	*Carapus* sp.	Papua New Guinea	In the respiratory tree	VandenSpiegel et al. (1992)
	Carapus mourlani	Indo-Pacific	In the coelomic cavity	Eeckhaut et al. (2004), Markle and Olney (1990)
	Encheliophis (Jornanicus) gracilis	Solomon Islands and elsewhere in West Pacific and Indian Oceans	In the coelomic cavity	Arnold (1956), Brasseur et al. (2016), Hamel and Mercier (1999), Jangoux (1987), Murdy and Cowan (1980)
	Encheliophis vermicularis	Galapagos Islands	In the coelomic cavity	Markle and Olney (1990), Murdy and Cowan (1980), Parmentier (2004)

[a]Species name may have changed since the references were published.

Fig. 18 Examples of symbiotic associates of *H. scabra*. (A) Scaleworm *Gastrolepidia clavigera*; (B) anterior region of the polychaete *Syllis unzima*; (C) ventral view of *H. scabra* showing *S. unzima* (red arrows) moving on the body wall of its host including one individual entering the mouth (right red arrow); (D) ventral view of a male harlequin crab *Lissocarcinus orbicularis*; (E) ventral view of a gravid female of *L. orbicularis*; (F) dorsal view of two individuals of *L. orbicularis* on the dorsal body wall of *H. scabra*; (G) dorsal view of a male pea crab *Holotheres semperi;* (H) dorsal view of the shrimp *Zenopontonia rex*; (I) a carapid *Encheliophis gracilis* between the tentacles of *Pearsonothuria graeffei*; (J) lateral anterior view of a carapid *E. gracilis.* The scale bars represent 0.3 cm in (A), 0.2 cm in (B), 2 cm in (C), 0.5 cm in (D and E), 0.7 cm in (F), 0.02 cm in (G), 0.4 cm in (H), 1.3 cm in (I) and 0.7 cm in (J). *Photo in (A) from Moorea Biocode, courtesy of G. Paulay, with permission; photos in (B) and (C) from Simon, C., San Martín, G., Robinson, G., 2014. Two new species of* Syllis *(Polychaeta: Syllidae) from South Africa, one of them viviparous, with remarks on larval development and vivipary. J. Mar. Biol. Assoc. U. K. 94, 729–746, with permission; photos in (D–F) courtesy of G. Caulier from Caulier, G., Lepoint, G., Van Nedervelde, F., Eeckhaut, I., 2014. The diet of the Harlequin crab* Lissocarcinus orbicularis*, an obligate symbiont of sea cucumbers (holothuroids) belonging to the genera* Thelenota, Bohadschia *and* Holothuria. *Symbiosis 62, 91–99, with permission; photo in (G) courtesy of J.-F. Hamel from Hamel, J.-F., Ng, P.K.L., McCurry, S., Mercier, A. 2019b. Note on the pea crab* Holotheres semperi *(Burger, 1895) parasiting the sea cucumber* Holothuria scabra *in Rempang (Indonesia). SPC Beche-de-mer Inf. Bull. 39, 36–38, with permission; photo in (H) courtesy of J. Poupin, with permission; photos in (I) and (J) courtesy of E. Parmentier, with permission.*

Syllis unzima (Fig. 18B, C) was found in high densities in aquacultured individuals of *H. scabra* in South Africa (sent from Madagascar). This species is characterized by a viviparious reproductive strategy involving large broods (more than 40 juveniles) that develop synchronously (Simon et al., 2014).

Among the crustacean symbionts of *H. scabra*, 5 species of crabs, 1 species of shrimp and 2 species of copepods were described (Table 13; Fig. 18D–H). *Hapalonotus pinnotheroides* [formerly *Hapalonotus reticulatus*] is an endosymbiotic crab belonging to the Pilumnidae family. Males and females live separately in different host individuals where they form a membranous cyst within the respiratory tree near the cloaca of *H. scabra* (Castro, 2015; Chia and Ng, 1999; VandenSpiegel et al., 1992). The harlequin crab *Lissocarcinus orbicularis* (Fig. 18D–F) is the commonest Indo-Pacific brachyuran colonizing the body wall or the buccal/cloacal cavities of several holothuroids belonging to the genera *Bohadschia, Thelenota* and *Holothuria*, including *H. scabra* (Caulier et al., 2012; Eeckhaut et al., 2004). This Portunidae lives on the host integument where it temporarily creates a slight depression in which it stands. Its chelipeds are smooth so that *L. orbicularis* can firmly attach to the dorsal papillae of its host without harming it. The colour of the crab varies from white to black, depending on the host body wall colour (Ayotte, 2005; Caulier et al., 2014). Three pea crabs belonging to the Pinnotheridae family can be found in *H. scabra*, i.e. *Buergeres deccanensis* [formerly *Pinnotheres deccanensis*], *Holotheres halingi* [also known as *Pinnotheres halingi*] and *Holotheres semperi* [formerly *Pinnotheres semperi*] (Fig. 18G). Pinnotheridae are exclusively comprised of symbiotic species (De Bruyn et al., 2016; Palacios-Theil et al., 2009). The three pea crabs are endosymbionts living in the respiratory tree or the posterior part of the intestine (i.e. often in the cloacal cavity). *B. deccanensis* may also be found in the perivisceral coelom of its host (Eeckhaut et al., 2004; Hamel et al., 1999, 2019a; James, 1995b; Ng and Manning, 2003). Hamel et al. (1999) observed atrophy of the respiratory tree and gonad reduction in *H. scabra* parasitized by the pea crab *Holotheres halingi* in the Solomon Islands. The infestation rate was highly variable depending on the investigated area and most crabs were found in pairs of opposite sex with young males being strongly attracted by the sea cucumbers hosting only one female. Copulation was observed within the female cyst and the embryonic development occurred 1 month after mating, leading to five zoeal pelagic larvae and the infestation occurring at the megalop stage after 2 months from mating. This new species was exclusively found in association with *H. scabra*. The emperor shrimp *Zenopontonia rex* [formerly

Periclimenes imperator] is a palaemonid that interacts as a generalist ectosymbiont on a large diversity of hosts (e.g. asteroids, nudibranchs, crinoids) including 11 different holothuroid species, among which *H. scabra* (Fransen and Hoeksema, 2014; Purcell et al., 2016b; VandenSpiegel et al., 1992). The copepods *Lichothuria mandibularis* and *Scambicornus idoneus* are two external symbionts living on the body wall of their host (Humes, 1980; Humes and Ho, 1969).

Three gastropod molluscs were observed on *H. scabra*. *Mucronalia* sp. [also known as *Eulima* sp.] and *Prostilifer* sp. are two Eulimidae snails while *Odostomia* sp. is classified into the Pyramidellidae. Both families are phylogenetically close and gather numerous ectoparasitic gastropods found on echinoderms but also annelid, crustacean and cnidarian hosts (Goodwill et al., 2009; Macnae and Kalk, 1962; Takano and Kano, 2014). The three gastropods can either live free on the body wall or embedded into it. Their boring proboscis allows them to feed from the coelomic cavity or the haemal system (Jangoux, 1984). They may induce the formation of a gall at the site of feeding (Bakus, 1973; James, 1998).

Three carapid fishes were exclusively found to live as endosymbionts in the coelomic cavity or the respiratory tree of *H. scabra* (Markle and Olney, 1990; Murdy and Cowan, 1980; VandenSpiegel et al., 1992), i.e. the pearlfishes *Carapus mourlani*, *Encheliophis* (*Jornanicus*) *gracilis* and *E. vermicularis*. They can enter the host through the cloaca and may perforate internal membranes to enter the general coelomic cavity. Symbiotic carapids have developed specific adaptations to their symbiotic life, including a general elongated shape and an important mucus production to cope with their host's chemical defences along with a resistance to anoxic conditions (Brasseur et al., 2016; Eeckhaut et al., 2015; Parmentier and Vandewalle, 2003). Moreover, pearlfishes communicate by exchanging acoustic information, allowing intraspecific interactions (Parmentier et al., 2003).

Even if sympatric holothuroid species do not always present exactly the same symbiotic community (Caulier et al., 2012; Queiroz et al., 2013), some investigators believe that at least two other taxa should be listed as symbionts of *H. scabra* because they are commonly found associated with sympatric holothuroids: protozoans like gregarines and ciliates (Doignon et al., 2001; Eeckhaut et al., 2004; Kroll and Jangoux, 1989) and umagillid platyhelminthes (Eeckhaut et al., 2004; Martens and De Clerck, 1994). However, this remains to be confirmed, especially since there are other symbionts that have been found on sympatric sea cucumbers but not on *H. scabra*.

19.2 Categories of symbionts

Numerous factors allow the classification of symbiotic associations as mutualistic, commensal or parasitic, also accounting for the fact that symbionts may switch from a category to another depending on the conditions. Therefore, symbioses should be considered as a broad continuum where the boundaries within categories are not well defined and are only based on the general impact on the host fitness (Parmentier and Michel, 2013). Details on the impacts that the symbionts can have on *H. scabra* are outlined in this section.

Symbiotic polychaetes are usually considered to be ectocommensals as they use *H. scabra* to shelter and do not represent an important symbiotic burden (Martin and Britayev, 1998). However, Britaev and Lyskin (2002) studied the diet of *Gastrolepidia clavigera* and found that it ingests its host integument, with holothuroid ossicles representing up to 75% of the polychaete's gut content, suggesting that they may be considered parasites. Similarly, some crustaceans (copepods and shrimp) can be described as ectocommensals of *H. scabra* (Eeckhaut et al., 2004; Humes, 1980; VandenSpiegel et al., 1992) while the symbiotic nature of brachyurans has been the subject of debates. The microscopic and genetic signature of the stomach content, along with its isotopic signature, suggested that the portunid crab *L. orbicularis* could potentially be considered as a moderate kleptoparasite, i.e. feeding directly on food present in the buccal cavity of its host (Caulier et al., 2014). The pinnotherids *Holotheres halingi* and *Holotheres semperi* seem to have more pronounced detrimental effects as they cause an atrophy of the right respiratory tree or the gonads (respectively) of their host *H. scabra* (Hamel et al., 1999, 2019a). Tresnati et al. (2021) suggested that *H. halingi* may switch from commensal to parasite when starving, being a potential threat for the natural broodstock of *H. scabra* in Indonesia. As for eulimid gastropods, they also present harmful effects as they use their proboscis to dwell into the body wall of *H. scabra* to feed from its internal fluids. Moreover, they may produce wounds that may be subjected to bacterial infection and thereby trigger the immune response of the host (Eeckhaut et al., 2004). Carapidae fishes are usually classified in two groups based on their stomach content and jaw morphology, with the gender *Carapus* being considered as commensal, using the holothuroid as a host but feeding on other food sources, and *Encheliophis* as a parasite that feeds on the internal organs of its host, specifically the gonads (Eeckhaut et al., 2004; Parmentier et al., 2016). In conclusion, with the exception of *Holotheres halingi, H. semperi* and *Encheliophis gracilis*, the known symbionts of *H. scabra* should be considered to have a generally neutral impact on their host, identifying them as commensals.

19.3 Host selection

One of the most important factors driving symbiotic associations concerns the symbiont recognition system towards its host. Indeed, communication between the associates is needed to ensure the maintenance of the symbiotic relationship through time (Caulier et al., 2013a; Parmentier and Michel, 2013). Since 1950, a few studies have shown that the secretion, transport and perception of semiochemicals in seawater through chemical sensing can trigger intraspecific (Hamel and Mercier, 1999) or interspecific (Davenport, 1950) interactions in holothuroids. Caulier et al. (2013a) revealed that the harlequin crab *L. orbicularis* can detect saponins to locate different holothuroid species, including *H. scabra* (Fig. 19).

In line with holothuroid species investigated so far, *H. scabra* produces triterpenoid saponins in its tissues (Caulier et al., 2013b; Mitu et al., 2017) and in the surrounding environment, which act as a chemical weapon to deter potential predators and generalist parasites (Caulier et al., 2011; Van Dyck et al., 2011). Symbionts living on/in *H. scabra* may benefit from the predator-deterrence properties of the saponins to increase their own fitness and thus may develop specific adaptations to resist such chemicals. This is notably the case of carapids that need to adapt in order to avoid gills damages because of saponins (Brasseur et al., 2016; Eeckhaut et al., 2015). It was also

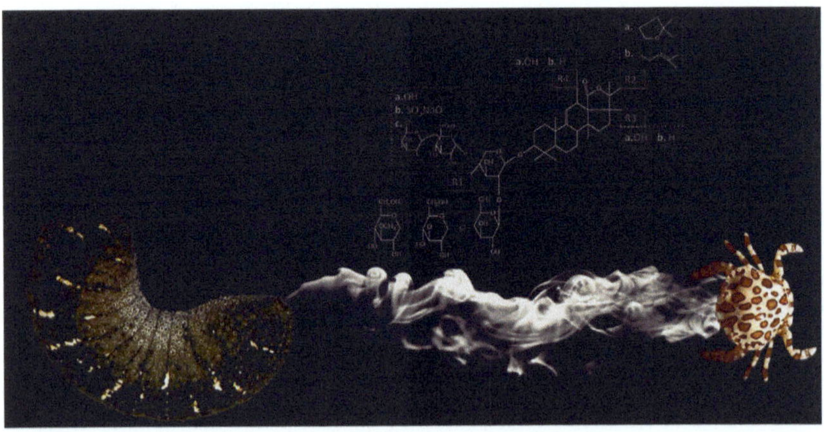

Fig. 19 Schematic representation (not to scale) of the role of saponins as mediators of chemical communication allowing the symbiont *Lissocarcinus orbicularis* to select its host *H. scabra*. The chemical structure represented in white corresponds to the summary of different saponins produced by *H. scabra* (i.e. desholothurin A, holothurins C and G, pervicosid C and scabrasides A and B). *Adapted from Caulier, G., Flammang, P., Rakotorisoa, P., Gerbaux, P., Demeyer, M., Eeckhaut, I. 2013b. Preservation of the bioactive saponins of* Holothuria scabra *through the processing of trepang. Cah. Biol. Mar. 54, 685–690.*

demonstrated that the pearlfish *E. vermicularis* could use saponins as a kairomone (i.e. a chemical cue allowing interspecific communication and providing a benefit to the organism detecting the molecules) to locate its host *H. leucospilota* (Miyazaki et al., 2014). It may therefore be suggested that most specific symbionts mentioned in this review use saponins detection, in combination with other modes of communication, to select their host.

19.4 Effect of anthropogenic activities on symbiotic community

The aquaculture of *H. scabra* raises the question of the potential relation existing between the symbiotic burden and sea cucumber farming. Even if most associations are considered as commensal in nature, i.e. neutral for the host, the high sea cucumber densities in farms can tip the natural balance by promoting proliferation of commensals, which may finally negatively impact the fitness of their host by switching to a parasitic mode of life because of a higher symbiotic burden (Eeckhaut et al., 2004; Houck, 1994). Future research may focus on the comparison between the symbionts of wild vs farmed individuals of *H. scabra* and on the indirect effects the symbionts may cause to cultivated holothuroids. One key question is whether specific commensal associations may favour the spread of diseases; for instance, wounds created by eulimid gastropods could be a point of entry for pathogenic microorganisms under high densities of sea cucumbers.

Inversely, the impact of farmed holothuroids on the symbiotic community would be an interesting topic to investigate in the future. As an example, the need for standardization of the size and body wall thickness of trepangs in aquaculture has probably a direct effect on the symbiotic community of *H. scabra*. Indeed, larger holothuroids display bigger symbiotic burden and a richer biodiversity (Lin, 2015) while the thickness of their body wall may have an impact on the ability of some parasites like eulimids to puncture their host with their proboscis (Queiroz et al., 2013).

Contrary to farming, overfishing of *H. scabra* has noticeably decreased the density of natural populations around the world, which could also negatively impact the general symbiotic diversity (i.e. coextinction) in benthic ecosystems where this species lives (Dunn et al., 2009). Surveys of the symbiotic community related to natural populations of *H. scabra* impacted by overfishing may be conducted at different intervals to assess the impact and the potential resilience of the symbiotic community of *H. scabra*.

H. scabra hosts a diversity of species from various phyla and thus may be considered a true micro-ecosystem and a unique ecological niche for numerous obligatory or opportunistic symbionts. This finding advocates for conservation plans not only to preserve the sea cucumbers themselves but the biodiversity that thrives because of them.

20. Fisheries

Fisheries for *H. scabra* are occurring in many countries of the Indo-Pacific where sea cucumbers are exploited, but their management is generally not specific. Since the original synthesis by Hamel et al. (2001), the interest for this marine resource has considerably increased over the last 20 years (Bruckner, 2006a; Conand and Muthiga, 2007; Friedman et al., 2008; Lovatelli et al., 2004; Muthiga and Conand, 2014; Purcell and Eriksson, 2016; Purcell et al., 2016b, 2018a; Rahman et al., 2020; Toral-Granda, 2007). However, much remains to be done with respect to management (Conand et al., 2018; Friedman et al., 2011). In fact, sharp declines in the stocks of *H. scabra* have earned this species the status of Endangered on the IUCN Red List of Threatened Species (Hamel et al., 2013).

Over the past two decades, sea cucumber fisheries have received increased attention; several international meetings were organized by the Food and Agriculture Organization of the United Nations (FAO) since 2003, which were followed by important publications (Conand, 2004, 2005; Lovatelli et al., 2004; Toral-Granda, 2007; Toral-Granda et al., 2008). Other global authorities, such as the Convention on International Trade in Endangered Species of Wild Fauna and Flora (CITES) (Bruckner, 2006a; Conand, 2006a,b) and the International Union for Conservation of Nature (IUCN) (see www.iucnredlist.org) have also invited specialists on working groups to help improve management practices (Conand et al., 2014b; Purcell, 2014a). The South Pacific Community (SPC) continues to publish the SPC Fisheries Newsletter (https://coastfish.spc.int/en/publications/bulletins/fisheries-newsletter) and SPC Beche-de-mer Information Bulletin (BDM) (https://coastfish.spc.int/en/publications/bulletins/beche-de-mer), which document the fisheries and management of sea cucumbers; the BDM started in 1990 with reports from the tropical Pacific but has since then expanded to cover world sea cucumber fisheries. The number and percentage of the different categories of articles on *H. scabra,* and on other holothuroids (e.g. on taxonomy, ecology, fisheries, socio-economics) are presented yearly, to show their relative importance (Conand, 2019, 2020; Conand et al., 2016, 2018).

Several books and syntheses have been published on the commercially important sea cucumbers, their stocks and management measures at world scale, by Purcell et al. (2012b, 2013). More recently, changes in sea cucumber fisheries worldwide have been analysed by Conand (2017, 2018a, 2019), Conand et al. (2016, 2018).

Considering that the regional scale is pertinent from the biological and socio-economic angles, meetings have been held in collaboration with both international and regional organizations. In the Indian Ocean, a project funded by WIOMSA (Western Indian Ocean Marine Science Association), involving five countries, was initiated in 2005 (Conand and Muthiga, 2007; Conand et al., 2006) and outputs were subsequently published (Muthiga and Conand, 2014; Muthiga et al., 2010). FAO organized a workshop in Zanzibar (FAO, 2013) and an ecosystem-based approach was then developed (Conand et al., 2013b; Eriksson et al., 2012a). In the tropical West Pacific, workshops were organized by various organizations (SPC, Worldfish, FAO, Australian Centre for International Agricultural Research/ACIAR), leading to a number of publications (Battaglene, 2000; Choo, 2008a,b; FAO, 2012; Friedman et al., 2008; Kinch et al., 2008a,b).

H. scabra is usually categorized as a first grade product on the international market (Arakawa, 1991; Baine and Choo, 1999; Conand, 1999a; Conand and Byrne, 1993; Purcell, 2014a; South Pacific Commission, 1994, 1995) where it fetches some of the highest prices (see below for details) (Adams, 1992; Conand and Sloan, 1989; Purcell, 2014b; Purcell et al., 2018a). Therefore *H. scabra* is among the most intensively exploited holothuroids in the Indo-Pacific (Bruce, 1983; Conand, 1997a, 1999a,b, 2018b; FAO, 2019; Gonzales, 1975; James and Ali Manikfan, 1994; McElroy, 1990; Preston, 1990; Purcell et al., 2012b, 2013; Sachithananthan, 1994a; Sloan, 1984; Taylor-Moore, 1994; Valayudhan and Santhanam, 1990; Van Eys, 1986, 1987). Data on world fisheries and trends were presented in Conand (1998a,b, 2001) and Jaquemet and Conand (1999), based on FAO compiled statistics and Hong Kong trade statistics (the main hub for the market). In 1996, approximately 23,000 metric tons (mt) of processed product were derived from the family Holothuriidae (including the genera *Actinopyga* Bronn, 1860; *Bohadschia* Jaeger, 1833; *Holothuria* Linnaeus, 1767; *Labidodemas* Selenka, 1867 and *Pearsonothuria* Levin et al., 1984). In more recent years, the international sea cucumber fisheries have been using the yearly statistics compiled by FAO, Hong Kong and exporting countries (Conand, 2004, 2006a, 2017, 2018b; Conand et al., 2014a, 2019; FAO, 2019; To et al., 2018; Toral-Granda et al., 2008). There has been an increase worldwide, with more countries now exploiting the resource (Anderson et al., 2011; Conand, 2017, 2018b; Eriksson et al., 2015c).

While it is generally possible to extrapolate the fishery data to a particular species where fisheries are more or less monospecific, as is the case for temperate areas (González-Wangüemert et al., 2018; Hamel and Mercier, 2008), it is not possible in most multispecies fisheries in the Indo-Pacific. Another difficulty arises from the change in species targeted by the fishers, which occurred over the past decades, when the first-grade species *Holothuria nobilis, H. whitmaei, H. fuscogilva*, and *T. ananas* became over-exploited in most areas (Anderson et al., 2011; Eriksson et al., 2015c; Purcell et al., 2013). The capture data are today mostly referring to a mix of the species, *H. scabra, H. lessoni, Bohadschia* spp. and *Actinopyga* spp.

Aside from being exported, *H. scabra* is also consumed locally by communities in Palau, Fiji and Papua New Guinea (Adams, 1992; Conand, 1990; Conand and Byrne, 1993; Kinch et al., 2008b), but this does not represent large harvested quantities.

20.1 History and price fluctuations

Sea cucumber has been recorded in the diet of Chinese since the Three Kingdoms between 220 and 280 AD, and its importance is shown by the fact that it is referred to as 'ginseng of the sea' in traditional as well as modern Chinese medicine (Yang et al., 2015a). According to Hornell (1917) and James and Baskar (1994) the Chinese have been trading sea cucumbers (as beche-de-mer) with Southern India and Sri Lanka for about one thousand years, anciently in exchange for porcelain, silk and sweetmeats. As *H. scabra* is one of the most common species along these coasts, it is likely that harvesting for this species dates back to this period. Carter et al. (1997) noted that Macassan sailors (Indigenous people from Indonesia) were fishing *H. scabra* in the 1600s. In Seychelles (Indian Ocean), the fishery for sea cucumbers dates back to the 18th century, with low quantities of several species harvested until the end of the 19th century (Aumeeruddy, 2007; Aumeeruddy and Conand, 2008). Today, *H. scabra* is exported to the Asian market via Hong Kong and Singapore (Conand, 1989; Conand and Byrne, 1993; Conand et al., 2014a; Sachithananthan, 1994a; To et al., 2018) and is probably the dominating tropical species, both in quantity and value (Conand, 2018b; Sachithananthan, 1994b).

Prices of dried products of *H. scabra* have fluctuated over the past decades. While it was considered a low-value species 35 years ago, its current trade value is among the highest on the market (Holland, 1994b; Purcell, 2014a). The selling price of *H. scabra* was 0.32 USD kg^{-1} in 1972 (Sachithananthan, 1972). James (1986b, 1989a) noted that *H. scabra* fetched no more than the fourth highest price on the market in 1986 when it was selling for 0.38–0.70 USD kg^{-1}. In the early 1990s, *H. scabra* was the most common medium-value beche-de-mer with prices around 14–20 USD kg^{-1} (Sommerville, 1993). Prices rose to 50–100 USD kg^{-1} in 1996 (Mercier and Hamel, 1997). In some areas, such as the Andaman and Nicobar Islands and elsewhere in India, fishermen devoted 90% of their effort to *H. scabra* at the expense of other valuable species (James, 1994a,b,c; James and James, 1994a). Due to this monospecific fishery, stocks of *H. scabra* started to decline rapidly (James and James, 1994b). For instance, in Fiji, the catch of *H. scabra* decreased by 80% between 1979 and 1993 (Stewart, 1993).

Additional data for several species, including *H. scabra* exported from the Pacific Islands, were presented in a detailed paper by Purcell (2014a). For Hong Kong, the average price for *H. scabra* in 2011 was 303 USD kg^{-1}, with a maximum of 1668 USD kg^{-1}, while in Guangzhou (China) it was 137 USD kg^{-1} on average and 200 USD kg^{-1} at most (Purcell, 2014b). A few years later, Purcell et al. (2018a) presented updated data on prices and relationships with product sizes; in 2016, the average and maximum prices in Hong Kong were between 369 and 1898 USD kg^{-1}, respectively, while in Guangzhou they were between 153 and 251 USD kg^{-1}, respectively. The product sizes and weights were both higher in Hong Kong than in Guangzhou, and the study determined that the price per individual increased exponentially with increasing length in *H. scabra* (Purcell et al., 2018a). The Chinese market shows resilience to changing situations (Eriksson and Clarke, 2015) and needs to be carefully followed.

20.2 Harvesting techniques

No important changes in the harvesting techniques for *H. scabra* have been documented recently (Lovatelli et al., 2004; Toral-Granda et al., 2008). When large numbers of *H. scabra* can be found in shallow water, on the sand or muddy sand flats, they are collected by hand while wading at low tide (MPEDA, 1989). This is related to the ecology of the species, which is mostly restricted to shallow waters. Snorkelling using face masks or goggles is a common collection method for deeper water sites (Fig. 20). In some places in India, women and children reportedly harvested *H. scabra* during

Fig. 20 Fishing and processing of *H. scabra*. (A) Large individual in muddy sea grasses in New Caledonia; (B) collecting in Papua New Guinea; (C) storage in bins after collection in New Caledonia; (D) boiling in New Caledonia; (E) gutted sea cucumber in New Caledonia; (F) drying on (F) racks and (G) tables in New Caledonia. *Photos in (A, D–F) courtesy of S. Purcell; photo in (B) courtesy of D. Vaïtilingon; photos in (C) and (G), courtesy of P. Chabre STHNC.*

low tide, while men focused on deeper waters by snorkelling (Radhakrishnan, 1994). In Sri Lanka and Tanzania, fishermen were known to use steel-pronged forks mounted on long handles (James, 1989a). SCUBA diving equipment was used when the only individuals available were in deeper water, such as in Sri Lanka, where fishermen dove to collect *H. scabra* at depths between 6 and 20 m (Anonymous, 1978b). Moreover, Dissanayake and Wijeyaratne (2007) indicated that, in Sri Lanka, the fishing crafts (with 8–10 divers aboard) left around 7h30–8h00 and came back between 14h30–15h30; some night fishing was also carried out. Fishermen in the Solomon Islands were reported to harvest deep individuals using a straightened fish hook inserted in a piece of lead of about 3 kg, attached to a fishing line (Baird and Sachithananthan, 1974), although these practices were not recommended as they damaged the skin and gave lower grade beche-de-mer products (Conand, 1999a). In India, *H. scabra* was sometimes also dredged by sail boats concurrently with fishes and prawns (MPEDA, 1989). In more recent publications, harvesting of *H. scabra* is often still done in shallow waters, by women and children gleaning at low tide, or by men snorkelling. It is briefly presented for many countries of the tropical Pacific and Indian Ocean, in Lovatelli et al. (2004), Toral-Granda et al. (2008) and Muthiga and Conand (2014).

Sachithananthan (1986) indicated that harvests of *H. scabra* kept out of water for long periods will develop skin lesions (especially in the fishing boat), and, if packed in irregular containers, their body wall will take the form of the container. They thus suggested that plastic boxes with a smooth interior should be used for stocking and transport. Sachithananthan (1986) also reported that freshly collected sea cucumbers will eject 2–3 cm of silt/sand from the cloacal opening, and this is a proper time to apply a pressure on the anal region to stimulate evisceration to get rid of most internal organs. This step is carried out in the fishing boat.

20.3 Processing into beche-de-mer

The traditional processing of holothuroids into the dried product known as beche-de-mer involves several steps, i.e. evisceration, boiling and drying, which slightly vary across species and *H. scabra* over the years. It was described and illustrated in several early publications (James, 1994f; Sachithananthan, 1979; South Pacific Commission, 1994), first updated in French (Conand, 1999a) and later more broadly (Purcell, 2014b). The traditional processing of *H. scabra* (Fig. 20) is similar to that of most tropical species, with an added step that involves removing the calcareous ossicles, which are very abundant in its body wall.

The process starts with small incision in the posterior area across the anus/cloaca (20 mm long), followed by removal of the internal organs (gonad, intestine, Polian vesicle, respiratory tree). The eviscerated body wall is then rinsed and undergoes a first boiling for 15–30 min, before being buried in the sand for about 18 h to facilitate the removal of the calcareous ossicles from the body wall. Individuals are sometimes frozen before the first boiling to have enough individuals for the processing (Yunizal et al., 1997). The sea cucumber is then cleaned with a rigid brush, to help remove any ossicle still attached to the body wall, and processed through a second boiling for 10–30 min, before it is finally dried in the sun and bagged (Adithiya, 1969; Conand, 1989, 1990, 1994, 1999a; Durairaj, 1982; Gurumani and Krishnamurthy, 1994; Hornell, 1917; James, 1986b, 1989a, 1994f; James and James, 1994a; Nair et al., 1994; Preston, 1993; Purcell, 2014b; Sachithananthan, 1994a; South Pacific Commission, 1994, 1995; Tebchalerm, 1984). Drying may take days to weeks, and it is sometimes preceded by smoking to suit the needs of specific buyers (South Pacific Commission, 1994, 1995).

In order to remove the calcareous ossicles remaining after boiling, Sachithananthan (1975) developed a motorized stainless steel apparatus referred to as a de-scummer in Sri Lanka. Moreover, Peranginangin et al. (1994) described the use of enzymatic treatment with papaya leaves to remove calcareous ossicles from smoke-dried sea cucumbers, while Purcell (2014b) reports using crushed papaya leaves in the water during the first boiling. These methods circumvent the necessity to bury the sea cucumbers overnight (see above for the classical method).

The traditional way of preparing the beche-de-mer from *H. scabra* has undergone slight changes over the last decades. Salt is often added immediately after evisceration, which preserves the quality of the tissues until enough individuals are ready for processing, and the burying method is no longer used in many locations (Purcell, 2014b).

A manual by Conand (1999a) illustrates the processing methods for *H. scabra* in the Indian Ocean using photos and cartoons; it is still in use, and updated information has been presented in Conand and Muthiga (2007) and Muthiga and Conand (2014). The recent manual for Pacific fishers by Purcell (2014b) details the steps for processing the main species harvested in the Pacific islands, including *H. scabra*. It is well illustrated and details the different steps involved in cutting, salting, cooking, smoking and drying. It is important to keep in mind that Asian consumers like dried sea cucumbers that are straight, without damage to the skin, have a dark colour (for dark species), mild odour and are not covered by salt. Nair et al. (1994) indicated that a good beche-de-mer should be properly dried, and free from fungal, insect and mite infestation. A recent project was conducted in the Philippines to develop alternative technology for processing *H. scabra* (Pangan et al., 2017): it includes the fabrication of a stainless preparation table, a stainless cooking vat with cover, a mechanical cleaning/scraping machine and a dryer using solar energy or biomass waste for heating. Sea cucumbers were subjected to different treatments during processing, using modified processing conditions, flavourings, and different packaging techniques.

Regardless of the method used, sea cucumbers decrease in size and weight considerably during processing (Conand, 1979, 1986, 1990; Preston, 1990; Vuki and Viala, 1991). Baskar and James (1989) measured a loss of approximately 42% in size and 91% in weight. Durairaj et al. (1984) stated that length shrinkage was between 50% and 60%. Shelley (1985a) found that final dry weight corresponded to 5% of original wet weight. SPC, in a Handbook for fishers (South Pacific Commission 1994, 1995) also noted that beche-de-mer from *H. scabra* corresponded to about 5% of the initial weight, e.g. from an initial total weight of 370 g, the dry product decreased to about 20 g. Holland (Holland, 1994a,b) also noted that up to 20 units of beche-de-mer produced with *H. scabra* were needed to obtain 1 kg of product in the Solomon Islands. Hair et al. (2018a) have obtained the recovery rates from wild harvested and cultured *H. scabra*: they were not statistically different for fresh total weight (5.0% vs 5.6% respectively), but not different for fresh length (52.3% vs 49.7%), eviscerated weight (8.8% vs 9.6%), collagen content 94.5 vs 104 $mg\,g^{-1}$ and ash content (15.2% vs 14.0%). Ram et al. (2014, 2016), following previous work by Conand (1979, 1990, 1999a) and by Purcell (2014b), have studied the influence of processing on the size of final products in several sea cucumber species, including *H. scabra*, which was collected in Fijian waters where it can be easily harvested or purchased from fishers, during a project aimed at raising awareness about the importance of sea cucumber processing in village communities. They found that extending salting times (by more than 48 h) affects the collagen composition and the final product quality.

As mentioned above, the processing methods used to produce beche-de-mer have a major influence on the price at all levels of the trade, from the harvesters to the retail markets (mostly in China). The main defects that can influence the quality (and price) of the final product are: insufficient drying, scratches on the body wall and unattractive incisions. Also, very small or non-uniform products, often associated with an inconsistent processing method, are less appreciated (Conand, 1999a; South Pacific Commission, 1994, 1995). Conversely, a sizable product with a proper appearance, no bad odour, an appropriate colour and a moisture rate lower than 20% are required for an optimal value. Sachithananthan (1994a) noted that the muscle bands should be preserved during evisceration to keep the beche-de-mer straight.

With better shipping and air freight options, a diversification of the products derived from holothuroids has occurred, as shown by the China fishery seafood exhibition of 2014 (Choo and Conand, 2015). Many new products with new packaging are now found in the markets (Purcell, 2014a). While the traditional dried products of *H. scabra* are found in many markets as shown in Fig. 21, additional options include dried product in gift boxes [see photo of *H. scabra* from shops in Sydney, Australia, in Purcell et al. (2014b)]. A variety of alternatives exist, such as frozen, vacuum-packed products, as well as extracts such as 'gamat water' produced by simmering the sea cucumber in water and 'gamat jelly' and soap, used as medicines or tonics. During the China Fisheries Seafood Expo 2014 held in Quindao (China), products from sea cucumbers coming from many countries where presented, including the dried forms of *H. scabra* cultured by the National Aquaculture Group of Saudi Arabia (Choo and Conand, 2015). This product diversification is also apparent in Hong Kong trade statistics; the codes have changed and imports have been presented in four different forms since 2012: imports for cultivation; live, fresh or kept chilled; frozen including those smoked; dried, salted or in brine including those smoked; prepared foodstuffs (Conand et al., 2014a). Unfortunately, the holothuroid species are not always specified, but *H. scabra* is probably very important in exports from the tropical Indo-Pacific countries to Hong Kong, and it is likely to be traded in most if not all forms.

Novel packaging for partially processed, cooked and treated with essential oils, and cooked and cold smoked as well as processed and reconstituted *H. scabra* were tested (Ram, 2017). Products were subject to modified atmospheric packaging (MAP), with a gas ratio of 60%CO_2/40%N_2 and 5%O_2/50%CO_2/45%N_2, and also vacuum packaging prior to shelf-life

Fig. 21 Various trade products of *H. scabra*. (A and B) In bags, Hong Kong market; (C) in containers, Singapore market; (D) in jars, Hong Kong market; (E) in boxes, Guangzhou market; (F) in bags, Guangzhou market. *Photos (A, D–F) courtesy of S. Purcell; photo in (B) courtesy of D. Vaïtilingon; photo in (C) courtesy of C. Conand.*

assessment tests. The cold smoked samples, packaged using MAP and vacuum-packing, recorded 0 CFU g^{-1} at day 0 for SPC, yeast and mould and this remained consistent throughout the shelf life. The cold smoked sandfish (MAP and vacuum-packed) recorded 0 CFU g^{-1} bacterial, yeast and mould counts until day 30. There were no differences between initial and 7 days samples for moisture, protein, lipid or NFE contents, across treatments. There were no differences in the levels of moisture, protein, lipid, ash and NFE between initial treatment samples. There were no differences in the levels of the metabolically important fatty acids C20:4n-6, 20:5n-3, 22:6n-3 between samples from either day, across all treatments after 7 days of storage. This was also true of the total level of n-3 polyunsaturated fatty acids (PUFA) in the samples. No differences were observed in collagen composition or texture, however, the tissue toughened with the storage time (Ram, 2017). Recent results on biological activities of sulphated polysaccharide show the need for adapted standardized processing methods (Li et al., 2021).

Aprianto et al. (2020) examined the quality of the processed forms of *H. scabra* harvested with different techniques in Indonesia (Fig. 22). The main causes of low value included elimination of the fumigation stage, storage in refrigerator, addition of boiling, overcooking, and using excessive papaya leaves. Although water, ash, protein, total plate count, *Salmonella*, lead and cadmium levels were in the acceptable range, the exporters still purchased beche-de-mer at lower prices because exporters have to undertake reprocessing to improve the quality of the final products.

Samples obtained from consolidators or processors show that the moisture content of thoroughly dried product is about 12% wet basis or 0.14 g g^{-1} dry matter (Yaptenco et al., 2013). Beche-de-mer is known as a hygroscopic material that can reabsorb moisture, leading to decay (Nair et al., 1994). The traditional drying process involves a combination of smoking and sun-drying until moisture content is 8–10%. However, sea cucumber is often poorly dried, leading to spoilage and mould growth during storage. To establish conditions for proper storage, Yaptenco (2017) developed moisture sorption isotherms (MSI) at 30, 45 and 60°C. The MSI of *H. scabra* followed a Type III curve, which was best fitted to a Peleg model. The best range of moisture content for storage was 0.12–0.18 g g^{-1} dry matter. Predicted shelf life for polyethylene (PE) and oriented polypropylene (OPP) film packs at 30°C were 114 and 372 days, respectively; at 35°C, the estimated shelf life decreased to 83 and 141 days, respectively.

Fig. 22 Processing of *H. scabra* into beche-de-mer (trepang) in South Sulawesi, Indonesia: (1) after weeding; (2) boiling; (3) softening of outer skin with papaya; (4–5) scraping of outer skin; (6) result of steps 4–5; (7) second weeding; (8–10) results after second weeding; (11 – 12) salting; (13–16) sun drying. *From Aprianto, R., Amir, N., Tresnati, J., Tuwo, A., 2020. Bycatch sea cucumber* Holothuria scabra *processing and the quality characteristics. In: IOP Conference Series: Earth and Environmental Science, vol. 473. IOP Publishing. p. 012001 under CC BY 3.0.*

Trondillo et al. (2018) also remarked that traditional methods of processing and drying *H. scabra* often result in poor quality and low value products. Smoking and sun drying can leave scorch marks and smoke residue, and inadequate drying, especially during poor weather also affect the product. A combination of hot air and solar drying was compared to the traditional process of smoke drying followed by sun drying to determine the effect on product quality and price (Trondillo et al., 2018). Results showed that important shrinkage occurred regardless of the drying treatment used; it was greater in samples subjected to the traditional process, although this was not statistically significant. Scanning electron microscopy revealed changes in muscle and collagen fibres between primary drying/Treatment I (hot air drying or solar drying) and secondary drying/Treatment II (solar drying or sun drying). Assessment of dried samples by a producer, consolidator, and exporter showed that Treatment II produced better quality products that could command higher prices by 25–89% compared to Treatment I.

20.4 Other products derived from *H. scabra*

Yenti et al. (2019) proposed to develop a drink from *H. scabra* and performed a proximate analysis of fresh meat, to obtain the best formulation in antioxidant and saponin for the functional drink. The best formulation based on organoleptic tests was the following: 20% sea cucumber, 20% lemongrass, 20% sugar, 20% rosella flower, 10% citrus, and 10% ginger. The antioxidant content was low because the IC_{50} was ~1838 ppm (Yenti et al., 2019).

Elfath et al. (2020) also explored the preparation of raw materials from *H. scabra* and the formulation of functional beverage products, with a focus on bioactive compounds such as saponins and antioxidants. The best scores in organoleptic tests were obtained for functional drinks with 30% *H. scabra*, 20% rosella, 10% ginger, 10% lemongrass, 10% kalamansi orange and 20% sugar. As in the study by Yenti et al. (2019), the antioxidant activity measured in the drinks was weak because the IC_{50} value was >200 ppm (1621 ppm). Saponins were identified by the formation of foam (Elfath et al., 2020).

Azam and Singh (2013) investigated value-added food products from *H. scabra*, especially to produce biscuits and jams. The two different types of biscuits (sweetened and salted) and four different types of jams were made with fruits like guava, banana, lemon and wi (wild apples or Polynesian plum). The proximate composition and appreciation of biscuits and jams were evaluated. High protein content of 11–12% was measured in both biscuits, higher than in normal biscuits (Oreo®: 1%). The four jams also had higher protein content (lemon sandfish jam: 11.27%, banana sandfish jam: 16.39%, guava sandfish jam: 15.37% and wi sandfish jam: 18.44%) than normal Fruto® jam (0.26%). However, there were no differences between the moisture, fat and protein content of biscuits and jams. Sensory evaluation indicated sweetened biscuits to be more acceptable in comparison to salted biscuits, and jams made of banana and wi fruit to be more acceptable than jams made of guava and lemon. However, there were no differences between the sensory characteristics of biscuits and jams (Azam and Singh, 2013).

20.5 Catch data

Data on holothuroid captures are abundant, although they mostly refer to a mix of commercial species, and only a few data specific to *H. scabra* are available. More than 30 years ago Conand (1989) indicated that a single diver, at the start of the exploitation in New Caledonia, could collect between 83 and 230 $kg h^{-1}$ of *H. scabra*. A few years later, a study noted that *H. scabra* represented approximately 28% of the sea cucumber catches in New Caledonia (Anonymous, 1993), with about 393 mt $year^{-1}$ (metric tons per year). An estimated 100–150 mt $year^{-1}$ were collected in Northwest Sri Lanka in the 1990s (James and Baskar, 1994). Anonymous (1996b) indicated that the catch of *H. scabra* in Tonga was 2089 kg in 1995. The total catch on the east coast of Queensland (Australia) was approximately 48 mt between July 1995 and June 1996 (Anonymous, 1997). In New Caledonia, commercial categories for the species were introduced into fisheries statistics (i.e. 1: high value, 2: medium value, 3: low value; with 2 + 3 categorized as other species) to obtain a better evaluation of the fishery (Conand, 1986, 1990). However, before 1998, *H. scabra* was classified as a medium-value species. Given the demand for this species, it is reasonable to assume that it dominated this medium-value category (possibly alongside *H. lessoni*). In those years in New Caledonia the catches were: 34.9 mt (on a total of 47.9 mt) in 1995, 13.3 mt (on a total of 33.6 mt) in 1996, 32.7 mt (on a total of 46.7 mt) in 1997 and 13.7 mt (on a total of 24 mt) in 1998 (Chantal Conand, unpublished data).

Depletion of stocks due to overfishing has been demonstrated for *H. scabra* in New Caledonia (Conand, 1989), and in Malaysia (Forbes and Ilias, 1999). In India, catches reached 91 mt in 1975 and dropped to

11 mt in 1985 (James, 1989a). Similarly, in Papua New Guinea, Lokani et al. (1995b) and Lokani (1996) indicated that harvests reached 192,000 individuals in 1991 and dropped to 39,000 individuals in 1993 as the yield estimate per hectare also decreased sharply between 1991 and 1993. Stocks of *H. scabra* declined similarly in Mozambique, with sea cucumber catches (not referred to species) dropping from 700 mt in 1993 to 54 mt in 199, as a result of intense exploitation (Abdula, 1998).

Publications in the first decade of 2000 provided data specifically on catches of *H. scabra* when populations were dense enough to warrant a fishery. They are presented in books of general interest on holothuroids published by FAO, such as in Lovatelli et al. (2004), which includes a chapter on the world sea cucumber resources and their utilization by Conand (2004), and in Toral-Granda et al. (2008). Purcell et al. (2012b) compiled 58 commercially important sea cucumbers, including *H. scabra,* with identifications, photos and general information on habitat and biology, exploitation and geographic distribution. Other publications, such Bruckner et al. (2003), Conand (2005) and Bruckner (2006a), present basic information on *H. scabra* and other species, including the taxonomy, distribution and conservation status (Conand, 2006b), the fishery and socio-economics (Conand, 2006a), the management (Bruckner, 2006b), the CITES status (Sant, 2006), and reports from 18 countries. Recommendations from the three working groups are given on: (1) National fisheries management, (2) Priorities on international Conservation and protection, (3) Potential CITES implementation issues.

Table 14 shows the total holothuroid captures, which include *H. scabra,* in the tropical Indian and Pacific Oceans, annually in mt, from recent FAO data (2020) in the four fishing areas, Western Indian Ocean (WIO), Eastern Indian Ocean (EIO), Western Central Pacific (WCP) and Eastern Central Pacific (ECP), for two periods (2010–13 and 2014–18), to show the trends of the mean captures year^{-1} and the percentage of the world captures. On the whole, 28 countries were analysed, including 12 from the Indian Ocean and 16 from the Pacific Ocean. The proportion of the catches from these areas (all sea cucumber species combined) decreased from 33.5% to 21.7% (of the world catch), while the global annual mean world catch increased from 40,196 mt in 2010–13 to 44,672 mt in 2014–18.

The overall status of the fisheries around *H. scabra* is difficult to estimate globally, given the changes occurring in management and conservation, with new countries entering the exploitation on a regular basis, and the changes in species fished, which affect the relative importance of *H. scabra.* Therefore, it is preferable to detail regional or national publications, to try understanding the

Table 14 Captures of sea cucumbers in the Indo-Pacific, in tonnes (fresh weight), during two periods. Σ 2010–13 and Σ 2014–18 are the sum for these periods, mean/Y is the mean per year in each period, and % is the percentage of the total world captures.

Fishing area	FAO code	Number of countries	Σ 2010–13	m/Y	%	Σ 2014–18	m/Y	%
Indian Ocean, Western	51	9	9950	2488	6.2	9063	1813	4.1
Indian Ocean, Eastern	57	3	16,767	4192	10.4	12,482	2496	5.6
Pacific, Western Central	71	13	25,653	2496	16.0	26,522	5304	11.9
Pacific, Eastern Central	77	3	1517	379	0.9	367	73	0.2
Total Tropical Indo-Pacific		28	53,887	13,472	33.5	48,434	9687	21.7
Total World Captures			160,785	40,196		223,363	44,672	

changes in fisheries. Recent publications, more specific to countries or regions of the tropical Indian and Central Pacific Oceans where *H. scabra* is exploited are presented below and Table 15 based on FAO data (2020). The captures by country for these areas between 2015 and 2018 are provided.

Sea cucumber catches from the tropical Pacific islands and Asia are provided in many contributions. The global review of fisheries and trade by Toral-Granda et al. (2008) includes all the regions where *H. scabra* is exploited. In the same global review, Kinch et al. (2008b) presented data from three geographic and cultural areas of the Western Central Pacific (WCP): Micronesia, Polynesia and Melanesia. The biological and population status, trade, and socio-economic importance of *H. scabra* in Asia was provided by Choo (2008a). Other contributions on holothuroids including *H. scabra* have been published by Purcell (2014a,b), Purcell et al. (2009a,b, 2013). Eriksson and Byrne (2015) reviewed the Australian fishery for holothuroids along the Great Barrier Reef Marine Park from 1991 to 2011 catches: they showed global patterns of serial exploitation (changes over time), from high-value species to less-value species, and highlighted key needs, such as gathering more precise data on the fisheries,

Table 15 Captures (in tonnes wet weight) of sea cucumber by country, in the Indian and Central Pacific Oceans, for years 2015 to 2018 (from FAO, 2020). The sum (Σ) for the Western (WIO) and Eastern (EIO) Indian Ocean, Pacific Western Central (WCP), and Pacific Eastern Central (ECP) areas are shown, along with the percentages (%) of the world captures they represent.

				Captures (metric tonnes, wet weight)			
Country	**Species**	**FAO fishing area**	**FAO code**	**2015**	**2016**	**2017**	**2018**
Egypt		I. O. Western	51	0	0	0	0
Kenya		I. O. Western	51	171	6	13	15
Madagascar		I. O. Western	51	1598	535	922	853
Maldives		I. O. Western	51	0	0	99	0
Mauritius		I. O. Western	51	0	0	0	0
Mozambique		I. O. Western	51	0	0	0	0
Seychelles	Black teatfish	I. O. Western	51	8	10	2	2
Seychelles	Prickly Red	I. O. Western	51	108	79	72	72
Seychelles	Sandfish	I. O. Western	51	0	0	0	0
Seychelles		I. O. Western	51	467	270	194	194
Seychelles	White teat	I. O. Western	51	108	81	66	66
Tanzania, U.R.		I. O. Western	51	0	0	0	0
Yemen		I. O. Western	51	32	27	23	23
Σ WIO		I. O Western	51	2492	1008	1391	1225

% of world capture			6.14%	2.20%	2.80%	2.42%
Indonesia	I.O. Eastern	57	309	477	805	1013
Malaysia	I.O. Eastern	57	0	0	0	0
Sri Lanka	I.O. Eastern	57	2560	1190	1090	1100
Σ EIO	I.O. Eastern	57	2869	1667	1890	2113
% of world captures			7.07%	3.64%	3.50%	4.17%
Σ IO	I. Ocean		5361	3683	4677	3338
% of world captures			13.2%	5.8%	6.4%	6.6%
Brunei D.	P.O. Western	71	0	0	0	2
Cambodia	P.O. Western	71	0	0	0	0
Fiji	P.O. Western	71	50	54	54	54
Indonesia	P.O. Western	71	3609	3424	3424	3689
Kiribati	P.O. Western	71	0	0	0	0
Malaysia	P.O. Western	71	0	0	462	432
New Caledonia	P.O. Western	71	462	496	320	192
Palau	P.O. Western	71	0	0	0	0
Papua N. G.	P.O. Western	71	0	0	0	0

Continued

Table 15 Captures (in tonnes wet weight) of sea cucumber by country, in the Indian and Central Pacific Oceans, for years 2015 to 2018 (from FAO, 2020). The sum (Σ) for the Western (WIO) and Eastern (EIO) Indian Ocean, Pacific Western Central (WCP), and Pacific Eastern Central (ECP) areas are shown, along with the percentages (%) of the world captures they represent.—cont'd

				Captures (metric tonnes, wet weight)			
Country	**Species**	**FAO fishing area**	**FAO code**	**2015**	**2016**	**2017**	**2018**
Philippines		P.O. Western	71	686	747	745	746
Solomon I.		P.O. Western	71	100	100	100	100
Vanuatu		P.O. Western	71	75	59	60	60
Wallis Futuna Is.		P.O. Western	71	0	0	0	0
Σ WCP		P.O.Western	71	4985	4880	5165	5275
% of world captures				12.3%	10.7%	10.3%	10.4%
FrenchPolynesia		P.O. Eastern	77	39	51	51	51
Tonga		P.O. Eastern	77	0	0	0	0
U.S.A.		P.O. Eastern	77	0	0	0	0
Σ ECP		P.O. Eastern	77	39	51	51	51
% of world captures				1%	1%	1%	1%
Σ CP		P. Ocean		5024	4931	5216	5326
% of world captures				12.4%	10.8%	*10.4%*	*10.5%*
Total World Captures				**40,544**	**45,773**	**45,773**	**50,661**

adopting a precautionary approach in harvests and extending closed periods in fishing zones. Recent issues of the SPC Beche-de-mer Information Bulletin and SPC Fisheries Newsletter also contain data on the fisheries and management of *H. scabra* in Melanesian countries where it is harvested (Bermudes, 2018; Govan, 2017a,b, 2019).

The Indonesian fishery, occurring in both the western central Pacific (WCP) and Indian oceans, is still the most important for tropical holothuroid captures, on the world scale (Choo, 2008a; Conand, 1990, 2004, 2018b; Tuwo, 2004). Based on the most recent (2018) data published by FAO (2020), including details of captures in the different FAO fishery areas, 1013 mt in area 57 (Pacific Western Central) and 3689 mt in area 71 (Indian Ocean Eastern) were captured, which makes a total of 4702 mt. Several case studies on fisheries from Indonesian villages exists for *H. scabra*: Purwati (2006a) tells about a visit in Lombok (village in central Indonesia) where harvesters collected large specimens of '*bantun beras*' (the local name of *H. scabra*) that were salted before further processing. Navarro et al. (2014) presented data from a small fishery in the village of Palau Misa, where harvesting was carried out by free divers during day trips, with women processing the catch on the same day. Conand and Tuwo (1996) had described this fishery in Barang Lompo (South Sulawesi) with 300 fishermen (30 boats) collecting several species, including *H. scabra*, that were kept in salt, during long trips, then transformed by 60 processors and dried in the sun on dirt roads. Some 20 years later, Yusuf et al. (2017) observed that the same number of boats brought back smaller catches to only 4 processors, but garnered higher prices.

The Philippines sea cucumber fishery has long been second in importance in the tropical Indo-Pacific (after Indonesia), but the catches declined after the 'boom' in 1985 (Choo, 2008b; Conand, 1990, 2004; Gamboa et al., 2004; Schoppe, 2000). From the data provided by FAO (2020) for the Philippines, only 746 mt were recorded for 2018. The species *H. scabra* is locally called '*putian*' or '*curtido*', and it was the most expensive species in 2000, at 35 USD kg^{-1}, for the best weight category (XL = 15 individuals kg^{-1}, or 65 g each) (Akamine, 2001). Dolorosa et al. (2017) recently observed the fishery in an island Wildlife sanctuary of Palawan archipelago, where the density of *H scabra* was 19 individuals ha^{-1} and the processed product very small, under 5 g per beche-de-mer.

Other fisheries for sea cucumbers from the WCP area, include Malaysia (Choo, 2012), Papua New Guinea (Hair et al., 2018a, 2020b; Kinch, 2002; Kinch et al., 2008a), Torres Strait islands, a protected zone, managed under cooperation between Australia and Papua New Guinea (D'Silva, 2001), Solomon Islands (Nash and Ramofafia, 2006), Fiji (Mangubhai et al., 2017) and the border between Cambodia and Vietnam (Otero-Villanueva and Ngoc Ut, 2007).

The eastern central Pacific (ECP) only contributes to 1% of the catches of holothuroids for recent years (Table 15). Overall, the central Pacific countries are only contributing to around 10% of the world sea cucumber captures for the last years. This is an important decrease, despite Indonesia and the Philippines declared relatively stable captures for the period 2015–18 (FAO, 2020).

Sea cucumber catches for the Indian Ocean have been presented in several publications by Conand (2008, 2017, 2018b) for both the Eastern and the Western regions, by FAO zones and in Table 14. They are detailed by country in Table 15 for recent years, 2015–18 (FAO, 2020).

For the eastern region of the Indian ocean (EIO), Indonesia (western section) and Sri Lanka are the most important producers (Choo, 2008a; Conand, 2008, 2017, 2018b). Asha et al. (2017) described the distribution of holothuroid species in India, with details about the fishery and trade that existed mainly in the Gulf of Mannar and the Palk Bay (south of the country), providing livelihoods to harvesters inhabiting the region. The fishing was mainly centred around two commercially species, *H. scabra* and *H. spinifera*. The indiscriminate exploitation led to decline of the wild population and forced a ban in 2001 (Asha et al., 2017). However, illegal fishery and trade continues, making the moratorium ineffective. A controlled regulatory mechanism of collection and trade of sea cucumbers involving all concerned stakeholders might be a preferred policy solution, as this would help in conservation as well as sustainable use of these resources (Asha et al., 2017). In Sri Lanka, the fishery was described in several publications, such as Kumara et al. (2005), Dissanayake and Wijeyaratne (2007), Dissanayake et al. (2010), Dissanayake and Stefansson (2012), and Nishanthan et al. (2019). Sri lanka now ranks first for the holothuroid catches in the Eastern Indian Ocean (Table 15). It is a seasonal fishery (due to monsoon winds), especially on the north coasts. The catches are mostly composed of low value species, with *H. scabra* as the only high value harvested, fetching mean fresh individual price of 6.7 USD. The CPUE (catch per unit effort) for *H. scabra* (number/person/day) is very different between regions, with 109 for scuba diving, 43 for snorkelling in the north-west, 3 for scuba diving, 10 for snorkelling and 14 for gleaning in the north, not reported in the north-east. Sri Lanka earned a total of 2.55×10^6 USD in 2016 from the export of 136 mt of beche-de-mer. More than 10,000 fishing families in the north, northwest and northeast coasts depend of that fishery for their livelihood to this day (all local species harvested confounded), based on the recent study by Nishanthan et al. (2019).

For the western region of the Indian Ocean (WIO), the recent statistics (FAO, 2020) are given in Table 15 and show that the area contributes to around 2.5% of the world sea cucumber catches. Conand and Muthiga (2007) and Muthiga and Conand (2014) provide details on the fisheries for *H. scabra* and other tropical species in the six countries who participated in a MASMA (Marine Science for Management) project funded by WIOMSA; three continental countries (Kenya, Tanzania and Mozambique) and three islands, La Reunion (an Overseas Department of France), Seychelles, and Madagascar present very contrasted socio-economic conditions. For example, the Human Condition Index (HCI) (Purcell et al., 2013) varied in 2013 from high values of 0.97 in La Reunion, where the reefs are well-managed reserves without holothuroid harvesting, and 0.84 in Seychelles, to much lower values of 0.47 in Kenya, 0.44 in Madagascar, 0.40 in Tanzania and 0.32 in Mozambique. Madagascar has a long-lasting holothuroid fishery (McVean et al., 2005; Rasolofonirina, 2007) and, despite the decline of catches, remains first for the WIO, followed by the Seychelles (Aumeeruddy, 2007; Aumeeruddy and Conand, 2007, 2008). The FAO statistics of 2020 for Seychelles are separated into five categories (the main four valuable species and others, including *H. scabra*), showing that it was not collected intensely in recent years. In Kenya and Seychelles, the harvesters were men, while in Zanzibar women represented 29%; in Madagascar women represented 13% and children 26% of the harvesters. In villages near Toliara in the south of Madagascar, the catches of *H. scabra* represented <10% of a total of 16 holothuroids species. The results from the MASMA project (Conand and Muthiga, 2007; Muthiga and Conand, 2014) gave an overview of the types of situations pertaining to holothuroid fisheries in the WIO region. Other publications have presented the national holothuroid fishery for Tanzania (Mmbaga, 2015) and Zanzibar (Eriksson et al., 2015a), including a comparison with Mayotte (an Overseas Department of France). Other holothuroid fisheries from the WIO (Table 15) have been reported for Egypt (Ahmed and Lawrence, 2007; Lawrence et al., 2004, 2006), Oman (Al-Rashdi and Claereboudt, 2010; Al-Rashdi et al., 2007b; Claereboudt and Al-Rashdi, 2011), Saudi Arabia (Hasan, 2008), Iran in the Persian Gulf

(Afkhami et al., 2012). In several WIO islands and countries, *H. scabra* is either rare or absent, such as in Mayotte (Conand et al., 2005; Eriksson et al., 2012a; Pouget, 2005), Mauritius (Lampe-Ramdoo et al., 2014), La Reunion (Conand et al., 2018), and the Eparses islands (Conand et al., 2013a,b, 2016; Mulochau et al., 2014). This scarcity is probably due to the non-reef ecology of the species rather than to overexploitation.

20.6 Management

Some 20 years ago, Hamel et al. (2001) showed that overexploitation of sea cucumbers was ongoing and that illegal captures were on the rise, resulting from the lack of adequate management in most fishing countries. Things have not improved much over the last two decades, even though holothuroids, like many other marine resources, have received increasing attention. Numerous cases of illegal fishing are reported recently (a few examples from different countries are presented in Appendix). In the tropical Pacific and Indian Oceans, management has been assessed in countries where *H. scabra* is present. This section presents an overview of these assessments, followed by more specific case studies at the regional level. The very dynamic Chinese market has shown resilience to overexploitation and changing situations (Eriksson and Clarke, 2015) and needs to be carefully followed.

20.6.1 Overview

Global reports were published following international meetings organized by FAO, where management was a key issue. The first one 'Advances in Sea Cucumber Aquaculture and Management' (ASCAM) was held in 2003, in Dalian, People's Republic of China and the results from different countries were presented as chapters (Lovatelli et al., 2004). Among them, information on *H. scabra* can be found, on the Great Barrier Reef (Uthicke, 2004), the Northern Territory of Australia (Shelley and Puig, 2004), Papua New Guinea (Polon, 2004), the Seychelles (Aumeeruddy and Payet, 2004) and the Solomon Islands (Ramofafia et al., 2004). A second global review by Toral-Granda et al. (2008) followed another FAO international meeting held in the Galapagos Islands in 2007. The approach of this review was regional and national management of holothuroids with comprehensive data related to *H. scabra* presented for the western central Pacific with a focus on Papua New Guinea (Kinch et al., 2008a,b), for Asia with a focus on the Philippines (Choo, 2008a,b), and for the Indian Ocean with a focus on the Seychelles (Aumeeruddy and Conand, 2008; Conand, 2008).

Based on the expertise gathered during this FAO meeting, an 'Ecosystem Approach' was advocated by Purcell et al. (2010). This document details the management principles, the regulatory measures, and the implementation of management and also provides useful practical annexes. Briefly, it was recommended that regulations at the different levels (harvesters, processors and traders) include size limits, gear limitation, marine reserves, access rights and licensing, monitoring and reporting along the market chain. Aspects to include in the plan of actions were also identified, including fishery-independent stock surveys, surveys of catch and effort, socio-economic surveys, education and communication with stakeholders, and training for processing. One conclusion was that while a rotational zoning system is used for the management of holothuroid resources in several developed countries with industrial-scale fisheries (Alaska, USA; British Columbia, Canada; Great Barrier Reef, Australia), it seems difficult to implement in less developed or small-scale fisheries of tropical islands that have a modest planning capacity and complex access rights to fishing grounds (Purcell et al., 2016a). Recently, concerns around climate change and how to limit its anticipated ecological impacts were presented in studies on the importance of fuel consumption by the harvesters (Purcell and Lalavanua, 2018; Purcell et al., 2018b).

Global conservation concerns were expressed at the CITES international workshop on sea cucumbers, with a publication edited by Bruckner (2006a) containing several background documents, including management and conservation strategies and practices (Bruckner, 2006b; Sant, 2006). Given the difficulties encountered to list holothuroids on CITES Appendices, only one species (*I. fuscus*) had previously been placed on Appendix III. The recent Conference of the Parties (CoP18), held in Geneva in 2019 adopted a resolution to place the teatfish group, i.e. *Holothuria* (*Microthele*) *nobilis, H. whithmaei* and *H. fuscogilva* on Appendix II, with a delay of 1 year, to allow the exporting countries to get a permit to prove NDF (Non Detriment Findings) (CITES, 2020; FAO, 2019). Will *H. scabra* be considered for CITES listing in the future? It remains to be seen, for two reasons: the processed product is not easily differentiated from other species and *H. scabra* is now grown more or less successfully in aquaculture facilities in several countries (see Section 21). However, based on five criteria, *H. scabra* was listed as Endangered on the IUCN Red list (Hamel et al., 2013), along with six other exploited species, including five tropical ones (Conand et al., 2014b; Purcell et al., 2014c).

20.6.2 Regional initiatives

Given the distribution of *H. scabra* throughout the tropical Indo-Pacific, its exploitation is not only examined at global scales, but several smaller regional initiatives have also tried to produce practical tools to improve its management. For example, following a meeting on conservation of sea cucumbers in Malaysia (Hashim et al., 1999), the Malaysian Network for Holothurian Conservation added several important points for management of holothuroids, such as establishing conservative management guidelines and baseline surveys prior to the start of a fishery. The management plans also proposed a ban on harvest during the breeding season, the introduction of quotas and minimum landing sizes, the establishment of permanent survey sites and preserved areas, the maintenance of records on harvesting data and the ban of SCUBA equipment.

During the following years, numerous countries increased interventions to protect the resource (Hashim et al., 1999). More recently, the Australian Centre for International Agricultural Research presented the 'manager's toolbox' (Friedman et al., 2008), with a section to help assess the health of the fishery using six indicators: (1) presence of breeding groups (areas with adult sea cucumbers), (2) fishing gear used, (3) global sea cucumber abundance, (4) ratio of species abundance, (5) size of sea cucumbers, (6) fishers profiting (benefits going to the fishing community); it is followed by a section on management practices. Studies were subsequently published on the pattern of vulnerability and recovery of sea cucumber stocks impacted by fishing in six countries, with different management regimes, showing that new precautionary approaches are needed (Friedman et al., 2011; Pakoa, 2013). In 2011, the FAO in collaboration with ACIAR, SPC and Southern Cross University organized a 4-day workshop in Fiji, called Sea Cucumber Ecosystem Approach to Management (SCEAM) Pacific (FAO, 2012). It included seminars delivered by the facilitators, pre-workshop fishery forms filled by the participants, short presentations by participants, as well as work-groups on stock health, on practical regulatory measures and on management actions to discuss the challenges and highlight many recommendations of further networking, research and training for sustainable use and management of sea cucumbers (FAO, 2012).

In the Western Indian Ocean (WIO), a regional concern was also raised during the Environmental Program of the Indian Ocean Commission (COI/EU) (Conand, 1999a). In Madagascar, for example, several actions have taken place, including the setting up of the National Trepang Traders Group (ONET). Several national meetings have favoured the exchange information between different participants in the 'Beche-de-Mer Fishery System', which includes harvesters, processors, traders, administrators and scientists involved in various projects (Conand, 1997a, 1998b). Another WIO project was funded by WIOMSA through the MASMA project on sea cucumbers, to build knowledge on management practices and identify challenges for several countries of this region (Conand and Muthiga, 2007; Muthiga and Conand, 2014; Muthiga et al., 2010). The same regional approach was more recently used for the entire Indian Ocean, during a SCEAM Indian Ocean workshop held in Zanzibar, Tanzania (Eriksson et al., 2012d; FAO, 2013). The need for improvement of the management capacity and planning was still recommended (Eriksson et al., 2015b; FAO, 2013).

At the country level, the main management measures recommend in a National Management Plan (Table 16), are: (*i*) establish a minimum size

Table 16 Plan of action for the management of *Holothuria scabra*.

Fishery system	**Management actions**
Natural resources in commercial holothurian species Local or national resources in *H. scabra*	Research on biology Stock assessment Hatcheries—production of juveniles Sea ranching—mariculture
Fishermen catches collected by wading, snorkelling, scuba diving	Respect of fishery legislation: size (ban on juveniles), periods, zones, national legislation Catches and CPUE Values
Processing by fishermen or women or processors	Improving the quality during all phases of processing Storage, grading Values Education
Fishery services and customs, national, regional and international trade International agencies	Communication between the agencies and stakeholders Storage, grading Standardized statistics on catches and values Access to information
Import and consumption	Information on market regulations and preferences Quantities and values

limit for capture, (*ii*) introduce strict quotas, including a total ban for a defined period when the stocks are decreasing, (*iii*) limit the number of export businesses, with strict quotas, (*iv*) alternate closed seasons, (*v*) ban SCUBA diving or specialized apparatus for harvesting, (*vi*) establish reserves, and (*vii*) promote stock enhancement (via aquaculture production). The involvement of all stakeholders, in a participatory manner, is essential.

20.7 Case studies

Case studies of management measures adopted at the national level will be presented below, with general examples applicable to tropical species of holothuroids, including *H. scabra*. Ideally, species-specific measures should be implemented in the future. The recent CITES decision regarding *Holothuria* teatfish species (Di Simone et al., 2020; FAO, 2019) will help support such initiative, but nothing has been developed with regards to *H. scabra* yet. For the time being, management systems in New Caledonia and Papua New Guinea provide relevant examples, since these countries have long established holothuroid fisheries that include *H. scabra* among the exploited species (Purcell et al., 2010; Toral-Granda et al., 2008).

In New Caledonia, several research programmes have been conducted on sea cucumbers, first by the Institut de Recherches pour le Développement (IRD) in the 1980s (Conand, 1989, 1990, 1993, 1998a, 2006a,b), then by Worldfish and the South Pacific Community (SPC) (Purcell et al., 2009a). More recently, recommendations for management were proposed and discussed, highlighting four main efficiency factors: spatial management and shared governance of fisheries; defining total allowable catches (TACs) per species and area, using stock biomass; implementing TAC-based operational spatial management; and the technical and financial capacities that managing departments need to acquire (Léopold et al., 2013a,b, 2015; Purcell, 2014b).

Papua New Guinea is representative of the Western Central Pacific (Toral-Granda et al., 2008), given its important sea cucumber fishery (Lokani et al., 1995b, 1996), which supplies international markets without any regulations (Kinch et al., 2008a; Purcell et al., 2010). In 1998, the National Fishery Authority (NFA) became responsible of the Management Plan, implementing permits, restricting the harvest season from January to September, province-specific TACs, and a minimal legal size set for *H. scabra* at 22 cm live and 10 cm dried. A long list of recommendations was made for different levels of scientific, economic, and sociologic studies. In 2006–07, socio-economic studies were conducted in several provinces of Papua New Guinea to refine the local management. Following the decreases in sea cucumber population densities, a precautionary approach was suggested (Friedman et al., 2011). The NFA imposed a moratorium on the fishery in October 2009. Recent publications by Hair et al. (2018b, 2019, 2020b) and Barclay et al. (2016, 2017, 2019) use this fishery as an example of 'Governability of High-Value Fisheries in Low-Income Contexts'. The ban lasted until 2017. The NFA revised the National Plan prior to the fishery re-opening and allocated provincial TACs based on estimated fishable stocks of sea cucumber (all species confounded). The results of the study in the New Ireland Province (Hair et al., 2020b) show 'the need to increase awareness of fisheries regulations (particularly species size limits to reduce the amount of undersized beche-de-mer that is brought in for sale), to strengthen the reporting requirements for companies and also to implement extension services to improve processing to lower the high rejection rates by traders of poorly processed beche-de-mer'. In 2017, the first year of the re-opening of the sea cucumber fishery, *H. scabra* represented 40.7% of the landings, CPUEs declined as the fishery progressed and the NFA's revised plan was not administered properly. Recommendations were made for improvements in future sea cucumber fishing seasons. A unified approach and initiation of the joint management bodies (government, community and industry) is advocated (Hair et al., 2018b, 2019, 2020b).

Sea cucumbers have long been fished in Tanzania without management or regulations, with a peak in 1998 (Conand and Muthiga, 2007; Mgaya and Mmbaga, 2007; Mmbaga, 2015; Mmbaga and Mgaya, 2004; Muthiga and Conand, 2014). Decrease of abundance was subsequently observed, and a ban was implemented in 2006 for mainland Tanzania, but not for the Zanzibar archipelago, where harvesting is still legal to this day, making evaluation of the origin of the catches difficult at the country level. Eriksson et al. (2010) first stated that a management reform was urgently needed. Later, Eriksson et al. (2012b) through interviews, discussions, participant observations, and boat monitoring, analysed the multifaceted scuba diving fishery, showing a 'roving bandit' behaviour on multiple spatial scales. A later comparison between two different small-scale fisheries (SSF) from the region, Zanzibar and Mayotte (France administration), allowed Eriksson et al. (2015b) to conclude that 'a governance model that distributes and shares management responsibility between government and community appears advantageous. This must be coupled with high institutional capacity to enforce rules and keep at bay the destructive responses that reinforce the trajectory of the fishery'. The spatial scale was later enlarged to 17 fisheries in the Indian Ocean (Eriksson et al., 2015c) showing that, notwithstanding the importance of a good inclusive governance, many other factors (inadequate regulations, weak enforcement, high number of harvesters) are involved. Participatory diagnosis for effective governance and management are useful tools to find solutions to local situations (Eriksson et al., 2016).

Aumeeruddy (2007) presented the situation in the Seychelles in the WIOMSA programme (Conand and Muthiga, 2007). This country was also the example for sea cucumber fisheries in Africa and Indian Ocean region (Aumeeruddy and Conand, 2008), as sea cucumber fishery is very important there, being second after the tuna fishery. Pinault and Conand (2007) visited some stakeholders and Aumeeruddy and Conand (2007) presented the length and weight distributions and length-weight correlations of the processed products from the species harvested. It was an old non-regulated fishery until 1999 when the Seychelles Fishery Agency (SFA) introduced licences for fishing and processing and required data on catch efforts and exports. A management plan was finalized in 2005, with a mix of input and output controls: 25 fishing licences and a TAC for each main species (*H. scabra* included) and 1707 mt as total eviscerated wet weight for all species. The implementation of the plan is managed by an Advisory Management Committee (AMC). The effects of the MPA was analysed by Cariglia et al. (2013). A project called SEACUSEY (Sea Cucumber Seychelles) is ongoing (Léopold and Govinden, 2018) to define and implement up-to-date operational management measures that are appropriate for the resource diversity, genetic structure, abundance, distribution and trends in the four main commercial species (not *H. scabra*). Despite the low importance of *H. scabra* in Seychelles, it is clear that the country is an example of active management, for a scuba dive-based sea cucumber fisheries on a very large spatial scale.

Although recent efforts were deployed to improve sea cucumber fisheries management through actions such as better evaluation of the research projects (Hiruy et al., 2019) and enforcement capacities, overfishing is ongoing (Hasan, 2019) and illegal fisheries have bloomed (Purcell and Eriksson, 2016). Conand (2018a) outlined many worldwide examples of illegal fisheries, in the years 2015–18, from several countries conducting traditional fisheries in the Western Tropical Pacific and the Indian Ocean, and from countries with more recent fisheries around the Galapagos and in the Atlantic Ocean. One clear example was published

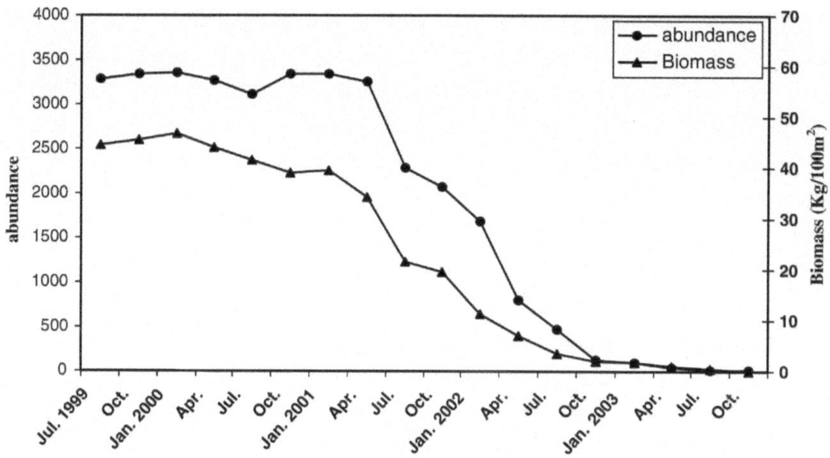

Fig. 23 Variations in the abundance and biomass of *H. scabra* at Abu Rhamada Island (Egypt) from July 1999 to October 2003. *From Hasan, M.H., 2005. Destruction of a* Holothuria scabra *population by overfishing at Abu Rhamada Island in the Red Sea. Mar. Environ. Res. 60, 489–511, with permission.*

by Hasan (2005), showing rapid destruction of a pristine population of *H. scabra* in the Red Sea between 1999 and 2003 (Fig. 23). The cases reported in recent years, from the Western Pacific concern Northern Australia where Indonesian and Vietnamese are poaching; the latter also fish illegally in Palau, Malaysia and New Caledonia on their 'blue boats'; in the Western and Eastern Indian Ocean, many examples come from India, Madagascar, Tanzania and Kenya. These illegal fisheries share the same characteristics: they are either practised by local people within MPAs, or mostly by foreign divers and poachers in remote sites, who remain mobile in order to escape controls and are thus referred to as 'roving bandits' (Eriksson et al., 2015b). These are more frequent in countries where there is a permanent ban, or weak regulations, or where remote fishing areas occur (Conand, 2018a). In an unpublished document by Sheppard and Price (unpublished data) mentioned that despite their isolation in the middle of the Indian Ocean, the Chagos atolls are subject to poaching. The number of boats captured from Sri Lanka averaged about 3 or 4 per year over the last 15 years. Sometimes the boats transported also sea cucumber fishers, collecting or dropping off groups of fishermen to live on the islands of the northern atolls and Great Chagos Bank. These sea cucumber fishermen glean from the shallow areas, use snorkelling and, sometimes, use scuba.

The catches are dried and salted and then, after a few weeks, the fishing boats return to pick them up along with their catches. However, the Chagos was established in April 2010 as a marine protected area and became a fully no-take marine reserve later that year (Sheppard et al., 2012). The archipelago is totally uninhabited apart from the island of Diego Garcia, which harbours a US military facility through an arrangement with the British Indian Ocean Territory (BIOT). Despite time and more surveillance being required, faint signs of recovery of the sea cucumber populations were noted in the following years (Price et al., 2013).

Mitigations to these problems are based on the general principles of management, including raising awareness at international, national and regional levels, increasing scientific input to monitor the changes in exploited species, providing fishery statistics at different levels and encouraging collaboration between the different stakeholders, administrations and scientists. These issues should be discussed broadly and be taught at an early age to help develop long-term solutions.

21. Aquaculture

21.1 Background

H. scabra has long been considered to be one of the best candidates for tropical sea cucumber aquaculture (Battaglene and Bell, 1999; Conand, 1999b; Mercier and Hamel, 2013). Research on aquaculture production was initiated in India by the Central Marine Fisheries Research Institute of the Indian Council of Agricultural Research (ICAR-CMFRI) in the 1980s (James, 1983, 1987b) leading to successful spawning induction, embryonic and larval development, as well as juvenile growth. D.B. James subsequently published several papers on the aquaculture of *H. scabra* (James, 1986b, 1987b, 1989a, 1994b,f; James and Baskar, 1994; James and James, 1994a; James et al., 1988, 1996); see also Asha et al. (2017) for a review of Indian publications on *H. scabra*. Many investigations related to aquaculture have since then been conducted to understand its life cycle, from fertilization to adult commercial size (Hair et al., 2012; Lovatelli et al., 2004; Purcell et al., 2010; Toral-Granda et al., 2008); see Pitt (2001), Giraspy and Walsalam (2010), Jimmy et al. (2012), Mills et al. (2012), Mercier and Hamel (2013), and Junus et al. (2018) for reviews. Studies were conducted in various countries at the start of aquaculture, but especially in the Solomon Islands (Battaglene, 1999a,b,c, 2000; Battaglene and Bell, 1999; Battaglene and Seymour, 1998;

Battaglene et al., 1998, 1999, 2002). Today, despite the large number of research programs and publications dedicated to *H. scabra*, few companies have developed cost-effective aquaculture operations. Comparing the aquaculture development of *H. scabra* in today's world to what was happening 30 years ago, the methods can be deemed successful. On the other hand, the number of profitable private aquaculture ventures that emerged from research projects remains limited, indicating that the aquaculture of *H. scabra* did not fully live up to the expectations and that it still cannot boast the large-scale success achieved with the temperate species *Apostichopus japonicus*. This mitigated success is partly due to aspects intrinsic to the biology of *H. scabra*, which are not immediately apparent during pilot-scale hatchery phases but emerge later during expansion towards mass production of this species.

21.2 Aquaculture research programs, pilot-scale and large-scale commercial enterprises

Since the last review on *H. scabra* published by Hamel et al. (2001), a number of aquaculture-related investigations have developed in several countries, which can be categorized broadly as: (*i*) research programs, (*ii*) pilot-scale aquaculture and (*iii*) large-scale aquaculture with commercial production. Research programs are mainly supported by academic and/or governmental institutions and aim to better understand and control the life cycle of *H scabra* from the fertilization to commercial-size adults. Pilot-scale enterprises are generally financed in whole or in part by universities and/or government and are sometimes supported by private funds. Such initiatives control the full life cycle under captive conditions, but annual yields generally do not exceed 500,000 hatchery-produced juveniles (i.e. seedlings). Large-scale commercial productions mainly finance themselves with private funds; their hatcheries can produce at least 500,000 juveniles annually and they export several tons of beche-de-mer each year.

Still today, aquaculture programs are most commonly at the research stage in most countries. When Purcell et al. (2012a) surveyed the number of hatchery-produced juveniles of *H. scabra* in the world, 11 countries were involved in producing them: Australia, Fiji, Federal States of Micronesia, India, Iran, Madagascar, Maldives, New Caledonia (France), Philippines, Saudi Arabia, Solomon Islands and Vietnam. With the exception of the Maldives, where the production by Masmeeru Pty Ltd. was said to reach 5 million juveniles year^{-1}, and Australia where Blue Fin Ltd. produced

Fig. 24 Map of the known geographic distribution of *H. scabra* highlighting the states where aquaculture has developed at a commercial scale (in dark blue, with the location of companies) and those where research projects or pilot-scale commercial production are taking place (in light blue). The yellow countries are those where no aquaculture project has been documented. South Africa is outside the map; however, research has been conducted there with *H. scabra* originating from Madagascar.

500,000 juveniles annually, the other countries did not produce more than 200,000 juveniles year^{-1} (except Madagascar, see below) and were generally in the research phase (Purcell et al., 2012a).

Fifty-seven states have a coastline in the geographic area corresponding to the natural distribution of *H. scabra* (Fig. 24). For 30 of them, fisheries past or present have been documented, but there are no scientific data related to aquaculture of *H. scabra*: Bangladesh, British Indian Ocean Territory, Brunei, Cambodia, Comoros, Cook Island, Djibouti, Egypt, Eritrea, Israel, Jordan, Mauritius, Mayotte (France), Mozambique, Myanmar, Nauru, Niue, Pakistan, Reunion, Samoa, Seychelles, Singapore, Somalia, Sudan, Taiwan, Tokalau, Tonga, Tuvalu, Wallis and Futuna and Yemen. The other 27 countries have conducted a number of research programs in recent years aimed at developing the aquaculture of *H. scabra*. However, ten of them either have pilot-scale (Indonesia and Malaysia) or commercial-scale companies (Australia, China, Madagascar, Maldives, New Caledonia, Philippines, Vietnam, and probably Saudi Arabia) whereby the private sector has funded or currently supports research projects (Fig. 24). For 15 of the 27 countries, scientific information shows a production of juveniles and sometimes adults but at a sub-industrial level (i.e. no established companies). In addition to India and Solomon Islands (mentioned above), information indicates that juvenile productions are now active in the Sultanate of Oman (Al-Rashdi et al., 2011, 2012, 2019), Sri Lanka (Chen et al., 2009; Haran et al., 2012; Kumara and Dissanayake, 2017; Kumara et al., 2013; Mallawaarachchi and Arulnanthan, 2014; Medagedara et al., 2019), Palau (Jimmy et al., 2012), Thailand (Vaitilingon et al., 2016), Papua New Guinea (Hair et al., 2016, 2020a,b, Jimmy et al., 2012), Fiji (Hair et al., 2011, 2012), Kiribati (Ito, 2014; Jimmy et al., 2012), Federate States of Micronesia (Cantero and Smith, 2016; Purcell et al., 2012a), and Vanuatu (Jimmy et al., 2012). Reproduction of *H. scabra* has been studied in Tanzania (Kithakeni and Ndaro, 2002) and Iran (Dabbagh and Sedaghat, 2012). Various programs are ongoing in Malaysia and Indonesia (Mills et al., 2012; Padang et al., 2016, 2017; Rizal and Muskita, 2018; Serang et al., 2016; Sugama et al., 2019; Tomatala and Rahantoknam, 2017; Tomatala et al., 2018a,b; Tomatala et al., 2018c). Small scale farming based on the collect of wild *H. scabra* has been studied in Zanzibar, Tanzania (Hacker, 2017) and Sri Lanka (Mahishalini et al., 2019). Laboratory experiments have been done on *H. scabra* juveniles in Okinawa (Japan) (Minami and Tamaki, 2013).

South Africa is not in the geographic area where *H. scabra* occurs naturally, but studies have been conducted with juveniles of the species imported from Madagascar. There is no scientific publication concerning the aquaculture of *H. scabra* in Mauritius but a sufficiently large population was recently discovered and fertilization and larval rearing trials have been carried out successfully (D. Vaïntiligon, Personal communication).

Hatchery production of *H. scabra* in India was developed over 30 years ago in the 1980s and 1990s (James, 1983, 1986c, 1987b, 1989a, 1994b,f, 1996a, 1999; James and Baskar, 1994; James and James, 1994b; James et al., 1988, 1994a,b). Knowledge is now used mainly towards sea ranching, but no large-scale production is documented in the literature. In 2001, success was achieved in the production of another commercial and over-exploited sea cucumber species, *H. spinifera* (Asha, 2005; Asha and Muthiah, 2002, 2005), which was the second major contributor to the beche-de-mer production from Indian waters. Through continuous research, several improvements were made in the hatchery and culture protocols of the two species. Determination of optimal spawning induction methods, larval feeding regimes, microalgal diets, environmental conditions, settlement preferences, and stocking densities at different stages of larval and early juvenile rearing have been investigated (Asha and Diwakar, 2013; Asha and Muthiah, 2005, 2006, 2007, 2008; Asha et al., 2004, 2011). Sea ranching of 11,335 juveniles of *H. scabra* (mean length of 23 mm) in different locations of the Gulf of Mannar was carried out during this period (Asha et al., 2007).

In China, *H. scabra* has been cultured since the 2010s in Xuwen County, near Zhanjiang City, on the western coast of Guangdong Province (Fig. 24), where the species naturally occurs according to Purcell and Wu (2017). The grow-out phase occurs in earthen seawater ponds on three farms, comprising more than 300 ha in surface area. Many of these ponds had previously been used for shrimp culture but were abandoned. The large ponds in Xuwen County are being used to grow sea cucumbers in a three-dimensional co-culture with pearl oysters *Pinctada martensii* and groupers (*Epinephelus* sp.). In 2015, approximately 4 million juveniles were produced. The starting stocking density for grow-out is about 2–3 ind. m^{-2}. From these juveniles, 1–2 million commercial-sized adults were reportedly harvested, with body lengths averaging ~20 cm and fresh body weights ranging from 300 to 500 g a piece (Purcell and Wu, 2017).

In Madagascar, the aquaculture of *H. scabra* is led by Indian Ocean Trepang (IOT), a company that was created in 2008 as an outcome of

Belgian-Malagasy research projects (Eeckhaut, 2021; Eeckhaut et al., 2008; Pascal and Robinson, 2011; Robinson and Pascal, 2009, 2012) (Fig. 24). It is currently one of the few industries to produce *H. scabra* with support from private funds exclusively, and it employs local fishermen as sea farmers. Hatchery production is over 2 million juveniles year^{-1} and the factory exports several tons of processed products to Hong Kong. The company, which is located in Toliara, has a hatchery for the production of seedlings, 10 ha of seawater ponds as nursery for grow-out and 400 ha of fenced enclosures in the ocean for further growth of individuals up to a marketable size (Eeckhaut, 2021). They work in collaboration with villages spread over 300 km of coastline.

In the south-east of New Caledonia, the 'Société d'Élevage Aquacole' (Ouenghi Aquaculture Farming Company; SEA) operates in the town of Boulouparis (Fig. 24), with a farm comprising a hatchery and 16 ha of ponds for raising sea cucumbers (*H. scabra* and *H. lessoni*). SEA began operating in 2010 after finalizing all phases of research concerning the reproduction, larval rearing and adult growth. A portion of the juveniles is released into the environment as part of lagoon reseeding programs, but the main portion is grown in ponds of a former shrimp farm. In 2019, SEA was able to produce sea cucumbers from 5 to 500 g. Its productivity is about 4–5 tons of beche-de-mer per year.

Australia's first commercial sea cucumber hatchery, Bluefin Seafoods Hatchery, officially started operations in 2003 in the Great Sandy Strait, Hervey Bay, on Queensland's mid-north coast Giraspy and Ivy (2005) (Fig. 24). In 2004, another company, Tasmanian Seafoods, started investigating the potential of propagation and juvenile production of *H. scabra*, with the objective of enhancing the wild fishery through sea ranching and exploring land-based grow-out (Fig. 24). It secured the lease on a prawn farm in 2009. The facility comprised eight 1-ha ponds, six 0.1-ha ponds, and one 2-ha reservoir. The 1-ha ponds were used for production, yielding 2.0–2.5 t wet weight per hectare.

In the Maldives, no activities are documented except for the information in Purcell et al. (2012a) concerning the production of 5 million juveniles by the Masmeeru Pty Ltd. company. Commercial aquaculture of sea cucumber is mainly limited to the hatchery production and farming at Shavayani Atoll in Kendhikulhudhoo lagoon (James, 2012) (Fig. 24). Other companies, such as BluBridge in Laamu atoll, have shown recent interest and success in initiating sea cucumber farming. The initial commercial hatchery and farming of sea cucumber relied on broodstock from India, brought around the year 2000 by Indian consultants. The hatchery breeding technology was refined over the years through trial and error, and the juveniles are now produced in local hatcheries and seeded in protected lagoons for grow-out (James, 2012). When the juveniles reach 2–3 cm in length, they are transferred to nearby company-owned atoll lagoons for further growth. In addition to local companies grow-out operations, considerable quantities of juveniles are grown, with similar grow-out periods, by contract growers and villagers from the nearby islands. When the sea cucumbers are fully grown, the local growers sell them back to the local companies and are paid a management fee according to the duration of care and quantity of sea cucumber produced.

Tuwo and Tresnati (2015) reviewed the sea cucumber farming in Malaysia, Philippines, Indonesia and Vietnam (Fig. 24). No large-scale industries are presently documented in Malaysia and Indonesia (Tuwo, 2004; Tuwo and Conand, 1992; Tuwo and Tresnati, 2015; Tuwo et al., 2012). In Vietnam, large-scale production has been conducted based on knowledge developed from research collaborations between the Vietnamese Research Institute for Aquaculture and ACIAR–WorldFish (Bell et al., 2007; Duy, 2005, 2010, 2012; Mills et al., 2008; Pitt and Duy, 2004b). Small juveniles are produced in hatcheries, grown in hapa nets described initially by Duy (2012) and Pitt and Duy (2004b), then grown in earthen ponds. The production is managed by local farmers and sold to local dealers and buyers in Ho Chi Minh City.

Accounts indicate that the Philippines are very close to mass-production of *H. scabra* (Gamboa and Juinio-Meñez, 2003; Gamboa et al., 2004; Juinio-Meñez et al., 2013, 2017). The South East Asian Fisheries Development Centre (SEAFDEC) inaugurated a sea cucumber hatchery in its main station in Tigbauan, Iloilo, in 2010. At least two companies currently produce *H. scabra* (Juinio-Meñez et al., 2017). Palawan Aquaculture Corporation (PAC) is located in Baquid Island, Coron in northern Palawan, while Alsons Aquaculture Corporation (AAC) is located in Alabel, Sarangani Province in south central Mindanao (Juinio-Meñez et al., 2017) (Fig. 24). PAC and AAC operate fish hatcheries, land-based ponds, and fish cages in leased sea areas. The PAC hatchery for *H. scabra* can rear up to 10 million oocytes per batch (Juinio-Meñez et al., 2017). PAC has a potential annual capacity to produce 450,000 juveniles from tanks, 225,180 juveniles from floating hapas and 71,316 juveniles from bottom-set hapas. The hatchery in AAC can rear up to 6 million oocytes per batch (Juinio-Meñez et al., 2017). AAC has a potential annual capacity to produce

525,720 juveniles from tanks, 183,984 juveniles from hapas in tanks and 104,316 juveniles from ponds (Juinio-Meñez et al., 2017).

According to Tuwo and Conand (1992) and Tuwo (1999, 2004), sea cucumber juveniles in Indonesia are mainly collected from the wild. However, to optimally utilize 10% of the potential farming area, Indonesia needs about 7.2–14.4 billion juveniles per year, suggesting that the development of hatchery-produced seedlings of *H. scabra* would be very important for the future of sustainable farming (Gamboa and Juinio-Meñez, 2003; Gamboa et al., 2004). Publications about aquaculture research on *H. scabra* in Indonesia can be found in Lantang and Paulangan (2015), Sembiring et al. (1996, 2004, 2015, 2016, 2017), Sembiring and Hutapea (2019), Tomatala et al. (2020) and Notowinarto and Putro (1992, 1993).

In Malaysia, a project for a large-scale industry started in 2014 under the business incubator programme of the Fisheries Research Institute (FRI) in Pulau Sayak, Kedah (Vaitilingon et al., 2016). It was the product of collaboration between the Malaysian Department of Fisheries, a spin-off from research conducted by an Australian-based company, Asia Aquaculture Development Co. Pty Ltd. and a local engineering company (Sesaga Engineering Snd Bhd). The project, led by the spin-off company, aimed at developing and commercializing a sea cucumber aquaculture system for producing premium quality beche-de-mer.

There is no published information on the production of *H. scabra* in Saudi Arabia, but an anonymous source estimated that one company produces hundreds of thousands of sea cucumbers annually (Fig. 24). Aquaculture is managed by the National Aquaculture Group (NAQUA). NAQUA is also among the largest, fully integrated shrimp farms in the world and covers an area of approximately 250 km^2 on the Red Sea coast. Born from a research project in 1982 and located 180 km south of the historic port of Jeddah, close to the town of Al Leith, NAQUA grows various organisms including *H. scabra*, which are hatchery-reared and grown-out in the Red Sea.

With funding from the Republic of Korea, FAO assisted the development of a multipurpose hatchery in Zanzibar to help local farmers and to reduce overfishing. The facility was completed and has been functional for a few years (Alessandro Lovatelli, Personal communication, 2022).

21.3 Hatchery production

The production of juveniles of *H. scabra*, i.e. of young post-metamorphic individuals of a few millimetres long, also known as seedlings or seeds, relies on proper control over four main stages: (i) broodstock collection and maintenance, (ii) spawning and fertilization, (iii) larval rearing and metamorphosis and (iv) settlement with post-settlement rearing.

21.3.1 Broodstock collection and maintenance

Broodstock collection, conditioning and spawning was most recently reviewed by Purcell et al. (2012a). Since the work of D.B. James in India and the first hatchery manual published on *H. scabra* (James et al., 1994b), the methods for broodstock collection and maintenance have remained very similar everywhere. Broodstock is collected from commercial catches meant for processing (Pitt and Duy, 2004b), from individuals in the wild (Ceccarelli et al., 2018; Kumara and Dissanayake, 2017; Morgan, 2000a,b; Purcell and Simutoga, 2008) or from aquaculture production (Pitt and Duy, 2004b). Batches of 20–30 (Asha et al., 2011; Battaglene, 1999a) to 30–45 individuals (Agudo, 2006; Pitt and Duy, 2004b) are typically required to induce a small proportion of animals to spawn. The average wet weight of broodstock individuals varies but is generally 250 g wet weight or more (Agudo, 2006). Transport of broodstock from the wild follows different procedures from one hatchery to another. Battaglene (1999a) mentioned that collection of broodstock needs to be performed with minimal temperature and salinity variations to avoid evisceration during transport. Very often, individuals must be placed individually in oxygen-filled bags with seawater (e.g. Abidin et al., 2016; Agudo, 2006; Hair, 2012; Ito, 2014; Juinio-Meñez et al., 2017; Kumara and Dissanayake, 2017). Ideally, the bags are placed in insulated containers during transport. However, IOT-Madagascar transports batches of tens of genitors from the wild to the hatchery in dried containers with seawater wetted towels (Igor Eeckhaut, unpublished observation). Tuwo et al. (2019) assessed the evisceration rate of *H. scabra* transported using open and closed transportation modes. Evisceration appeared to be triggered by the presence of dead and decaying individuals in the bag, mostly those that had eviscerated before transport. Individuals of *H. scabra* that survive transportation generally recover almost immediately after being placed in a pre-prepared holding tank.

Broodstock are maintained in tanks, earthen ponds or sea pens at low density (Hair, 2012; Ivy and Giraspy, 2006; James et al., 1996; Kumara and Dissanayake, 2017; Militz et al., 2018; Morgan, 2000a,b; Purcell et al., 2012a). Ponds and pens are generally used for long-term holding, and genitors are transferred to tanks when they are almost ready to spawn. Tanks of variable volumes are used, and their bottom is generally covered with marine sediment. Food is added if the genitors are not induced to spawn immediately after collection or transfer. Various food types are used, such as extract of *Sargassum* sp. and live unicellular algae (*Chaetoceros* sp. and *Isochrysis* sp.) (Dabbagh et al., 2011, 2012), seaweed and mud (Mazlan and Hashim, 2015), a mixture of *Sargassum* and *Enhalus* (Abidin et al., 2016), a mixture of *Sargassum, Gracillaria* and *Ulva* (Vaitilingon et al., 2016) or powdered *Spirulina* [1 $g m^{-2}$ day^{-1} (Militz et al., 2018)]. Broodstock must be stocked at a low density; i.e. a density of 15–30 individuals per 1000 L in static aerated water kept between 27 and 30°C was proposed early by James (1996a).

To optimize production, obtaining a maximum of competent (fertilizable) oocytes is necessary. This requires that the females used as broodstock have the most mature ovaries possible at the time of spawning induction (or artificial oocyte maturation; see below) and fertilization. The development of the gonads of *H. scabra* was investigated by Demeuldre and Eeckhaut (2012) who observed the histological appearance of the first germ cells in individuals 9–10 months old. The study of the reproductive cycle (gametogenesis) of *H. scabra*, especially in females (oogenesis), is essential in the specific area where production occurs, as this cycle may vary across geographic locations. Broadly speaking, *H. scabra* is distributed between the two tropics in the Indian and the western Pacific Oceans. It is generally accepted that the closer populations occur to the equator, the more homogeneous the gonads are (i.e. reproduction may occur monthly without any discernable peak season), whereas in populations closer to the tropics, gonad maturity shows seasonality due to less homogeneous environmental conditions (especially temperature) over the year. This is partially demonstrated by the results of certain analyses. For example, in evaluating the gonad maturity with histological data and spawning induction tests, Morgan (2000a,b) noted that the reproductive cycle of males and females in an Australian population of *H. scabra* at Stradbroke Island (27°S) peaked in November of 1996 and 1997. At a similar latitude in the Indian Ocean, the reproductive cycle was investigated in a population at Toliara (23°S), south-west of Madagascar with similar analytical methods, i.e. gonad histology and spawning tests (Rasolofonirina et al., 2005). Surveys from November 1998 to April 2000 revealed that the highest gonad index and spawning occur between December and March. At the opposite end of the latitudinal distribution, reproduction of *H. scabra* was investigated in the Solomon Islands close to the equator (8°S), over a 3-year period (1996–98) (Ramofafia et al., 2003a). The months showing the highest gonad maturity in females were different from year to year and Ramofafia et al. (2003a) suggested that reproduction was continuous in this population. In agreement with their findings, the investigators concluded that spawning could be induced in *H. scabra* year-round with the greatest response obtained from September to December. Similarly, histological data show that mature gonads occur year-round in *H. scabra* from Kunak (4°N; Sabah) in Malaysia, also suggesting a potential for extended gamete release (Arsad et al., 2017). However, the tendency for a longer (even continuous) reproductive period near the equator is far from a universal rule in *H. scabra*. Serial analyses of gonad indices, whether or not coupled with histological analyses, show that the gonads in *H. scabra* may exhibit annual, biannual, seasonal or non-periodic cycles of gonad maturation (the latter often associated with post-spawning ovaries found throughout the year). Annual reproduction is observed at Moreton Bay, Australia [27°S; (Morgan, 2000a, b)], in Toliara, Madagascar (23°S; (Rasolofonirina et al., 2005)), in Kudat, Malaysia [6°N; (Arsad et al., 2017)] and Mahout Bay, Sultanat of Oman [20°N; (Al-Rashdi et al., 2019)]. Biannual reproduction is observed in New Caledonia [22°S; (Conand, 1993)], in Sulawesi, Indonesia [5°S; (Tuwo, 1999)] and in the Gulf of Manar, India [9°N; (Krishnaswamy and Krishnan, 1967)]. So-called continuous reproduction are reported in Calatagan, Philippines [13°N; (Ong Che and Gomez, 1985)], Solomon Islands [8°S; (Ramofafia et al., 2003a)], in Kunak, Malaysia [4°N; (Arsad et al., 2017)] and in Indonesia at Lampung Bay [5°S; (Darsono et al., 1995; Purwati, 2006b)], Saugi Island [0.8°N; (Purwati, 2006b; Tuwo, 1999)] and Ambon Bay [3°S; Purwati, 2006b)]. It should be noted that, very often, when reproduction is said to be continuous, peaks of intensity in spawning exist: e.g. two peaks in spawning exist in Indonesia (Purwati, 2006b) and one in Kunak, Malaysia (Arsad et al., 2017) and in Solomon Islands (Ramofafia et al., 2003a).

Knowing the period during which females of *H. scabra* harbour mature gametes ensures a reliable source of oocytes throughout the year in regions near the equator and seasonally in tropical regions (see details in Section 8.6). Without background knowledge of the reproductive cycle of *H. scabra*, hatchery productivity would be unreliable and inconsistent over the annual cycle. For a hatchery based in a location where *H. scabra* spawns seasonally (only during a portion of the year), the cost effectiveness of the enterprise may be uncertain. In this context, conditioning the broodstock to obtain mature gonads before induction of fertilization is essential. Currently, there are no clear or definitive reliable methods available to ensure that gonads, especially ovaries, are induced to mature in captivity. Fully mature broodstock has sometimes been collected from the wild or bought from dealers and induced to spawn almost immediately (Pitt and Duy, 2004b) but, in recent years, broodstock that is not fully mature is often conditioned in tanks in the hatchery for a period of about a month to allow gonad to reach full maturity (Duy, 2012). This is the case in Vietnam, where a fine shrimp starter diet is added to the broodstock tanks at concentrations of $1\,g\,m^{-3}$ twice a day (Duy, 2012). In addition, broodstock is co-cultured with the Babylon snail (*Babylonia areolata*), which is presumed to provide a supplement of protein-rich feed that would contribute to sea cucumber growth and gonad conditioning (Duy, 2012). Broodstock might lose weight during this month-long conditioning but weight loss of <20% has been considered acceptable (Duy, 2012). A supplement of vitamin E in the diet was also shown to have a positive effect on gonad development in broodstock from Indonesia (Sembiring and Hutapea, 2019). These authors found that when food (mainly *Sargassum* and *Ulva*) contain $300\,mg\,kg^{-1}$ of vitamin E, the gonad index was significantly higher, with more mature oocytes in the ovaries (36 individuals tested over 90 days).

21.3.2 Spawning induction and fertilization

There are essentially three broad methods used to obtain fertilized oocytes in sea cucumber aquaculture: natural spawning, induced spawning and in vitro fertilization. The first method is difficult to use with *H. scabra* when the optimal spawning period is not easily determined. Unforced natural spawning is efficient on other species such as *I. fuscus*, in which spawning events are very well correlated with the lunar cycle so that fertilized oocytes can be obtained on predictable days within the reproductive period (Mercier et al., 2007).

Thermal shocks are the most widespread spawning induction technique used in sea cucumber aquaculture. It was successful used with *H. scabra* in Iran (Dabbagh et al., 2012), India (James et al., 1994b), Maldives (James, 2012), Philippines (Gamboa and Juinio-Meñez, 2003), Vietnam (Pitt and Duy, 2004b), Australia (Ivy and Giraspy, 2006; Morgan, 2000b), Solomon Islands (Battaglene et al., 2002), Fiji (Hair et al., 2011), New Caledonia (Agudo, 2006), Tanzania (G. Robinson, Personal communication), Indonesia (Sitoresmi and Pursetyo), Sri Lanka (Kumara and Dissanayake, 2017) and Malaysia (Mazlan and Hashim, 2015; Vaitilingon et al., 2016). Thermal shocks were also successful in Mauritius but on other species, i.e. *Bohadschia marmorata* and *Holothuria atra* (Laxminarayana, 2005). Thermal shocks involve placing genitors into baths of different temperatures, which may vary depending on species and locations. For *H. scabra*, the water temperature is generally raised by 3–5°C for 1 h, either by adding warmed seawater to the spawning tank or using aquarium heaters (Agudo, 2006; James et al., 1994a). The water temperatures are ideally kept within the range of 28–32°C. If the ambient water temperature is >30°C, it is recommended to give a cold shock treatment for 1 h before the heat shock. Sealed plastic bags containing ice are added to the tank to quickly lower the water temperature by 5°C (Agudo, 2006). Thermal shocks have some drawbacks as success is inconsistent and unpredictable over the annual cycle, which can impede continuous larval and juvenile production with impacts on financial profitability. Water pressure and dry treatment are often used in combination with thermal shocks or when thermal shocks are unsuccessful. The broodstock is left exposed to air in a tank in the shade or in a few centimetres of seawater, for 30–45 min (i.e. desiccation) before subjecting them to a powerful jet of seawater at ambient seawater temperature for a few minutes (Agudo, 2006). The broodstock is then returned to the spawning tank at ambient water temperature.

In some cases, the addition of food as stimulant is used to induce sea cucumbers to spawn. Dried algae (*Spirulina*) (Abidin et al., 2016; Cabacaba and Campo, 2019; Campo et al., 2019; Duy et al., 2016b; Militz et al., 2018) or the commercial formula Algamac 2000 at a concentration of $0.1\,\mathrm{g\,L^{-1}}$ or rice bran, soy powder and *Sargassum* sp. powder (4:1:2) at a ratio of $50\,\mathrm{g}\ 500\,\mathrm{L^{-1}}$ is added to the tank containing broodstock (Agudo, 2006). After 1 h, the water is removed from the tank and replaced with clean seawater at ambient temperature. In the Philippines, broodstock was successfully induced to spawn within 3–4 days using a combined treatment of desiccation, thermal shock and food addition (*Spirulina*) (Juinio-Meñez et al., 2012a).

The number of males and females used during spawning trials varies from a few males and females in the case of research programs to several dozen males and females in the case of large-scale hatcheries. In Madagascar, between September 2014 and December 2015, the number of spawning inductions was 180 (all successful) with a total of 1862 males and 2052 females used. A total of 876,525,200 oocytes were obtained, producing 140,677,900 young auricularia (Igor Eeckhaut, Personal observation). In Vietnam, from March 2002 to August 2003, 38 spawning inductions were successful out of 71 assays, producing some 275 million oocytes (Pitt and Duy, 2004b). It is important to use several males and females for each spawning trial and, ideally, a sex ratio of 1:1 (i.e. at least 10 of each sex on average), to avoid a possible genetic drift after several generations.

In vitro fertilization (IVF) is a process by which oocytes are surgically removed from the female genital tracts and fertilized externally. The problem with in vitro fertilization in sea cucumber is that the development of oocytes in the ovaries is arrested at the prophase I stage of meiosis and that fertilization naturally occurs only at the end of the meiosis. Maturation naturally concludes in sea cucumbers immediately before or during spawning, resulting in mature oocytes that are competent, i.e. ready to be fertilized (also called ootids). Consequently, oocytes that are surgically extracted from ovaries cannot be fertilized. IVF was used early in India (James et al., 1994b) but the low fertilization rate obtained did not encourage the use of this method. At present, four molecules acting on oocyte maturation have been identified in sea cucumbers including three in *H. scabra*. The four molecules are (*i*) an unnamed peptide isolated from the radial nerve of *A. japonicus*, (*ii*) cubifrin, produced by the radial nerve of holothuroids, (*iii*) a neurohormone relaxin-like gonad-stimulating peptide (RGP) discovered in *H. scabra* and (*iv*) a thioredoxin, isolated from sea urchin ovaries and acting on the oocytes maturation of many holothuroid species. Of these four molecules, the first three are not yet commonly used in aquaculture but are promising. The fourth (thioredoxin) is currently used routinely by the IOT company in Madagascar.

The unnamed peptide isolated from the radial nerve of *A. japonicus* is 4.8 kDa and induces germinal vesicle breakdown (GVBD) in 80% of oocytes (Katow et al., 2004). At the same time, Kato et al. (2009) purified and identified another peptide named Cubifrin-I (NGIWYamide) also produced by the radial nerve of *A. japonicus*. Cubifrin-I induces oocyte maturation, ovulation and spawning. A synthetic derivative, cubifrin-L (NGLWYamide), was shown to be even more active than the natural product (Fujiwara et al., 2010). The cubifrin-I peptide has also been discovered in the transcriptomes of *H. scabra* (Suwansa-Ard et al., 2018; Turner, 2016) and *Holothuria glaberrima* (Suwansa-Ard et al., 2018). Chaiyamoon et al. (2020b) characterized the cubifrin in *H. scabra* and further investigated its effect on spawning. The deduced amino acid sequence of the neuropeptide precursor transcript provided five copies of two different mature sequences and were designated as cubifrin-Y and cubifrin-F. Physiological bioassays demonstrated that the newly identified cubifrin-F was more potent in stimulating spawning, with about 70% success rate, than cubifrin-Y, which could stimulate only 10% spawning, compared with the control group. The in situ hybridization showed intense expression of cubifrin transcript within the wall of the ovarian tubule at late ovarian stages, suggesting that cubifrin was produced locally and probably stimulated the spawning by causing contraction of the wall of the ovarian tubule.

Chieu et al. (2019) successfully produced a recombinant of a neurohormone relaxin-like gonad-stimulating peptide of *H. scabra* from the yeast

Pichia pastoris. To assess bioactivity, they tested four levels of purification of the recombinant using in vitro bioassays of germinal vesicle breakdown (GVBD). The purest form induced 98.56% of GVBD in *H. scabra* (Chieu et al., 2019). Cruder levels of purification still resulted in some GVBD (6.62–7.57%). Upon single injection into females of *H. scabra*, the recombinant RGP induced the spawning posture (head waving behaviour) followed by spawning within 90–170 min. Spawned oocytes were fertilized successfully, larvae settled and developed into juveniles.

Léonet et al. (2009) and Eeckhaut et al. (2012) described an in vitro fertilization (IVF) method used routinely in Madagascar and its effects on the oocyte maturation of *H. scabra* and other holothuroid species. The oocyte maturation inducer (OMI) was extracted from sea urchin ovaries. The bioactive molecule was further identified by mass spectrometry as thioredoxin and the active site of the protein as WCNPCK (Léonet et al., 2019). As this site is a redox catalytic site, the thioredoxin induces oocyte maturation, at least partly, by acting on the redox potential. Delroisse et al. (2021) used pharmacological agents known to have activating or inhibiting actions on oocyte maturation to study the intracellular reaction cascade induced by thioredoxin on oocytes. As a redox-active protein, thioredoxin reduces target proteins within the oocyte membrane and initiates an intracellular reaction cascade that lifts the block on oocyte meiosis. Thioredoxin-induced oocyte maturation is cAMP independent, but dependent of the presence of calcium in the seawater. The OMI always enhances oocyte maturation (92.2% on average) compared to controls made of seawater (15.8%) (Léonet et al., 2009) although the maturation effect, always superior to controls, depends on the season (Igor Eeckhaut, Personal observation). In general, it allows the fertilization of more than 90% of vitellogenic oocytes exposed. Other OMIs described in literature like 1-methyladenine, dithiothreitol (DTT), dimercaptopropanol (DMP) and L-cysteine never exceed 40% success and often lead to abnormal development (Léonet, 2010; Léonet et al., 2009). In 2018, half of the 2.5 million juveniles from the hatchery in IOT-Madagascar were produced by IVF; the other half was obtained by natural spawning and thermal shocks (Igor Eeckhaut, Personal observation).

21.3.3 Larval rearing and metamorphosis

The number of early embryos obtained from hatchery production in *H. scabra* reportedly varies from 20,000 to 22.23 million per successful trial (Table 17), the variation owing to the number of spawning individuals used and to timing of the trials relative to the natural breeding period (see above).

Table 17 Characteristics of embryonic and larval development of *Holothuria scabra* in hatcheries worldwide, based on various studies, showing the country, number of early embryos obtained initially (million), time to blastula or gastrula, time to larva (auricularia/doliolaria/pentactula) and percent survival (%) from fertilized oocytes (eggs) to seeds (i.e. juveniles destined for nurseries).

Country	Embryos (million)	Time to blastula/ gastrula	Time to larva (days)	Survival (%)	References
Australia	NM	NM	NM	0.4–20 (to settlement)	Bowman (2012)
Fiji	NM	NM	NM	0.009	Hair (2012)
Indonesia	0.378	NM	NM/13/17	NM	Sitoresmi and Pursetyo (2020)
Indonesia	0.76 and 2,6 (2 females)	NM	NM	NM	Sembiring et al. (2018)
Iran	NM	NM	2/14/20	NM	Dabbagh and Sedaghat (2012)
Kiribati	0.05–0.225	NM	1/7/9	NM	Ito (2014)
Madagascar	0.02–2 (per female)	6 h / 24 h	3/15/17	3	Rasolofonirina and Jangoux (2005) Eeckhaut (Personal observation)
Malaysia	0.154–0.534	NM	2/11/18	4.2 (to metamorphosis, 18 d)	Mazlan and Hashim (2015)
Malaysia	0.7	NM / 24 h	2/21/26	NM	Abidin et al. (2016)
Malaysia	1 (per female)	NM	NM/12/15	0.09	Vaitilingon et al. (2016)
New Caledonia	several	40 min–3 h/24 h	2/10/12	NM	Agudo (2006)
Oman	NM	12 h/16 h	2/18/22	NM	Al-Rashdi et al. (2012)
Papua New Guinea	NM	NM	NM/13/17	1.12	Militz et al. (2018)

Continued

Table 17 Characteristics of embryonic and larval development of *Holothuria scabra* in hatcheries worldwide, based on various studies, showing the country, number of early embryos obtained initially (million), time to blastula or gastrula, time to larva (auricularia/doliolaria/pentactula) and percent survival (%) from fertilized oocytes (eggs) to seeds (i.e. juveniles destined for nurseries).—cont'd

Country	Embryos (million)	Time to blastula/ gastrula	Time to larva (days)	Survival (%)	References
Philippines	7–22.23	NM	NM/10/12	0.11–067 (to juveniles 10–30 mm)	Juinio-Meñez et al. (2012b)
Philippines	1.9–6.8	NM	2/10/NM	0.02–2.2 (to juveniles 3 mm)	Gamboa et al. (2012)
Philippines	NM	NM	2/10/15	NM	Nievales et al. (2015)
Solomon	NM	NM	2/10/12	<1 to 35 (to settlement)	Battaglene (1999a)
Sri Lanka	0.24–2.6	NM	2/14/16	0.2–5.2 (from early auricularia)	Kumara and Dissanayake (2017)

NM = not mentioned.

General trends in the early development of *H. scabra* are uniform throughout the existing literature; however, the tempo following which the developmental stages develop may vary (Table 17), which is most likely due to variations in environmental conditions, especially temperature and food supply. Post fertilization, the eggs undergo first cleavage to produce an embryo with two blastomeres, generally after about 2 h (Fig. 25A, B). Subsequent cleavages generate a blastula, generally after 12 h, and a gastrula after 12–24 h post fertilization (Figs 25C, 26A, B). The gastrula is characterized by the presence of a blastopore and a developing archenteron (Figs 25D, E, 26A). It is ciliated and able to move. The planktotrophic development continues through a pelagic feeding larval stage, the auricularia (Figs 25F–I, 27C–I), a transitory non-feeding metamorphic stage, the doliolaria (Figs 25J, 26J, K), and a near-benthic stage, the pentactula, which metamorphoses into the benthic juvenile (Figs 26K–M). Generally, under common aquaculture conditions (Table 17), the majority of auricularia larvae appear between the 2nd and the 3rd day, the doliolaria stage develops from the 7th to the 18th day and the pentacula from the 9th to the 22th day. Metamorphosis in *H. scabra* begins in the water column,

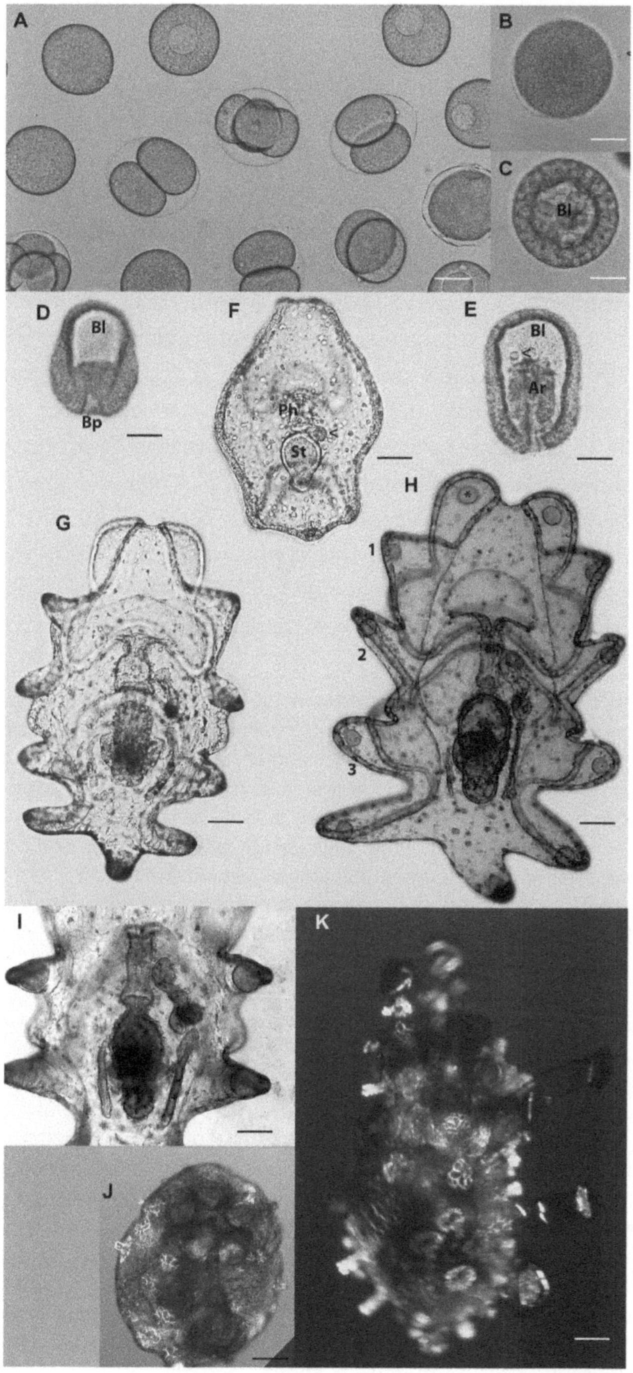

Fig. 25 Light microscopy view of embryonic and larval development in *H. scabra*. (A) Photo taken 2 h after fertilization, showing unfertilized oocytes with intact germinal *(Continued)*

before settlement to the substrate, which coincide with the transition from the pelagic to the benthic state.

Descriptions of the morphology and size of developmental stages are quite homogeneous and follow the original descriptions published in India (James et al., 1988, 1994a,b). The gastrula measures 190–250 μm. The auricularia stage is fully formed after 5 or 6 days under typical conditions. Three sub-stages of auricularia are often considered and named early, mid- and late auricularia. The early auricularia measures about 560 μm and the late auricularia about 1100 μm in length, when fully developed. Auricularia larvae are ear-shaped, transparent, and possess a preoral loop anteriorly and an anal loop posteriorly. The digestive tract consists of a mouth, an elongated pharynx and a sacciform stomach with an anus, the whole enabling the larvae to feed on planktonic matter. Various types of calcareous ossicles appear in the auricularia of *H. scabra* and they have been reviewed by Rasolofonirina and Jangoux (2005). The late auricularia (Fig. 25H) is the competent larva that is physiologically ready to metamorphose if properly stimulated. Battaglene (1999a) indicated that the presence of distinct hyaline spheres at the late auricularia stage appeared to be a good

Fig. 25—Cont'd vesicle, fertilized oocytes without germinal vesicle (and some polar bodies visible depending on orientation), embryos with two or four blastomeres. (B) Two hour-old fertilized oocyte with polar bodies (arrow). (C) Blastula, generally appearing 6 h after fertilization. This stage begins to spin on itself. (D) Early and (E) late gastrula, which can be observed ~10 h after fertilization. The blastopore is present and the archenteron is growing. Mesenchymal cells detach from the blind end of the archenteron into the blastocoel. (F) Early auricularia observed 2 days after fertilization. The digestive system is composed of a pharynx and a stomach with a mouth and anus. Cells separating from the digestive tract on the right will form the right part of the coelomic cavity (arrow). (G) Mid auricularia generally present between days 3 and 10 after fertilization. The three posterior lobes are formed as well as the preoral and postoral ventral lobes. The larva is now very active and feeds on phytoplankton. (H) Late auricularia observed 10 days after fertilization under optimal rearing conditions. The larva is very mobile and actively feeds. It has hyaline/lipid spheres (asterisk) and its coelomic attributes are clearly visible in (I); they include the axo-, hydro- and somatocoel on the right as well as the only left somatocoel. (J) The doliolaria is a barrel-shaped larva that appears 12 days post fertilization. This larva is less mobile and repeatedly touches the substrate in the pre-settlement phase. It includes an anterior vestibule containing five tentacles, which are not yet extensible. This larva also has dermal ossicles observable under a phase contrast. (K) Early juvenile, immediately following the pentactula stage, with five extensible tentacles and one to two primary podia. The individual shown has two primary podia plus a secondary podium. Pentactulae and early juveniles appear around day 15 post fertilization in healthy cultures. Scale bars: 100 μm in (A, H, J); 50 μm in (B–D, K); 80 μm in (E, F); 70 μm in (G, I). *Photos courtesy of R. Rasolofonirina.*

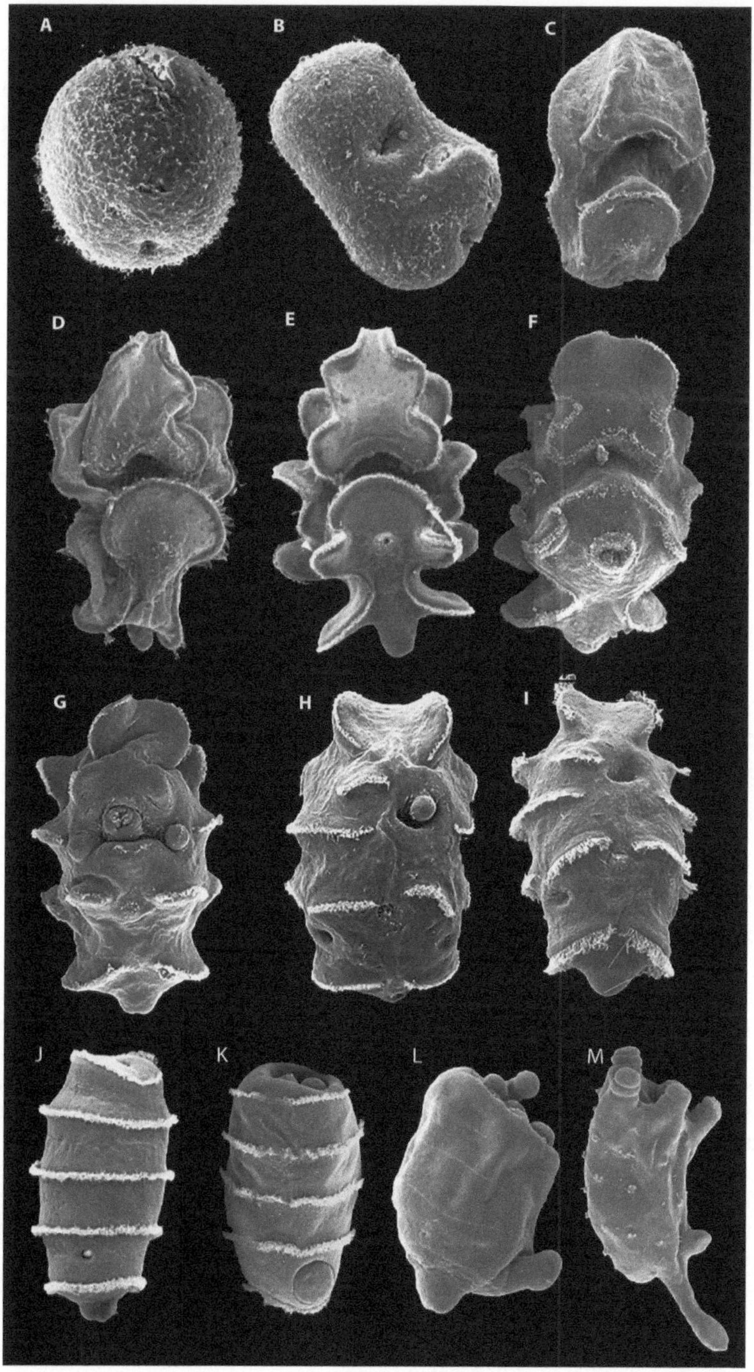

Fig. 26 Scanning electron microscopy view of larval development and metamorphosis in *H. scabra*. (A) Gastrula stage (10 h old; 200 μm long) with the blastopore and a uniform

(Continued)

indication of larval competency for metamorphosis. Unlike the settlement phase, metamorphosis is usually not stimulated but instead occurs naturally. Metamorphosis of late auricularia, however, is not something that happens radically at a fixed time post-fertilization; it can be delayed for up to 30 days, probably because the larvae do not get adequate stimulation or because their development is anomalous, which decreases the profitability of aquaculture. To date, no research on the initiation of metamorphosis in late auricularia larvae of *H. scabra* has been published. The auricularia transforms into the doliolaria (Figs 25J, 26I–K), which is a barrel-shaped larva with five ciliated bands around the body and five tentacles present in a vestibular cavity (i.e. visible through the transparent body but not extended outside of the body) (Fig. 26J, K). This stage measures between 460 and 620 μm in length. The doliolaria spends less time swimming, intermittently touching the bottom of the culture environment in an attempt to find an appropriate substrate to settle on. If the stimulation is adequate, the doliolaria transforms into the pentactula, which is the stage that clings to the substrate thanks to the five tentacles (that have become fully expendable) and one to two primary podia (also called terminal tentacle). The pentactula is 600–700 μm long. At this stage, the young individual has five ciliary bands still faintly visible (Fig. 26L) while older individuals possess cilia tufts (Fig. 26M). The colour of the body is greenish brown. Formally, the pentactula becomes a juvenile following the acquisition of a third podium and/or other tentacles (Fig. 25K). Settlement is fully completed when the individual is a juvenile.

Fig. 26—Cont'd ciliation. (B) Intermediate stage between the gastrula and early auricularia (6 h old), showing the appearance of the mouth in addition to the blastopore (anus) as well as the outline of the developing ventral lobes. (C) Early (2 days old; 400 μm long), (D) mid (6 days old; 800 μm long) and (E) late auricularia (10 days old; 1000 μm long) showing the formation of the three posterior lobes in addition to the development of the preoral and postoral lobe. (F–I) Metamorphosis of auricularia into doliolaria. The diameter of flattened auricularia hitherto increases by pushing the preoral lobe upwards and the postoral lobe downwards. The anus closes while the mouth is integrated into the vestibule, the ventral opening of which migrates to the apex of the anterior part. The ciliary bands, present along the different lobes in the late auricularia, are reorganized to form five ciliated bands surrounding the body of the doliolaria. (J) Early and (K) late doliolaria (both 12 days old; 500 μm long) which are barrel-shaped with five well-developed transverse ciliary bands and an anterior vestibule with five non extensible tentacles. The doliolaria in (K) shows a developing primary podia. (L) Early and (M) late pentactula (both 15 days old; 700 μm long). In (L), five faint ciliary bands are still visible and there is one primary podia. In (M), the body is covered with cilia tufts and two primary podia are visible. *Photos courtesy of R. Rasolofonirina.*

Fig. 27 (A) Wet room with larval hatchery tanks; (B) hapa nets in ponds used as nurseries; nursery ponds showing (C) preparation before a growing cycle and (D) aerial view; (E) juveniles released in ponds and (F) after growth, ready to be transferred in sea pens. *All photos from IOT in Madagascar, except photo in (B) from Philippines, courtesy of J.-F. Hamel.*

Data on the development of *H. scabra* in southern China were published by Yan et al. (2012) who mentioned that fertilized oocytes required 4 h to reach blastula, 5 h to gastrula, and 19 h to Auricularia at the average temperature of 29°C and salinity of 34 psu. It took about 7 days to develop into doliolaria and 15 days to reach the juvenile stage under the same conditions.

The optimal conditions for adequate larval rearing are reviewed in contributions by James et al. (1994b), Agudo (2006), Al-Rashdi et al. (2012), Mercier and Hamel (2013) and Ito (2014). The main conditions that must be controlled are larval density, water renewal, and auricularia nutrition. The normal culture density of embryos should be 0.3–0.7 per mL (James et al., 1994b). The water renewal varies from one hatchery to another but, irrespective of the protocol, it must ensure constancy in temperature, oxygenation, and pH as well as a low concentration of nitrogenous substances (nitrite/nitrate) in the water column. Agudo (2006) suggests not to change the water in the culture tanks until 2 days post fertilization and then to change either 30% of the water each day, 100% of the water every second day, or 30% of the water with addition of erythromycin at a rate of $2\,g\,m^{-3}$ after each water change, until the pentactula stage.

Auricularia larvae need to be fed with pelagic microalgae to help them complete their development. The feeding mechanism of the auricularia consists of conveying unicellular algae and suspended fragments of organic matter into the alimentary canal by ciliary movement (James and James, 1994b; James et al., 1988, 1989, 1994a,b; MPEDA, 1989). Food accepted by the larvae is $<40\,\mu m$ and up to $80\,\mu m$ in the early and late auricularia, respectively (James and James, 1994b; James et al., 1994b). Several microalgae are used in hatcheries: *Chaetoceros* sp. (Agudo, 2006; James et al., 1994b; Sitoresmi and Pursetyo, 2020), *C. calcitrans* (Abidin et al., 2019; Battaglene, 1999a,b; Cabacaba and Campo, 2019; Campo et al., 2019; Knauer, 2011), *C. mulleri* (Battaglene, 1999c; Ito, 2014; Knauer, 2011), *C. gracilis* (Cabacaba and Campo, 2019; Campo et al., 2019), *C. ceratosporum* (Makatutu et al., 1993), *Isochrysis* sp. (Duy et al., 2016a; Safitri et al., 2020), *Isochrysis galbana* (Agudo, 2006; James et al., 1994b; Knauer, 2011), *Nannochloropsis* sp. (Abidin et al., 2019; Al-Rashdi et al., 2012; Safitri et al., 2020), *Nannochloropsis oculata* (Morgan, 1999), *Nitzschia acicularis* (Mazlan and Hashim, 2015), *Navicula* sp. (Mazlan and Hashim, 2015), *Pavlova* sp. (Duy et al., 2016a; Safitri et al., 2020), *Pavlova lutheri* (Ito, 2014), *Pavlova salina* (Knauer, 2011), *Phaeodactylum tricornutum* (Al-Rashdi et al., 2012), *Rhodomonas maculata* (Ito, 2014), *R. salina* (Agudo, 2006; Battaglene, 1999c), *Tetraselmis* sp. (Agudo, 2006), *Tetraselmis chuii* (Morgan, 1999), *Thalassiosira pseudonana* (Todinanahary et al., 2018) and *Thalassiosira weissflogii* (Duy et al., 2016a). A mixture of algae is often deemed more efficient than a single species (Agudo, 2006; Ito, 2014).

Some studies compared the effect of different strains of microalgae on the growth and survival of larvae in *H. scabra*. Makatutu et al. (1993) indicated that larvae grew more rapidly when fed with *Chaetoceros ceratosporum*.

When larvae were fed with different algae, Morgan (1999) observed that the growth declined as follows: *C. cerastosporum* > *Tetraselmis chuii* > *Nannochloropsis oculata*. Battaglene (1999c) indicated that larvae survived and grew best on single diets of the red alga *Rhodomonas salina* and *C. muelleri*. They grew poorly on diets of *I. galbana* and *Tetraselmis chuii* and to a lesser extent when fed *C. calcitrans* (Battaglene, 1999c). Safitri et al. (2020) compared the growth rate and survival rate of *H. scabra* larvae when fed with *Nanocloropsis* sp., *Isochrysis* sp. or *Pavlova* sp. The best result was obtained with *Isochrysis* sp. Campo et al. (2019) measured the growth performance, development, and survivorship of *H. scabra* larvae fed with *C. gracilis* (Cgr) and *C. calcitrans* (Cc) that were administered singly and in combination. Growth was highest in combined diet with mean final length of 2088 μm, followed by Cc with 1855 μm and Cgr with 1800 μm. Survival rates among treatments were not statistically different (Campo et al., 2019). Some investigators have also reported successful replacement of live cultured micro-algae with commercially available micro-algae concentrates as a food source (Duy, 2017; Duy et al., 2016a,b, 2017; Militz et al., 2018). Micro-algae concentrates are readily ingested and their cell walls efficiently digested by larvae (Duy et al., 2015), supporting high rates of growth and survival through settlement when used as the sole food source (Duy et al., 2016a,b). The use of commercially available micro-algae concentrates as a replacement for live micro-algae in hatcheries supports development of a cheaper, simpler post-settlement rearing protocols for this species. Knauer (2011) assessed the nutritional value of four species of microalgae for larvae of *H. scabra*: single diatoms (*C. muelleri, C. calcitrans*), flagellates (*Isochrysis* aff. *Galbana, P. salina*) or a ternary microalgal diet (TMD) comprised of 40% *C. muelleri*, 40% *I. galbana*, and 20% *P. salina* for 12 days. This study showed *C. muelleri* to be the most effective of the four microalgae tested as a single microalgal diet for larval *H. scabra* and, as such, the author proposed *C. muelleri* as a reference diet for *H. scabra* larvae. Duy et al. (2016a) confirmed that the presence and size of hyaline spheres at the end of the auricularia stage is a reliable indicator of subsequent performances (proportion of competent doliolariae, post-settlement size and post-settlement survival). In their experiments, 2-day old larvae were fed one of three commercially available micro-algae diets that varied in their nutrient compositions: (*i*) *Isochrysis* sp., (*ii*) *Pavlova* sp. or (*iii*) *T. weissflogii* or a combination of the three. The results were the best with *T. weissflogii*.

Variable microalgal concentrations have been reported by different hatcheries. In India, unicellular algae were fed twice in a day, but the quantity given each time depended on the particular larval stage

(James et al., 1994b). In general, 20,000–30,000 cells mL^{-1} in the rearing tank water was maintained (James et al., 1994b). In the Philippines, a mixture of *I. galbana* and *Chaetoceros calcitrans* at a density of 10,000–15,000 cells mL^{-1} was fed to the larvae twice a day (Juinio-Meñez et al., 2012a). Agudo (2006) and Al-Rashdi et al. (2012) recommended 20,000–40,000 cells mL^{-1} from early to late auricularia; beyond this concentration, auricularia survival decreased (Agudo, 2006). In Kiribati, Ito (2014) suggested starting with 500 cells mL^{-1} on day 1, increasing to 5000 cells mL^{-1} on day 2, gradually increasing to 10,000 cells mL^{-1} on day 9, and then gradually decreasing to 500 cells mL^{-1}. In Bali (Indonesia), auricularia larvae were fed with *Chaetoceros* sp. at a density of 20,000 cells mL^{-1} (Notowinarto and Putro, 1992; Sembiring et al., 1996, 2015). The concentration used by IOT-Madagascar is reportedly from 20,000–40,000 cells mL^{-1} a day until the appearance of the doliolaria, after which the concentration is decreased to 2000 cells mL^{-1} [Igor Eeckhaut, unpublished data; similar concentrations have been used by others (Knauer, 2011)]. In their experiments showing the importance of the diet on the appearance of hyaline spheres, Duy et al. (2016a) used from 10,000 to 31,000 cells mL^{-1} per day (concentration increasing from day 2 to 22) for *Isochrysis* and *Pavlova* and from 3900 to 12,000 cells mL^{-1} for *T. weissflogii*. Todinanahary et al. (2018) reported on the influence of phytoplankton food density on the time of appearance of doliolaria and pentactula stages. Two different low-food daily concentrations were used during the larval rearing: ~1500 cells mL^{-1} and ~ 120 cells mL^{-1} and the results showed that the doliolaria larvae appears 7 day earlier when the auricularia larvae were fed at the highest concentration.

21.3.4 Settlement and post-settlement rearing

Settlement and post-settlement processes in *H. scabra* have been studied in the laboratory where independent and paired choice experiments revealed that several substrata could induce metamorphosis of doliolaria into pentactula, but that specific substrata such as seagrasses and seagrass extracts favoured settlement (Mercier et al., 2000a); see Section 10 above for details.

Based on biological and ecological knowledge, settlement in hatcheries can be conditioned using different techniques, although it is also possible to achieve settlement on the side and bottom of the culture vessels without any specific preparation or addition of conditioned plates. Some substances can be added in the tanks such as (*i*) extracts of finely filtered *Sargassum* (James and James, 1994b; James et al., 1994b), (*ii*) the commercial feed Algamac 2000 at a concentration of 0.25–0.5 $g\,m^{-3}$ day^{-1} (Agudo, 2006) and (*iii*) seagrass leaves that promote a distinctive biofouling layer (Agudo, 2006).

Settlement is chiefly stimulated by adding different conditioned substrates to the tanks around the 12th day of culture. Suitable settlement surfaces include plastic sheets (PVC, polythene or polypropylene), fibreglass plates, mesh screens, and rough surface tiles suspended in the water. The settlement surfaces can be conditioned by (*i*) immersion of plates in cultures of diatoms (*Nitzschia* sp., *Navicula* sp., or *Platymonas* sp.) for a few days; (*ii*) addition of extracts of filtered seaweed *Sargassum* sp. or seagrass (*T. hemprichii, E. acoroides*) over a period of 4–5 days to form a fine coating on the settlement surfaces; (*iii*) painting of the plates with dried *Spirulina*; and (*iv*) immersion of the plates for 4–10 days in outdoor tanks under partial shade (50–75%) in running 1-μm filtered seawater to promote natural conditioning with diatoms (Agudo, 2006; Duy, 2010, 2012; Mills et al., 2012). In Kiribati, a thick solution of the commercial brand Algamac is also used to paint the surface of plastic substrates, which are dried overnight and submerged into the settlement tank the following day (Ito, 2014). In the Philippines, the use of *Spirulina*-coated plates to induce settlement, and the extended use of *Chaetoceros calcitrans* combined with *Navicula ramosissima* to feed the new settled juveniles increased the survival (Gamboa et al., 2012; Juinio-Meñez et al., 2012a). The effects of four different species of seagrass leaves (*E. acoroides, Syringodium isoetifolium, Cymodocea serrulata* and *Cymodocea rotundata*) in the settlement of larvae were investigated by Indriana et al. (2017). The initial number of tested larvae was 1000 individuals per assay. *E. acoroides* showed the best result with 0.26 ind. cm^{-2} settlement preference and 10.66% survival rate.

Pentactula larvae turn into juveniles within a few hours after the onset of settlement. Those juveniles (Fig. 25K) measure from 0.3 to 1.5 mm long and must remain 30–45 days in culture tanks before being transferred to a nursery. During this time, juveniles must be fed to grow properly. A fine paste of sieved (40–80 μm) seaweeds (*Sargassum* sp., *Halimeda* sp.) can be used (Agudo, 2006; Al-Rashdi et al., 2012; Ito, 2014; James et al., 1994b) sometimes with seagrass, e.g. *Siryngodium isoetifolium* (Agudo, 2006). This filtered extract is given daily to the juveniles, in the morning and in the evening. Juveniles can also be fed with cultured diatoms (*Nitzschia* sp., *Navicula* sp.) daily for the first month, with the addition of a fertilizer (Aquasol, 7 $g\,m^{-3}$) weekly to promote growth of diatoms directly in the tanks (Agudo, 2006). A supplement of commercial dried algae (Algamac 2000, *Spirulina*) is sometimes added (Agudo, 2006; Ito, 2014). Other foods, such as a fine-grade shrimp starter pellet, may also be used at a daily rate of 1–1.5 $g\,m^{-3}$ (Agudo, 2006).

Lavitra et al. (2009b) tested the effects of seven different diets on the survival and growth rates of newly settled juveniles 0.5 mm long. The diets were an organic biofilm (tanks conditioned by unfiltered seawater for 3 days) or various extracts: (*i*) of the seagrass *T. hemprichii*, (*ii*) *T. hemprichii* with 3% *Spirulina*, (*iii*) of the seagrass *Thalassodendron ciliatum*, (*iv*) of the seagrass *S. isoetifolium*, (*v*) of the brown alga *Sargassum latifolium* or (*vi*) of *S. latifolium* with 3% *Spirulina*. The results showed that extracts of *S. latifolium* with or without *Spirulina* gave the best growth and highest survival rates (60% and 68%, respectively, after 8 weeks). Rearing densities of 150, 300, 450 and 600 individuals m^{-2} were also tested (Lavitra et al., 2009b). The best survival rates (>60%) were observed when juveniles had an initial density of 450 individuals m^{-2} or less. The best growth rate was observed at a rearing density of 150 individuals m^{-2}. The survival rates recorded for early juveniles (8 weeks post-settlement) were 34–66% depending on the type of food, and 54–72% depending on the rearing density (Lavitra et al., 2009b). Nievales et al. (2015) also compared various post-larval rearing methods to determine which scheme would give the bet yield of newly settled juveniles (>1 mm body length). Five types of postlarval feeding methods were tested: (*i*) pelagic *Chaetoceros calcitrans* only, (*ii*) benthic diatom *Navicula*, (*iii*) *Navicula* combined with *C. calcitrans*, (*iv*) *Spirulina* paste on settling plate combined with *C. calcitrans*, and (*v*) *Spirulina* paste on settling plates combined with *Navicula*. After 3 days of rearing, transparent but visible early settled juveniles were observed and assessed. Mean percent juvenile yield in week 1 was highest in treatment (*i*) at 17%, followed by treatment (*iv*) at 14%. Yield increased and peaked in week 2, especially for rearing methods with *Navicula*, while those without it declined dramatically by week 3 (Nievales et al., 2015).

Reported survival rates vary across studies. It was 34% after 1 month of rearing on plates, with bacterial and diatoms film, at an unknown density of juveniles in Battaglene (1999a), and 41–75% depending on the temperature regime in Rasolofonirina (2004). Survival up to 50% with a growth rate of 0.2–0.8 mm day^{-1} was reportedly obtained after 1 month at a density of 500 juveniles m^{-2} (Agudo, 2006). The effects of light (ambient vs shaded) on the growth performance of newly settled juveniles 4 mm long were determined by Gorospe et al. (2019) in a 30-day field experiment. Average length and absolute growth rate of juveniles in the shaded treatment were significantly higher by day 15, but differences between light treatments were not significant by day 30. Juvenile survival was higher in the shaded ($56.9 \pm 20.7\%$) treatment compared to the ambient light ($34.4 \pm 22.9\%$) treatment.

However, the authors indicated that rather than light reduction, the presence of net enclosures placed in the systems to reduce light might have lowered mechanical stress (i.e. due to high wave actions) and possibly the recruitment of potential predators, which resulted globally in higher survival of juveniles reared in shaded than ambient light treatment (Gorospe et al., 2019).

Survival and growth rates are two essential parameters to estimate the profitability of a hatchery. The first will determine the number of juveniles that will be transferred to the nursery. This survival rate is very variable in the scientific literature because it is calculated differently: the initial value is based either on the number of embryos post-fertilization or the number of young auricularia larvae, and the end value either on the number of juveniles recently settled or on the number of juveniles leaving the hatchery. For example, from fertilized oocytes to juveniles <10 mm long, the survival rate was of 0.76–2.08% in the Philippines (Juinio-Meñez et al., 2012a). At Tasmanian Sea Food (Australia), the settlement rate was highly variable between tanks, often ranging from 0.4% to over 20%, but post-metamorphic survival is not known (Bowman, 2012). Sembiring et al. (2018) mentioned 1.35 and 3.28 million eggs laid by two females and hatching rates of 57.4% and 78.6%. Four rates can be calculated to understand the evolution of hatchery survival, i.e. the hatching rate (HR), the settlement rate (SR), the juvenile survival rate (JSR) and the rate of overall hatchery survival rate (OHSR). The HR corresponds to the number of young auricularia larvae (after 48 h) that survive compared to the number of embryos (obtained 12 h after fertilization). The SR is the number of pentactulae and young juveniles settled (difficult to distinguish from each other in practice) compared to the number of early auricularia larvae (after 48 h). The JSR is the number of older juveniles ready to leave the hatchery (average 10 mm long; 0.3 g in wet weight) which survive relative to the number of pentactulae and young juveniles found on the substrate. The OHSR is the product of the previous three rates ($OHSR = HR \times SR \times JSR$). As an example, 29 spawning trials were carried out in IOT-Madagascar from January 2009 to February 2010, starting with 9,192,743 embryos produced. Values for HR, SR and JSR (expressed in %) were 36.5, 5.3 and 58.3, respectively, with an overall hatchery survival rate (OHSR) of 1.1%. At IOT (Madagascar), the usual OHSR varies from 1% to 3% per spawning trial, generating 20,000–50,000 juveniles per week (Igor Eeckhaut, unpublished data).

21.3.5 Nursery systems

Nurseries (Fig. 27B–F) are contained areas where young juveniles coming from the hatchery (Fig. 27A, E) are placed at high density before being transferred to larger areas at lower density for long-term growth (Fig. 27F). Hatchery-produced juveniles proved hardy for transport to nurseries in high densities (100 and 200 juveniles per bag) in seawater held at cool, constant temperature (Purcell et al., 2006a). The first nurseries of *H. scabra,* that were created in India in the 1990s consisted of rectangular iron cages placed directly in the sea, larger tanks fixed to the bottom at 1.5 m depth or sea pens made of bamboo screens (James et al., 1994b). The type of nursery selected was based on the number of juveniles obtained from the hatchery (James et al., 1994b). At the exception of a few experiments that were made in indoor or outdoor tanks, two main types of nursery systems exist today (Fig. 28): (*i*) benthic systems, mainly ponds but also large tanks, bottom set trays (at sea) and bottom set hapas (at sea or in large ponds) where sea cucumber juveniles are reared on sediment in semi-controlled conditions; and (*ii*) floating hapas (at sea, in channels or in ponds) whereby the hatchery-reared juveniles are placed in floating nets of very fine mesh, which are themselves placed in ponds or at sea in calm areas. The type of nursery and the time juveniles spend there both vary from place to place.

Survival and growth are again two of the essential parameters monitored in the nursery (and grow-out) phase, which will determine whether the production of *H. scabra* is profitable. These two parameters are calculated and reported in various ways in the literature. Survival is generally expressed as survival rate (SR), and growth is either expressed as growth rate (GR) or specific growth rate (SGR). An indication of the overall final biomass (FB) obtained at the end of this phase is sometimes also provided. These indicators are calculated as follows:

$$\mathrm{SR} = (\mathrm{Nt}/\mathrm{Nt}_0) \times 100\%$$

where SR is the survival rate expressed in %, Nt and Nt_0 are the final and initial numbers of juveniles;

$$\mathrm{GR} = (\mathrm{Wt}/\mathrm{Wt}_0)/\mathrm{t}$$

where GR is the growth rate expressed in g day^{-1}, Wt_0 is the initial wet weight (g), Wt is the final wet weight (g), and t is the length of the culturing period (days);

$$\mathrm{SGR} = 100\ (\ln\ \mathrm{Wt}\ \ln\ \mathrm{Wt}_0)/\mathrm{t}$$

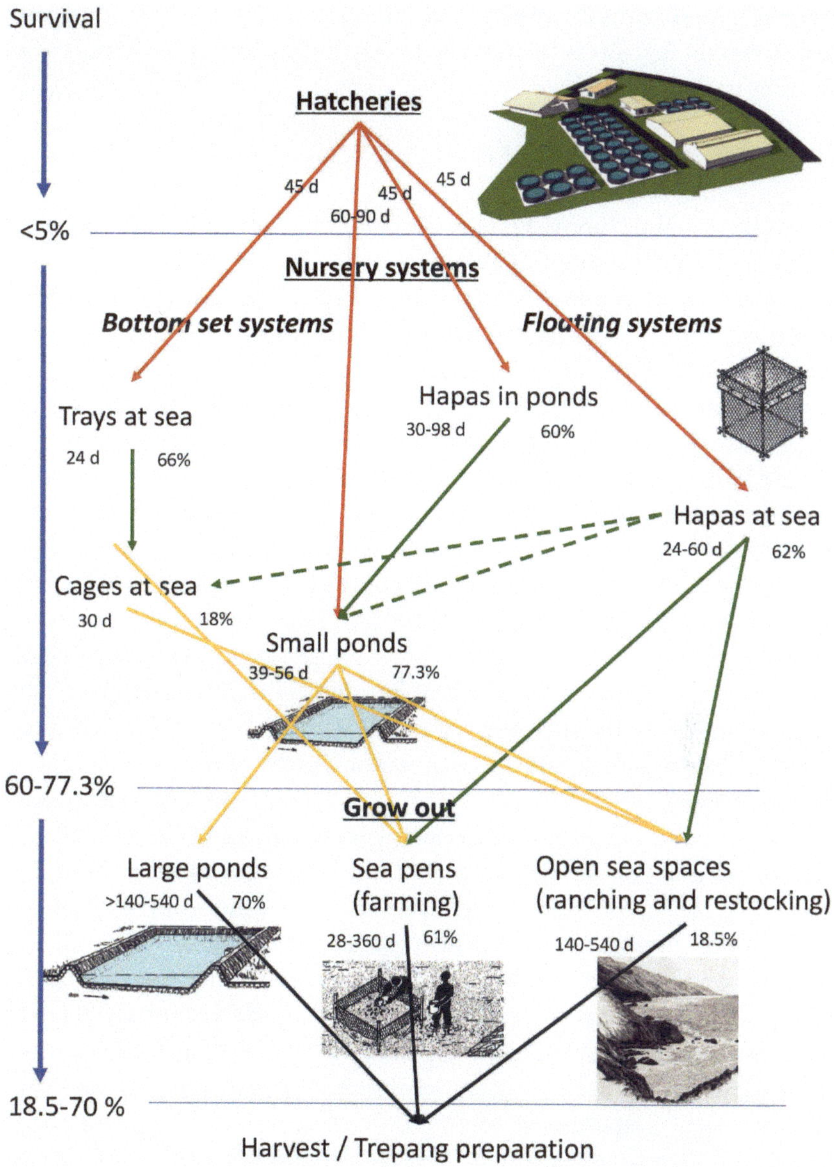

Fig. 28 Typical stages in the aquaculture of *H. scabra* and corresponding survival (%). Three steps are universal: larval and early post-metamorphic development in hatcheries, development of juveniles (sensu aquaculture) in nurseries and grow-out phase. The hatchery extends from 45 to 90 days, yielding juveniles <1 g (ww). They are then transferred to one of two main nurseries systems: bottom set or floating systems, where they grow to a max. of ~21 g in various systems (ponds, trays or cages at sea, floating hapas in ponds, floating hapas at sea); after a period in floating hapas, juveniles are often redirected ponds or cages at sea. As soon as the juveniles are ~15 g, they are placed in grow-out either in ponds, in sea fences (farming) or in protected areas at sea (sea ranching, restocking) where they will grow to the minimum legal weight (or size) for harvest.

where SGR is the specific growth rate expressed in % day^{-1}, Wt_0 is the initial wet weight (g), Wt is the final wet weight (g), and t is the length of the culturing period (days);

$$FB = Wt \times Nt$$

where FB is the final biomass (expressed in g), Wt is the final wet weight (g), and Nt is the final number of juveniles.

Irrespective of the type of nursery, survival and growth of juveniles depend on various conditions, among which salinity, temperature, food type, food concentration and juvenile density are the most influential. Meirinawati et al. (2020) showed that temperature and salinity are two crucial physical parameters for juvenile rearing systems. Furthermore, they identified phosphate, nitrate, and ammonia as essential chemical parameters. Dissolved oxygen, particulate organic matter, and pH may be monitored as well (Meirinawati et al., 2020). Temperature and salinity should be kept as steady as possible and are more easily controlled in nursery pond systems (Lavitra et al., 2010b) than in floating hapas. For small nursery ponds (e.g. 32–100 m^2) the use of shading and greenhouse systems can help stabilize the temperature (Lavitra et al., 2010b) but these correctives are challenging and costly to apply in large nursery ponds (e.g. 1000 m^2). Damage due to rainfall or hypersalinity due to evaporation in the warm season must be avoided (Lavitra et al., 2009a). Some studies demonstrated that juveniles of *H. scabra* show enhanced metabolic plasticity allowing them to adapt to a certain degree of temperature variation. In that context, Kühnhold et al. (2019b) compared the effects of an exposure of juveniles of *H. scabra* with those of the temperate species *Holothuria forskali* to identical cold- and warm-acclimation stress using key metabolic parameters such as the respiration rate, enzyme activity (electron transport system, lactate dehydrogenase, isocitrate dehydrogenase) and energy reserve fractions (lipid, carbohydrate and protein). Their results show much broader respiratory adjustments, in response to temperature change, in *H. scabra* than in *H. forskali*. Moreover, the tropical species exhibits a marked up and down regulation of metabolic enzymes and shifts in energy reserves, due to thermal acclimation, while the same metabolic indicators remain consistent in the temperate species. Kühnhold et al. (2017) suggested that a rearing temperature of 33°C might be possible, which could improve aquaculture production of *H. scabra* by enhancing growth. In another study Kühnhold et al. (2019a) showed that a good temperature range is essential during the rearing of juveniles of *H. scabra* (which is probably also the case for adults) and that they cannot survive temperatures below 22 or above 38°C. Within the thermal window of 22–38°C, *H. scabra* sustained positive aerobic capacity, with assumed optimal performance range between 29 and 31.5°C (Kühnhold et al., 2019a). Between 39 and 41°C, *H. scabra* decreased respiration progressively, while gene expression levels of Hsp70 increased at 41°C, indicating prioritization of heat shock response and homeostatic disruption (Kühnhold et al., 2019a). At the cold end (17–22°C), homeostatic disruption was visible through incrementally rising energetic expenses to fuel basal maintenance costs (Kühnhold et al., 2019a). Kamyab et al. (2017) determined the effects of temperature (i.e. cold and warm) on the immune and oxidative stress responses of juveniles of *H. scabra,* using the biochemical biomarkers involved in each process. Juveniles displayed increased superoxide dismutase and phenoloxidase activities when temperature was elevated, and low levels of superoxide radicals and damage in both cold and warm treatments throughout the experiment, confirming that they were able to mitigate thermal stress (Kamyab et al., 2017). This multi-biomarker approach may be useful to monitor the health of *H. scabra* and to gain insight concerning its use in global-scale aquaculture across regions submitted to different temperature regimes (Kamyab et al., 2017).

The juveniles of *H. scabra* are sometimes also released into sea pens directly from the hatchery, without passing through a nursery phase (Fig. 28). Even though this procedure often gives very poor survival results (Fig. 28), it is used in some instances to understand the important parameters allowing better yields in areas where farming or ranching occurs. In Ceccarelli et al. (2018), optimal microhabitat features of seagrass meadows were identified for juveniles of *H. scabra* in a coastal bay of New Caledonia at sites with varying biotic and physical features. Released juveniles were 2–8 g and the survival after 8–10 days ranged from 0% to 100% and averaged 77%. Survival was significantly higher at shallow depths with an intermediate seagrass cover of 42% (Ceccarelli et al., 2018).

The post-metamorphic growth patterns of *H. scabra* in nursery tanks and in sea pens have been examined Lavitra (2008). Polynomial regressions provided a good fit in both situations (Fig. 29).

21.3.5.1 Benthic systems

Earthen ponds (Figs 27C, 28) for growing juveniles have been tested principally in Indonesia, New Caledonia, Philippines and Madagascar (Table 18). In Madagascar, earthen ponds have become the regular

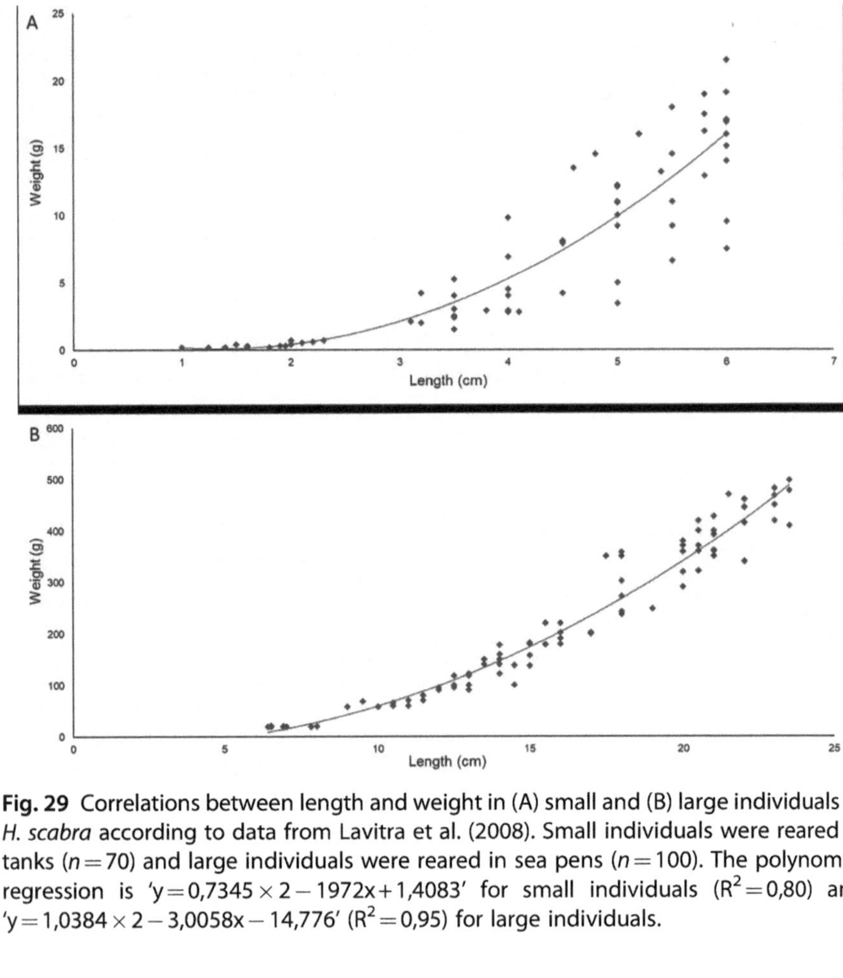

Fig. 29 Correlations between length and weight in (A) small and (B) large individuals of *H. scabra* according to data from Lavitra et al. (2008). Small individuals were reared in tanks ($n=70$) and large individuals were reared in sea pens ($n=100$). The polynomial regression is '$y=0{,}7345\times 2-1972x+1{,}4083$' for small individuals ($R^2=0{,}80$) and '$y=1{,}0384\times 2-3{,}0058x-14{,}776$' ($R^2=0{,}95$) for large individuals.

intermediate phase between hatchery and grow-out. It is therefore in Madagascar that most of the optimization tests for growth in nursery ponds were conducted. Juveniles are generally transferred at the weight of 1 g from hatcheries to earthen ponds [with one assay done on juveniles of 3–13 g in Philippines; (Gamboa et al., 2012)] (Table 18). The growth and survival of juveniles are mainly influenced by the water temperature and the stocking density (Table 18). Salinity is also an important parameter; i.e. if salinity fluctuations are important, they can lead to mortality (Lavitra et al., 2009a). A lot of trials have been done around the world and, all trials combined, the growth rate varies from 0.01 to 1.0 g day^{-1} and the survival from 18% to >95%. Juveniles may spend about 21–98 days in earthen ponds

Table 18 Characteristics of nurseries and grow-out systems worldwide for the aquaculture of *Holothuria scabra*, showing type of system, country, growth period in days (d), initial weight of the sea cucumber in grammes (g), initial stocking density per square metre (ind. m^{-2}), final weight (g), growth (g day^{-1}), survival in percent of initial stock (%), and references.

Type	Country	Growth period (d)	Initial weight (g)	Initial stocking density (ind. m^{-2})[a]	Final weight (g)	Growth (g day^{-1})	Survival (%)	References
NURSERIES								
Indoor tanks	Indonesia	170	<1	500	15[b]	0.08	5.2–8.7	Sembiring et al. (2018)
Indoor tanks	Madagascar	141	0.8	4	6.5–7.74	0.013–0.040	94.5	Rasolofonirina (2004)
Outdoor tanks[c]	Vietnam	70	2.1	20–21	6–22	0.055–0.284	NM	Pitt and Duy (2004a)
Outdoor tanks	Vietnam	39	0.94	36–72	0.7–17.2	−0.01 to 0.59	9–100	Pitt et al. (2004)
Earthen ponds[d] (32 m^2)	Madagascar	39	<1	22	9.94	0.294	89.43	Lavitra et al. (2010a)
Earthen ponds[e] (32 m^2)	Madagascar	39	<1	22	3.74	0.095	93.43	Lavitra et al. (2010a)
Earthen ponds[f] (32 m^2)	Madagascar	52	<1	20	4–9	0.076–0.17	41–65	Rakotonjanahary et al. (2016)
Earthen ponds[g] (32 m^2)	Madagascar	56	0.24	10–40	5–18	0.07–0.32	79->95	Lavitra et al. (2010a)

Continued

Table 18 Characteristics of nurseries and grow-out systems worldwide for the aquaculture of *Holothuria scabra*, showing type of system, country, growth period in days (d), initial weight of the sea cucumber in grammes (g), initial stocking density per square metre (ind. m^{-2}), final weight (g), growth (g day^{-1}), survival in percent of initial stock (%), and references.—cont'd

Type	Country	Growth period (d)	Initial weight (g)	Initial stocking density (ind. m^{-2})[a]	Final weight (g)	Growth (g day^{-1})	Survival (%)	References
Enclosures (hapas) in earthen ponds ($1 \times 1 \times 0.5\,m^2$)	Indonesia	42 98	4.21 4.21	15 15	13.2 7.09	0.21 0.03	93 31	Firdaus and Indriana (2019)
Enclosures (trays) in earthen ponds ($0.4 \times 0.31 \times 0.11\,m^2$)	Indonesia	24	<1	250	1.2	0.04	66	Juinio-Meñez et al. (2017)
Enclosures (cages) in earthen ponds ($1 \times 1 \times 1\,m^2$)	Indonesia	30	<1	750	0.4	0.02	18	Juinio-Meñez et al. (2012b)
Enclosures in earthen ponds[h]	New Caledonia	21	1	40	3 (maximum)	0.095	40–100	Bell et al. (2007)
Enclosures (bag nets) in earthen ponds ($1 \times 1 \times 1\,m^2$)	New Caledonia	21	1–2	150	2.68–4.1	0.08–1	NM	Agudo (2006)
Enclosures (hapas) in earthen ponds ($2 \times 1 \times 1\,m^2$)	New Caledonia	23	0.5–1	150	2.8–3.8	0.1	97	Agudo (2006)
Enclosures in earthen ponds	Philippines	30	<3–13	50	13–21.3	0.77–0.42	58–80	Gamboa et al. (2012)
Enclosures (hapas) in ponds ($2 \times 1 \times 1\,m^2$)	Philippines	30	<1	1000	0.6	0.01	57	Juinio-Meñez et al. (2012a)

Enclosures (hapas) in ponds (2 × 2 × 1 m)	Vietnam	41	<1	20	1	0.02	56	Pitt and Duy (2004a)
Sea pens (1 m^2)	New Caledonia	35	0.38	100	0.59–0.77	0.006–0.011	69–89	Purcell and Agudo (2013)
Sea pens (100 m^2)	Philippines	90	>2	200	0	–	0	Juinio-Meñez et al. (2017)
Sea pens (100 m^2)	PNG	40	3–20	NM	70–120	NM	40–95	Hair et al. (2016)
Floating hapas in ponds (1 × 1 × 0.5 m)	Indonesia	84	1.42–7.7	15	12.34–20.50	0.13–0.15	77.78–93.33	Indriana and Firdaus (2020)
Floating hapas in ponds (1 × 1 × 0.5 m)	Indonesia	42 98	4.21	15	13.69 7.49	0.22 0.03	95.56 75.56	Firdaus and Indriana (2019)
Floating hapas in channels (2 × 1 × 1 m)	Philippines	30	<1	400	2.1–5.45	0.07–0.18	42–96	Gamboa et al. (2012)
Floating hapas in ocean (1 × 1 × 1.2 m^2)	Philippines	24–60	<1	1000	0.4–1.23	0.01–0.03	14–69	Juinio-Meñez et al. (2017)
Floating hapas in ocean (unknown dimension)	Philippines	42	<1	250–750[i]	1.5[b]	0.066–0.137	27. 8–58.2	Damaso and Argente (2019)
Floating hapas in ocean (2 × 2 × 1 m)[j]	Philippines	60	0.02–0.06	150–2000	0.79–2.49	0.04–0.083	67.6–97.1	Altamirano and Noran-Baylon (2020)

Continued

Table 18 Characteristics of nurseries and grow-out systems worldwide for the aquaculture of *Holothuria scabra*, showing type of system, country, growth period in days (d), initial weight of the sea cucumber in grammes (g), initial stocking density per square metre (ind. m^{-2}), final weight (g), growth (g day^{-1}), survival in percent of initial stock (%), and references.—cont'd

Type	Country	Growth period (d)	Initial weight (g)	Initial stocking density (ind. m^{-2})[a]	Final weight (g)	Growth (g day^{-1})	Survival (%)	References
Floating hapas in ocean (2 × 2 × 1 m)[j]	Philippines	60	0.02–0.06	150–2000	1.93–3.08	0.01–0.024	69.1–88.9	Altamirano and Noran-Baylon (2020)
Floating hapas in ocean (2 × 2 × 1 m)	Philippines	45	0.05–3	250–324	2.49–4.54	0.114–0.134	45.6–85.3	Altamirano and Noran-Baylon (2020)
GROW-OUT SYSTEMS								
Tanks	Sri Lanka	308	11	2	77	0.21	95	Kumara and Dissanayake (2017)
Outdoor tanks	Vietnam	113	2.3–11.6	1	119–182	1,03–1,59	NM	Pitt and Duy (2004a)
Indoor and outdoor tanks	Indonesia	140	14	45	11.81–27.41	0.1 (maximum)	28.21–48.89	Indriana et al. (2017)
Sea ranches	Australia	140	3–15	–	80[b]	0.46–0.55	14.7	Taylor et al. (2016)
Sea ranches[k]	Philippines	540	5 7	0.016–0.04	500–630	0.9–1.15	2–39	Juinio-Meñez et al. (2012b)
Sea pens (3.51 m^2)	Maldives	124	20.17–21.2	10	146.2–147.8	1.56–1.61	>90	Ahmed et al. (2018)

Sea pens (2 m^2)	Tanzania	70	67.18	5	116.83–118.89	0.731–0.734	70–76	Hamad et al. (2019)
Sea pens (2 m^2)[l]	Tanzania	70	67.18	5	132.18	0.89	80	Hamad et al. (2019)
Sea pens (2 m^2)[m]	Tanzania	70	67.18	5	140.76	1.038	60	Hamad et al. (2019)
Sea pens (2, 25 m^2)[n]	Tanzania (Zanzibar)	84	106–115	3–7	118–163	0.14–0.8	75–83	Namukose et al. (2016)
Sea pens (500 m^2)[o]	New Caledonia	360	1–10	6	150	0.41	2.6–11.9	Purcell and Simutoga (2008)
Sea pens (9 m^2)	Madagascar	270	15	3–12	73–198	0.21–0.68	>75	Lavitra et al. (2010b)
Sea pens (60–625 m^2)	Madagascar	60–300	15	0.5–1.9	150–410.6	0.5–1.8	11–99	Robinson and Pascal (2012)
Sea pens (unknown size)	Indonesia	28	30–50	15–25	NM	0.035–0.607	92–100	Serang et al. (2016)
Sea pens (500 m^2)	Fiji	330	1–10	NM	165–167	0.5	28 (after 6 months)	Hair (2012)
Sea pens (500 m^2)	Sri Lanka	308	11	2	267	0.83	83	Kumara and Dissanayake (2017)
Sea pens (9 m^2)	Philippines	60	40	100	0	–	0	Juinio-Meñez et al. (2017)

Continued

Table 18 Characteristics of nurseries and grow-out systems worldwide for the aquaculture of *Holothuria scabra*, showing type of system, country, growth period in days (d), initial weight of the sea cucumber in grammes (g), initial stocking density per square metre (ind. m^{-2}), final weight (g), growth (g day^{-1}), survival in percent of initial stock (%), and references.—cont'd

Type	Country	Growth period (d)	Initial weight (g)	Initial stocking density (ind. m^{-2})[a]	Final weight (g)	Growth (g day^{-1})	Survival (%)	References
Earthen ponds (2000 m^2)	New Caledonia	540	0.9–11.7	1.6	325–425	0.60–0.79	>70	Agudo (2012), Bell et al. (2007)
Earthen ponds (600 m^2)	Sri Lanka	308	11	2	153	0.46	3	Lavitra et al. (2010a)
Earthen ponds	Solomon	140	15[b]	NM	42–98	0.3–0.7	93	Battaglene (1999a)
Earthen ponds (14,000 m^2)	Vietnam	420	2	1	350	0.83	80	Duy (2012)[p]
Earthen ponds (10,000 m^2)	Vietnam	305/274	10/20	1	310/300	0.97/0.93	85/87	Duy (2012)[p]

[a]Stocking density in individuals per square metre except in hapas where the dimension is indicated in the first column
[b]Calculated with curve from Lavitra (2008) showing the relation between weight and length.
[c]Various trials with individuals fed or unfed, with light or no light
[d]Water temperature: 30.8°C.
[e]Water temperature: 27.7°C.
[f]Various experiments with juveniles on normal sediment, fed with provender, fed with diatoms or cultured with tilapia; temperature from 21 to 34°C and salinity from 32.6‰ to 34.8‰.
[g]Reared on sediment from mangrove, sediment from sea grass bed or on sediment from flat reef.
[h]Various experiments with or without the shrimp *Lytopenaeus stylirostri*.
[i]Indicated in the publication as ind. m^{-3}.
[j]First line referring to the first 30 days, 2nd line referring to the last 30 days.
[k]Mentioned densities are those observed at the end with maximal weight observed (not means) and maximal growth rates.
[l]With *Euchema denticulatum*.
[m]With *Kappaphycus alvarezii*.
[n]With *Euchema denticulatum*.
[o]Various assays, only the large scale assay detailed here.
[p]With some information coming from Juinio-Meñez et al. (2017).
NM = not mentioned.

(Table 18) and growth rates are variable from place to place (Table 18). For example, growth rates of 0.28 g day^{-1} were measured in Vietnam at a rearing density of 20 ind. m^{-2} with shrimp food (Pitt and Duy, 2004b) and 0.07–0.32 g day^{-1} in Madagascar in experiments under various conditions (Lavitra et al., 2010a). These experiments have been done in small ponds <40 m^2 or in small enclosures of 1–2 m^2 inside large ponds (1310–11,400 m^2 in New Caledonia). Currently, at IOT-Madagascar, yield with hatchery juveniles released in large nursery ponds of 1000 m^2 is estimated at about 50–75% survival to 15 g individuals after 60 days, with an average growth rate of 0.24 g day^{-1} (Igor Eeckhaut, Personal observation). The bottom of these large ponds is covered with a few centimetres of sediments coming from quaries (Igor Eeckhaut, Personal observation) (Fig. 27C, D). The growth of captive juveniles of *H. scabra* in nursery systems has also been analysed by Watanabe et al. (2014) who determined the best-fit growth model to estimate age- and size-specific daily growth rates. The Gompertz model was considered optimal to express growth, both as length and weight.

The effects of the addition of extra-food other than sediment is often not obvious on the growth and survival of juveniles in ponds. Lavitra et al. (2009b, 2010a) tested three types of sediment coming from micro-atolls, mangroves and seagrass beds, and the results showed that the nature of the sediment did not affect the survival or growth of juveniles. High survival and normal growth rates were observed during 8 weeks of rearing using the three type of substrate described above (Lavitra et al., 2010b). In another study, Mondal and Kunzmann (2018) measured carbon stable isotopes in algal food sources and juveniles to study the assimilation of *Sargassum, Fucus, Spirulina* or *Ulva* separatedly. The poor performance of selected food sources towards the growth of *H. scabra* was highlighted. In contrast, Giri et al. (2018) indicated that juveniles could utilize a *Sargassum*-based diet at the nursery stage.

Robinson et al. (2013) carried out an experiment in South Africa to confirm the reported positive effect of sand on juvenile growth, relative to using bare tanks. The results confirmed the benefit of using sand as a substrate; however, sand in diets did not promote growth in the same way, indicating that inclusion of sand in formulated diets is unlikely to compensate for lack of sand as a substrate (Robinson et al., 2013). Rakotonjanahary et al. (2016) tested the effects of three types of enrichment to optimize the growth of *H. scabra* in nursery ponds: (*i*) marine sediment and provender (i.e. food for livestock and farmyard animals) based on fisheries by-products, (*ii*) marine sediment and green water (i.e. water where diatom production has been boosted), including tropical local diatoms and (*iii*) marine sediment with tilapia (*Oreochromis niloticus*) in co-culture. The stocking density at the beginning of the experiment was 20 ind. m^{-2} and, at the end of the experiment, no significant differences could be detected between the survival rates, mean weights or biomasses obtained for each type of enrichment tested (Rakotonjanahary et al., 2016).

Lavitra et al. (2010b) tested juvenile stocking densities of 10, 20, 30 and 40 ind. m^{-2} (juveniles size: 0.24–15 g). Average growth rates were always >0.22 g day^{-1} across treatments, but stocking density had a clear influence. At the lowest stocking density of 10 individuals m^{-2}, the best growth rate recorded over 10 weeks was 0.53 g day^{-1}, obtained between the fourth and the sixth week of rearing. The survival was >95% after 2 months (Lavitra et al., 2010b). The strong negative relationship between stocking density and growth rate observed by Lavitra et al. (2010b) was however not observed in another experiment conducted in Thailand where the negative correlation was lower [C.M. Raison, unpublished data in Mills et al. (2012)]. This experiment compared growth and survival over 36 days among treatments comprising two densities of early juvenile (average weight 1.74 g at densities of about 50 and 85 g m^{-2}), and five feeding rates (0.05–0.25 g day^{-1} of commercial shrimp starter feed). Results indicated a strong positive relationship between feeding rate and growth rate, and a weaker negative relationship between stocking density and growth rate.

Besides earthen ponds, other bottom set systems (called hapas, trays or cages; Figs 27B, 28) that are placed either in large ponds or at sea have been mainly tested in Indonesia, New Caledonia, the Philippines and Vietnam (Table 18). The general principle of these systems is identical to the ponds (the juveniles feed on the organic matter of the sediment) but they allow a control easier and less expensive than if these animals were scattered on a large surface or if ponds had to be built. The nets are of 1 or 2 m^2 and juveniles stay there between 21 and 93 days. The survival is generally good (from 34% to 55%) and juveniles are at the end transferred in larger area for final growth. Gorospe et al. (2017) suggested that the bottom-set tray with artificial sediment can be a practical alternative ocean nursery unit for rearing early sandfish juveniles particularly when the sea surface condition is rough. In Philippines, juveniles coming from the hatcheries are sometimes first reared in trays (mesh size < 1 mm) at sea then in cages before being released in larger areas (Juinio-Meñez et al., 2017). In China, juveniles of *H. scabra* with an average initial body weight of 0.04 g were reared in small benthic net cages at a site for shellfish culture in the open sea for 95 days (Chen et al., 2015). The results showed that juveniles had an average final body weight of 0.84 g with a daily weight gain of 0.008 g and a survival of 57%. Sea pens enriched with *Sargassum* spp. make viable ocean nurseries and significantly increase growth and survival of juveniles (Sinsona and Juinio-Meñez, 2018). Because conditions vary across locations and the carrying capacity of the grow-out area changes over time, regular monitoring of growth and survivorship is recommended (Juinio-Meñez et al., 2014). In addition, transfer to another sea pen when growth starts to decline can help optimize economic returns (Juinio-Meñez et al., 2014).

21.3.5.2 Floating systems

To overcome space limitations in hatchery tanks or in nursery ponds, juveniles can be transferred from hatcheries to floating hapas (Fig. 28) (Duy, 2012; Gamboa et al., 2012; Indriana and Firdaus, 2020; Pitt and Duy, 2004b), which can either be placed in earthen ponds (Firdaus and Indriana, 2019; Indriana and Firdaus, 2020; Juinio-Meñez et al., 2012b; Pitt and Duy, 2004b), in man-made canals (Gamboa et al., 2012) or in the ocean (Altamirano and Noran-Baylon, 2020; Damaso and Argente, 2019; Juinio-Meñez et al., 2017). Floating hapas are made of white fine mesh polyethylene (PE) net of 0.66–1 mm in diameter (Agudo, 2006; Altamirano and Noran-Baylon, 2020; Duy, 2012; Gamboa et al., 2012; Pitt and Duy, 2004b) with a dimension of 1 × 1 × 1 m or 2 × 1 × 1 m (length × width × depth). The bag is suspended from the water surface by various floats [e.g. a rectangular PVC pipe (Altamirano and Noran-Baylon, 2020)]. Weights made of plastic bottles filled with sand are attached to the four corners of the bottom of the hapa net, while a rectangular perforated PVC pipe frame (2 cm diameter) is placed inside the hapas to shape their floor (Altamirano and Noran-Baylon, 2020). Following usual nursery culture protocols (Gorospe et al., 2019; Sinsona and Juinio-Meñez, 2019), the nets are typically pre-conditioned by deploying out to sea prior to stocking them with juveniles (see below). Juveniles may spend from 24 to 60 days in floating hapas before being transferred into larger rearing areas (Agudo, 2006; Edullantes and Juinio-Meñez, 2009; Gamboa et al., 2012, 2017; Juinio-Meñez et al., 2012b; Pitt and Duy, 2004b) (Table 18). The weight of juveniles transferred to floating hapas varies from 1 to 2 g (Agudo, 2006), 2 to 5 g (Mills et al., 2012) or 1.4 to 7.8 g (Indriana and Firdaus, 2020). In the Philippines, it has proven difficult to find ponds with adequate seawater exchange rates to maintain the salinities required for juvenile growth. Consequently, efforts have been directed at developing floating hapa systems that can be deployed in sheltered marine embayments (Juinio-Meñez et al., 2012a, 2017; Mills et al., 2008) (Table 18).

Food addition during the nursery phase is normally not necessary in hapas; however, when ocean productivity is low, better growth may be obtained by adding ground shrimp pellets (Agudo, 2006). Hapas can be conditioned for 3–5 days to allow a substantial biofilm to grow on them before adding the juveniles (Gamboa et al., 2012). The algal mat reportedly serves as natural food for the growing juveniles and eliminates (or reduces) the manual task of feeding them (Gamboa et al., 2012). The results of a recent study of Sinsona and Juinio-Meñez (2019) suggest a shift in the early juvenile's feeding mode and diet from primarily grazing on diatoms when they are 5 mm long to feeding on detritus along with associated bacteria and organic matter, once larger (>1 g wet weight). It has also been suggested that juveniles that were previously reared in hatchery tanks with sediment had thicker body walls and grew faster in floating hapas than those reared previously in tanks without sediment (Dumalan et al., 2019; Schagerström, 2003). Accordingly, Juinio-Meñez et al. (2012a) suggested that, even if juveniles can be reared in floating hapas to a release size of >3 g, sand conditioning prior to release may be necessary to improve their survival in the wild.

Estimated growth rates of *H. scabra* in floating hapas range from 0.01 to 0.22 g day^{-1} and the survival rates vary from 15% to 97% (Table 18). The baseline density of juveniles had the greatest effect on both metrics. With densities of 150 juveniles m^{-2} (Agudo, 2006), survival averaged 96% over 23 days and growth rate was of 0.1 g day^{-1}. The growth performance and survival of early juveniles reared in floating hapas at 250, 500, and 750 ind. m^{-3}, were tested by Damaso and Argente (2019). Results revealed that the treatment with the lowest density yielded the highest growth and survival at the end of the 42-day culture. Altamirano and Noran-Baylon (2020) also studied the effects of initial stocking density, size grading and net replacement frequency during the culture of early juvenile *H. scabra* in sea-based floating hapas (2 × 1 × 1 m). Testing different initial stocking density treatments ranging from 150 to 2000 early juveniles (4–10 mm; 0.02–0.06 g) per floating hapa, they confirmed significantly faster growth (0.06–0.08 g day^{-1}) and higher survival (72–97%) in low densities (≤500 $hapa^{-1}$), compared to high densities (≥700 $hapa^{-1}$: 0.02–0.04 g day^{-1}; 67–80%).

21.4 Grow-out and adult production

The minimum legal length of *H. scabra* to be exported varies from country to country (e.g. 18 cm long for live *H. scabra* in Madagascar). Correspondence between length and weight at various ontogenetic stages and under various conditions have been established (Table 3; Fig. 29). To reach marketable size, juveniles coming from nurseries are typically raised in three different ways: sea ranching, sea farming and earthen pond growing. The three practices have been well documented in various publications such as Lovatelli et al. (2004), Toral-Granda et al. (2008), Purcell et al. (2012a), Hair et al. (2012) and Mercier and Hamel (2013). 'Sea cucumber stocking' is also a general term used in scientific literature to mean the release of juveniles into open areas with the expectation that they will grow to larger sizes (Purcell et al., 2012a). Bell and Nash (2004), Bell et al. (2005, 2008) and Bartley and Bell (2008) defined different types of sea cucumber stocking interventions, including sea ranching, restocking and stock enhancement (redefined in Purcell, 2012). The three types of stocking start either with the release of juveniles into open habitats for harvesting, into open habitats to build nucleus of breeding population or into a broader fishery area, respectively. The main criteria for determining a suitable grow-out site are highlighted in Agudo (2006), Giraspy and Walsalam (2010), Pascal and Robinson (2011), Purcell (2012) and Rizqi (2019). There are bioecological, environmental, and technical criteria; subcriteria include presence/absence of seagrass, presence/absence of *H. scabra*, substrate and water quality, tidal depth, transportation facilities, safety and labour/community support facilities. Geographic Information System (GIS) can help analyse the data and determine the suitable cultivation areas (Marizal et al., 2012; Mulyani et al., 2017, 2019).

Sea ranching pursues the objective of stock enhancement or restocking of overexploited wild populations through release of hatchery-reared individuals into No-Take Zones (NTZs) (Battaglene, 1999a,b; Battaglene and Bell, 1999; Purcell, 2004a; Robinson, 2013). Alternate stocking methods include earthen pond culture and sea pen farming, which revolve around human-created and human-delimited habitats, respectively. Both methods make use of nursery-reared juveniles having reached a weight that varies according to the rearing sites (Fig. 28). Earthen ponds and sea pens have their own sets of advantages and disadvantages. Earthen ponds (Figs 28, 30D, E) are more manageable from a 'diseases and predators' standpoint but they are expensive to maintain and time consuming. Their construction requires large areas on land which are not always readily available. Sea pens (Figs 28, 30F) are installed at sea and are therefore more susceptible to diseases, predation and poaching.

Regardless of the grow-out location (sea ranching or sea farming in open habitats, earthen pond), juveniles of *H. scabra* exhibit cyclical burrowing

Fig. 30 (A) A 350-g sea cucumber after growth in sea pens; (B) outcome after sea farming in Madagascar; (C) processing at IOT, in Madagascar; (D) final drying phase of beche-de-mer at the IOT factory; (E) view of the marine fences at IOT with watchtower where guards are posted to prevent smuggling; (F) sea pens in a village in Madagascar. *Panel (C) photo courtesy of B. Pascal.*

patterns, depending on their age (Battaglene, 1999a; Mercier et al., 1999, 2000b; Uthicke, 2001; Yamanouchi, 1939, 1956) (see earlier section dedicated to burrowing behaviour). Burying behaviour of cultured and wild juveniles of *H. scabra* was compared over a 48-h period in the natural habitat with or without seagrass (Hair et al., 2020a). Results showed that cultured juveniles were slower to burrow in the substrate after release, less likely to be burrowed at most times, and more likely to be burrowed in substrate where seagrass was present. However, they became better synchronized with their wild counterparts after 30 h (Hair et al., 2020a). Likely due to predation pressure, these patterns follow diel rhythms, burrowing at sunrise and emerging at sunset (Mercier et al., 1999, 2000b). Under laboratory conditions (i.e. animals kept in aquaria under a constant light regime of 14 h light and 10 h darkness with temperature reduced by 1°C each day from 24 to 17°C), sea cucumbers showed a distinct diel burrowing and feeding cycle, with most individuals exposed and feeding between 13:00 and 20:00 h, and mostly burrowed and inactive between 1:00 and 9:00 h (Wolkenhauer, 2008a,b). Periods over which individuals were burrowed increased with decreasing temperature from 6.7 h per day at 24°C, to 14.5 h per day at 17°C (Wolkenhauer, 2008a,b). Feeding activity decreased from 9.8 h a day at 24°C, to 0.8 h per day at 17°C. Temperature was positively correlated with both feeding and burrowing behaviour (Wolkenhauer, 2008a,b).

Alongside burrowing behaviour, *H. scabra* shows dispersal behaviour which can be an obstacle to the profitability of the grow-out phase in the natural environment. Indeed, thousands of individuals were seen displaying active buoyancy adjustment in farms in Madagascar, whereby they either floated near the sea surface at low tide (in 0.2–0.7 m depth) or tumbled on the sediment, and got carried away by the current (Hamel et al., 2019c). The speed of floating *H. scabra* was estimated to be 0.1 $m\,s^{-1}$, equivalent to 6 $m\,min^{-1}$, whereas tumbling speed was around 0.02 $m\,s^{-1}$. Between 33 and 73 tumbling/floating individuals of *H. scabra* were recorded by each observer during rounds conducted at the full or new moons every month for over 2 years, which was extrapolated to constitute between 0.5% and 4% of the entire surveyed population on any occasion. Instances of tumbling/floating occurred at unequal frequencies in the different enclosures monitored (mean conspecific density of 1.5 ind. m^{-1}). The individuals housed in the offshore enclosures (35–36 psu) did not exhibit the behaviour, whereas those in the six focal enclosures closest to land (18–27 psu) displayed the tumbling/floating behaviour (Hamel et al., 2019c).

The quality of the sediment is important as it forms the habitat where *H. scabra* burrows and also provides a source of nutrients. Tsiresy et al. (2011) and Plotieau et al. (2014b) respectively reported the mineral and organic features of sediments in three and four farms where *H. scabra* is reared by villagers in Madagascar. The sediments were composed of 0.4–2.0% organic carbon (TOC measured by carbonization) and 98–99.6% of mineral matter (relative to the total dry weight) (Plotieau et al., 2014b). This organic fraction was sufficient to support a growth rate of 0.5–2.1 g per day. The particle content <500 μm varied from 43% to 85% depending on the village (Plotieau et al., 2014b). The protein content was 0.0265 $\mu g L^{-1}$, the bacteria count was 1.1×10^9 bacteria g^{-1} (count by DAPI staining), and the chlorophyll *a* concentration was 4 $mg g^{-1}$. The proportion of carbonate to silica varied from 35:65 to 94:6 (Plotieau et al., 2014b).

Plotieau et al. (2014a) compared the importance of the assimilation of three ^{15}N-labelled food components (bacteria, dissolved alanine and microphytobenthos) in the tissues of *H. scabra*. The results showed that these three markers were effectively assimilated by *H. scabra* (Plotieau et al., 2014a). Alanine was rapidly taken up in its dissolved form, but the investigators indicated that its incorporation may also be mediated by bacteria. Two other teams studied the assimilation of diet components in *H. scabra* using ^{15}N and ^{13}C stable isotopes (Indriana et al., 2018; Watanabe et al., 2013). The first provided no direct conclusion concerning aquaculture but suggested that faeces may be used to determine ingested organic matter from the wild (Watanabe et al., 2013). The second indicated that the assimilation rate of digested food is low and the authors consequently proposed that added food in aquaculture should be homogenized and mixed with sediment (Indriana et al., 2018).

Bacterial diversity of sediments transiting trough the gut of *H. scabra* was analysed in Plotieau (2012) and Plotieau et al. (2013). The number of bacteria significantly increased (by a factor of 1.5) in the foregut as compared to the external sediments. The number of bacteria decreased in the midgut and stayed stable across the faeces. Some γ-Proteobacteria, especially *Vibrio*, were less affected by digestion than the other bacteria. The season had an impact on the bacterial diversity of the sediment transiting through the gut: in the dry season, gamma-Proteobacteria were the most abundant while in the rainy season it was the α-Proteobacteria (Plotieau et al., 2013).

21.4.1 Sea ranching

Sea ranching is currently one of the most extensive forms of sea cucumber aquaculture, making use of large areas of seagrass beds (~5 ha); it is typified by relatively low survival (2–39%) and yields (58–220 kg ha^{-1}) from a commercial perspective (Hair et al., 2021; Juinio-Meñez et al., 2012a,b; Robinson, 2013). Sea ranching, first explained in James (1990) and James et al. (1994b) is mainly documented from Australia (Taylor et al., 2016) and Philippines (Juinio-Meñez et al., 2012a,b, 2017; Olavides et al., 2011). The release of juvenile nursery in sea fences has also been analysed beforehand for a future sea ranching in Fiji (Hair, 2012) and Papua New Guinea (Hair et al., 2016, 2021). An overview of sea ranching in South East Asian regions is presented in Mills et al. (2012) and the progress and problems of sea farming on the tropics is reviewed in Purcell et al. (2012a). Growth rates range from 0.46 to 1.15 g day^{-1} when nursery juveniles are released in open spaces (not in fences) (Table 18) and can be highly variable, depending on the quality of the site and micro-scale changes in habitat (Juinio-Meñez et al., 2012b; Purcell and Simutoga, 2008; Robinson, 2013). Three pilot sea-ranching sites (from 1 to 4 ha) were established in north-western Luzon (Philippines) where multiple releases of cultured juveniles produced from local wild broodstock were conducted (Juinio-Meñez et al., 2012b). Within 7–10 months, effective spawning populations had established; growth and survival rates were variable among sites with the maximum estimated overall density reaching 1119 ind. ha^{-1}, and an estimated survival rate of 39% after 19 months (Juinio-Meñez et al., 2012b). In comparison, the Bolinao–Anda reef system supported a wild population density of 6 ind. ha^{-1} (Olavides et al., 2010), whereas density of a relatively unexploited natural population of *H. scabra* in Papua New Guinea was reported (Shelley, 1985a) to be 2.9 ind. ha^{-1}. After 19 months of sea ranching, synchronous spontaneous spawning was also observed in the Bolinao and Anda sites (Olavides et al., 2011). In another study in the Philippines, a total of 14,300 fluorochrome-stained juveniles (>3 g) were released in 8 batches over a 15-month period in a 5-ha pilot communal sea ranch (Juinio-Meñez et al., 2013). Apparent survival over the study period was estimated at 20–30%. Incidences of in situ spawning were observed within a year, and estimated densities of reproductively mature sea cucumbers increased from 37 ind. ha^{-1} after 7 months to 249 ind. ha^{-1} after 15 months. Juveniles without fluorochrome-stained ossicles were found during most monitoring periods, indicating the presence of wild recruits. Recently, a 5-ha trial sea ranch has been investigated by Hair et al. (2021) in Papua New Guinea. Over 10 months, 5655 fluorochrome-marked juveniles of *H. scabra* were surveyed with line-transects prior to stocking the sea ranch and at the end of the study. All juveniles were weighed and checked for fluorochrome marks to determine cultured or wild origin. Only an estimated 3.7% of cultured juveniles were recoverable in the post-stocking survey.

One of the major problems with sea ranching is that it is very difficult to estimate the survival rate of a moving, transiently burrowed, animal across an undefined area. Purcell and Kirby (2006) measured short-term movement paths of released individuals (1–105 g) and wild adults (130–690 g) of *H. scabra* in a seagrass bed in New Caledonia. They developed an individual-based model to predict long-term dispersal of released juveniles (1–16 g) at 1 individual m^{-2} within a 1-ha area, drawing on distributions of speed and directionality and the relationship between speed and individual weight from field data. The models determined that 6–12% of surviving animals were predicted to remain in the original 1-ha release site after 10 years. To protect surviving individuals as breeding populations for 10 years, and assuming 10% spillover, a surface of 19–40 ha would reportedly be needed (Purcell and Kirby, 2006).

Population genetics analyses should be considered when releasing hatchery-produced juveniles of *H. scabra* when spawners from non-local populations are used (Purcell et al., 2012a). In New Caledonia, within-region genetic heterogeneity of *H. scabra* populations (examined through allozyme electrophoresis) indicated high gene flow (Uthicke and Purcell, 2004). Inclusion of *H. scabra* from Bali, Australia and Solomon Islands showed that populations were different. The rapid increase in genetic distance over the first few hundred kilometres supports the view that the spatial extent of any translocation needs to be carefully considered (Uthicke and Purcell, 2004).

21.4.2 Sea farming

The sea farming of *H. scabra* began to develop in Madagascar under the impetus of projects bringing together Malagasy and Belgian scientists (Eeckhaut et al., 2008; Robinson, 2011; Robinson and Pascal, 2009, 2012). Since 1998, when the first hatchery was built there, sea farming has grown, with the total area of sea pens reaching 153 ha in 2018. Two types of enclosures were developed: village pens and industrial pens. Village pens (Fig. 30F) measure between 625 and 900 m^2, are constructed close to villages and stocked with batches of hatchery-reared juveniles at a density <2 ind. m^{-2}. In 2018, 26.2 ha of village pens involved 255 households and 175 farm-enclosures across 6 villages. According to the latest estimates, sea cucumbers reach a minimum size of 300 g after 4–12 months of growth (Fig. 30A, B) when they are harvested by farmers and sold to the commercial partner, Indian Ocean Trepang (IOT), for processing and export (Fig. 30C). Industrial pens in Madagascar are fully managed by the company IOT, measure 1–10ha each and covered a total of 127ha in 2018. Growth rates recorded during sea farming were variable within and between sites due to effects of stocking density and seasons, indicating that spatio-temporal variation is important (Robinson and Pascal, 2012). The average recorded growth rates in targeted villages varied from 0.5 to 2.1g day^{-1} (Plotieau et al., 2014b; Robinson and Pascal, 2012) and the survival varied greatly from 1% to 80% (Robinson and Pascal, 2012). Trials of sea farming (or release in fences for future ranching) have also been conducted in various other regions of the world: in Fiji (Hair, 2012), Indonesia (Serang et al., 2016). Maldives (Ahmed et al., 2018), New Caledonia (Purcell and Simutoga, 2008), Philippines (Juinio-Meñez et al., 2017), Sri Lanka (Kumara and Dissanayake, 2017), Tanzania (Hamad et al., 2019; Namukose et al., 2016) (Table 18). In Fiji, where juveniles were released into 100 m^2 sea pens, the survival was around 28% after 6 months and the average size was 166g after 11 months (Hair, 2012). In bottom-set cages deployed in a lagoon in the Maldives, the mean survival was 97.14% at the end of the 124-day culture period with a specific growth rate of 1.58% day^{-1} (Hamad et al., 2019). Survival rates in sea farming vary in the scientific literature between 2.6% and 100% for all periods of growth combined (Table 18). These rates however rarely exceed 75% after 6 months of sea farming. There is no data in the scientific literature reporting the survival rate in an industrial-type farm after several culture cycles. In the industrial-type farm IOT-Madagascar, after 5 culture cycles, the survival rate of *H. scabra* in sea pens was only 25% (Igor Eeckhaut, Personal observation). This low rate is due to poor control over four variables: predation, diseases and theft, and possibly to sea cucumbers undergoing active buoyancy adjustment (Hamel et al., 2019c).

The farming of *H. scabra* can have a positive impact on seagrass beds (Wolkenhauer et al., 2010). These authors measured seagrass productivity, seagrass and benthic microalgae (BMA) biomasses and organic matter (OM) during two experiments conducted using in situ cages over 2 months. Density of *H. scabra* was manipulated in caged exclusions (near-zero density, 'EX'), caged controls (natural densities, 'CC') and uncaged controls (natural density, 'NC'). Seagrass growth was lower when holothuroids were excluded. Seagrass biomass decreased in all treatments, but reduction was greater in EX than in controls. Both BMA biomass and OM increased in EX compared to NC and CC. Arnull (2018) and Arnull et al., 2021 investigated the impacts of varying densities of sea cucumbers on aspects of seagrass growth rate and ecosystem status at an aquaculture farm in southwest Madagascar. A key result is that the growth-rate of seagrass (measured as linear leaf extension rate) was higher in the presence of *H. scabra* than in the control sites without *H. scabra*. This apparent effect of *H scabra* on seagrass growth rate was seen even at moderate sea cucumber stocking densities of ~200 $g\,m^{-2}$, where seagrass leaves grew ~60–70% faster than in the control treatment.

Plotieau et al. (2013) observed various impacts on the sediments where intensive farming of *H. scabra* has occurred. Comparing various parameters in the upper layer of the sediment inside and outside sea pens in two farms in the south-west of Madagascar, they showed that after 2 years of intense farming: (*i*) the proportion of the finest grain size fraction (<250 μm) decreased from 5% to 14%, (*ii*) the global carbonate proportion decreased by 5%, (*iii*) aragonite was the carbonate type most affected by holothuroids, followed by calcite and magnesian calcite, (*iv*) the total organic matter did not vary significantly, (*v*) the number of bacteria decreased by up to 50%, and (*vi*) the concentration of photosynthetic microorganisms fell by up to 22%, within the pens. Based on these results, they suggest that fallow practice (i.e. temporarily leaving the sediment to rest) should be considered in sea cucumber farming (Plotieau et al., 2013). On the other hand, de la Torre-de la Cruz et al. (2018) determined the effect of *H. scabra* on the sediment components and particle composition after a year of rearing. With 10.3% of the stocked sea cucumbers recovered from the experimental pens, the results indicated that sea ranching at densities of two sea cucumber per m^2 can be made without causing sediment degradation.

21.4.3 Earthen pond culture

H. scabra is amenable to pond conditions because it is an inshore shallow-water species often found in silty habitats (Fig. 30D, E) (Agudo, 2012; Hamel et al., 2001; Purcell, 2004b; Purcell et al., 2012a; Raison, 2008). Survival to an adult size tends to be higher in ponds than in the sea, usually around 50–85% (Agudo, 2012; Bell et al., 2007; Duy, 2012; Gamboa et al., 2012; Purcell et al., 2012a). Comparison of the most recent data comes from the work of Purcell et al. (2012a) and it shows a strong correspondence of growth rates of *H. scabra* in earthen ponds among trials conducted in New Caledonia, Vietnam and Australia. Growth rates of individuals of 5–50 g tend to be slow in the first few months in ponds (0.1–0.9 g day^{-1}), with an increase (e.g. 1.0–1.7 g day^{-1}) once they reach a size at which they can ingest larger quantities of food (Agudo, 2012; Pitt and Duy, 2004b; Pitt et al., 2001; Purcell and Kirby, 2006). In Vietnam, the commercial monoculture of *H. scabra* in rotational cycles with shrimp was investigated (Pitt and Duy, 2004b), and it yielded 2–3 t wet weight ha^{-1} $year^{-1}$ (Duy, 2012; Pitt et al., 2001).

21.4.4 Co-culture with other organisms and bioremediation enhanced by H. scabra

A variety of studies have examined the feasibility of rearing *H. scabra* in co-cultures (with other animals or algae) and examined the role of deposit-feeding sea cucumbers (including *H. scabra*) inside integrated multitrophic aquaculture (IMTA; reviewed by Zamora et al., 2018). The co-culture of *H. scabra* with other organisms is currently documented in around 20 scientific articles. In these, *H. scabra* is found in co-culture with macroalgae (four–five species), shrimp (three species) and fishes (five species) as main taxa (Table 19). Trials are also mentioned with sea urchins, lobsters, mussels, oysters and snails. The majority of studies showed a neutral effect of other species on *H. scabra*; some were deemed to be deleterious and only one study, done with *Tilapia* fish, reported a strong positive effect.

Some projects with seaweeds (macrophytes) were performed in Indonesia, where wild-caught transplanted juveniles of *H. scabra* grew in the presence of *Eucheuma cottonii* (Madeali et al., 1993; Mulian, 1993), *Eucheuma* sp. (Daud et al., 1993; Rachmansyah et al., 1992; Tangko et al., 1993b) or *Gracilaria* sp. (Daud et al., 1991; Tangko et al., 1993a; Tomatala et al., 2019). Wild-caught juveniles of *H. scabra* were also co-cultured with the red seaweed *Kappaphycus striatum* in Zanzibar, Tanzania (Beltran-Gutierrez et al., 2016). Results indicate that *H. scabra* is a viable candidate species for lagoon co-culture with seaweed: growth performance for seaweed and sea cucumbers did not differ between monoculture and co-culture treatments, co-culture offering a more efficient use of limited coastal space over monoculture. Similarly, an integrated mariculture system with *H. scabra* and the seaweed *Eucheuma denticulatum* was tested in Zanzibar (Tanzania) to investigate the effect of stocking density on the growth and survival of cultured species, including the total organic matter and total organic carbon content in the sediment (Namukose et al., 2016). The study demonstrated that seaweed cultured under high stocking density of *H. scabra* exhibited a higher specific growth rate than that cultured at the medium or low densities of sea cucumbers. The effect on sea cucumbers was not evident, as individuals of *H. scabra* cultured at low stocking density displayed a higher mean growth rate

Table 19 Species associated with *Holothuria scabra* in co-culture, showing associated species, country and effect on sea cucumbers.

Associated species	Region/ country	Effect on *H. scabra*	Remark	References
Seaweed				
Eucheuma cottonii/*Eucheuma* sp.	Indonesia	Neutral		Daud et al. (1993), Madeali et al. (1993), Mulian (1993), Rachmansyah et al. (1992), Tangko et al. (1993b)
Gracilaria sp.	Indonesia	Neutral		Daud et al. (1991), Tangko et al. (1993a)
Kappaphycus striatum	Zanzibar (Tanzania)	Neutral	Survival 100% Growth: 0.9–1.6 g day^{-1}	Beltran-Gutierrez et al. (2016)
Eucheuma denticulatum	Zanzibar (Tanzania)	Neutral	Sea cucumbers cultured at low stocking density had a higher mean growth rate of 0.80 g day^{-1} compared to those cultured at medium or high densities	Namukose et al. (2016)
Sea grapes (*Caulerpa lentillifera*)	Vietnam	Neutral to positive	Survival was reduced in co-culture treatments (77.6% vs 97.2%); Mean weight gain and absolute growth rate were around 37% greater in coculture	Dobson et al. (2020c)

Shrimp				
Penaeus monodon	Vietnam	Negative in most cases	Survival: 0–100% Growth: −0.0017 to +0.067 g day^{-1}	Duy (2012), Pitt et al. (2004)
Litopenaeus vannamei	Vietnam	Unknown	Authors indicated that ‘trial was set up under parameters that should have greatly favoured sea cucumbers’	[Mills and Duy (unpublished data) in Mills et al. (2012)]
Litopenaeus stylirostris	New Caledonia	Neutral at most	Sea cucumbers visible on the surface did not differ significantly between monoculture and co-culture[a]; Death and morbidity of *H. scabra*[b]; Trials showed some initial promise however, larger scale trials proved unsuccessful[c]	Purcell et al. (2006b) Bell et al. (2007) Mills et al. (2012)
Fish				
Unidentified juvenile of Lethrinidae, *Balistapus undulatus*, *Cheilinus fasciatus*, *Halichoeres hartzfeldii*, *Lethrinus harak*, *Monotaxis grandoculus*, *Pentapodus caninus*, *Scolopsis trilineatus*, Unidentified adult of Scaridae	Solomon Islands	Negative	Small bites or ingestion	Dance et al. (2003)

Continued

Table 19 Species associated with *Holothuria scabra* in co-culture, showing associated species, country and effect on sea cucumbers.—cont'd

Associated species	Region/ country	Effect on *H. scabra*	Remark	References
Chanos; *Trachinotus blochii* and *Lates calcarifer*	Philippines	Negative	Small juveniles experienced mass mortality; Survival of larger juveniles was high but not significantly different from the seagrass (63%)	Dumalan et al. (2019), Zamora et al. (2018)
Red tilapia (*Oreochromis aureus*)	Thailand	Positive	Survival: up to 95% (1.3 × better than in monoculture) Growth: $0.68\,g\,day^{-1}$ (2.5 × better than in monoculture)	Sithisak et al. (2013)
Groupers (*Epinephelus* sp.)	China	Unknown		Purcell and Wu (2017)
Others				
Tripneustes gratilla	Australia	Unknown		[Carton, unpublished data in Zamora et al. (2018)]
Lobster and mussels	Vietnam	Unknown	Individuals placed to reduce the organic matter charge	Du et al. (2004), Zamora et al. (2018)
Pearl oysters (*Pinctada martensii*)	China	Unknown		Purcell and Wu (2017)
Babylon snail (*Babylonia areolata*)	Vietnam	Neutral to positive	Survival was reduced in coculture treatments (77.6% vs 97.2%). Mean weight gain and absolute growth rate were around 37% greater in co-culture	Dobson et al. (2020a,b, 2021), Dobson et al. (2020c)

[a] Purcell et al. (2006b)
[b] Bell et al. (2007)
[c] Mills et al. (2012)

compared to those cultured at medium or high densities, which is also the result found in monoculture. Another experiment was conducted in Zanzibar to assess the survival and growth performance of *H. scabra* in co-culture systems with *E. denticulatum* and *K. alvarezii* (Hamad et al., 2019). The growth of *H. scabra* was higher (1.038 g day^{-1}) in pen systems co-cultured with *K. alvarezii* compared to 0.898 g day^{-1} in pen systems co-culture with *E. denticulatum* (Hamad et al., 2019). Survival rate of *H. scabra* was however higher (76%) with *E. denticulatum* than with *K. alvarezii* (70%) (Hamad et al., 2019). Dobson et al. (2020c) recently assessed integration of sea grapes (*Caulerpa lentillifera*) into a co-culture system with *H. scabra* and Babylon snail (*Babylonia areolata*) in Vietnam. Sea cucumber survival was not affected by culture with *Caulerpa*, however mean individual weight gain was greater without the seaweed. In turn, co-culture with the Babylon snail had a positive effect: *H. scabra* grew from 18.2 to 119.0 g in 92 days in the presence of *Bactrocera areolata*, although growth in the latter was not increased relative to its monoculture (Dobson et al., 2020a,b, 2021).

Co-culture with various shrimp species, mainly *Penaeus monodon* and *Litopenaeus stylirostris*, was investigated in different places. Early in the aquaculture history of *H. scabra*, Battaglene (1999a) and Battaglene and Bell (1999) suggested that there may be scope for rearing juveniles of *H. scabra* in penaeid shrimp ponds as a way of mass producing them for restocking and stock enhancement. Later, trials were undertaken involving the co-culture of *H. scabra* and the shrimp *P. monodon* (Pitt et al., 2004). No negative impacts of sea cucumber on shrimp growth or survival were noted; however, the effect of shrimp on sea cucumber was more complex. Survival and growth of *H. scabra* was often as good with shrimp as without, and sometimes apparently better; however, shrimp also caused mortalities in sea cucumbers in several trials (Pitt and Duy, 2004b). Lebata-Ramos et al. (2012) investigated the feasibility of co-culture with the black tiger shrimp *P. monodon* in mangroves. Results showed that growth and survival of *P. monodon* and *H. scabra* grown together or separately did not differ. In general, the salinity of a shrimp pond can be a problem for the growth and survival of *H. scabra*. In a study conducted in Sri Lanka with *P. monodon*, at salinities ranging from 27 to 52 psu, Kumara et al. (2013) found that *H. scabra* grew well up to 38 psu, although growth slowed (to 0.33 g day^{-1}), and that stress symptoms (swollen body, damaged epidermis, evisceration of internal organs) appeared at levels above 45 psu.

Purcell et al. (2006b) tested the viability of tank co-cultures at the early juvenile stage of both *H. scabra* and the shrimp L. *stylirostris* for 3 weeks in

New Caledonia. Survival of shrimp and sea cucumber was high in all treatments. Growth of shrimp did not differ between monoculture and co-culture, but sea cucumbers grew slower in co-culture, although their cyclic burrowing behaviours were apparently unaffected (Purcell et al., 2006b). Jiang et al. (2017) studied the effect of polyculture of *H. scabra* with *Litopenaeus vannamei* on some nutrients in the laboratory. Their results indicated that nutrients in the culture system was influenced by the sea cucumber: the phosphate and nitrate concentrations were enhanced in the polyculture system relative to the monoculture system, but by contrast, the ammonium concentration remained low in the polyculture. Sabilu et al. (2020) evaluated the performance and physiological responses of *H. scabra* fed with shrimp pond sediment (*L. vannamei*). Five substrate treatments were prepared with different proportions of sediment, while the control consisted in *H. scabra* held in two tanks and fed with *Sargassum* and seagrass powder. The highest production performance was observed for sea cucumbers fed with 40% shrimp pond sediment, with no difference in health status relative to the control (Sabilu et al., 2020). Bell et al. (2007) conducted experiments using hapa nets within shrimp ponds to assess any deleterious effects of sea cucumber on shrimp, and vice versa. This work showed that a high stocking density of juvenile sea cucumbers had no effects on growth and survival of shrimp (Bell et al., 2007). However, survival and growth of sea cucumbers reared with shrimp for 3 weeks were lower than for sea cucumbers reared alone (Bell et al., 2007). To obtain basic information for co-culture methods of black tiger shrimp (*P. monodon*) and *H. scabra* for possible mitigation of shrimp-pond eutrophication and prevention of disease outbreaks, laboratory experiments were conducted by Watanabe et al. (2012b) in the Philippines. A feeding trial of juvenile sea cucumbers showed that they did not grow well with fresh shrimp feed on hard substrate (Watanabe et al., 2012b). However, another trial indicated that sand substrate enhances the growth of juvenile sea cucumbers fed with shrimp feed (Watanabe et al., 2012a,c). A feeding trial using shrimp detritus, shrimp faeces and the benthic diatom *N. ramosissima* as food sources showed that *H. scabra* grew fastest with the faeces, followed by detritus and *N. ramosissima* (Watanabe et al., 2012b).

Purcell (2004b) estimated that grow-out of *H. scabra* in shrimp ponds is technically feasible. Sea cucumbers could act as bio-remediators for shrimp pond aquaculture by eating unused organic deposits, but conversion to soluble nitrogenous excreta should be considered. Bioturbation by sea cucumbers could thus benefit shrimp culture (Purcell, 2004b), a conclusion also shared by Mathieu-Resuge et al. (2020) who characterized the assimilation of organic matter from shrimp (*Penaeus stylirostris*) farming by *H. scabra* in an experimental culture system in New Caledonia. Their results suggest that culture of *H. scabra* as the potential to reduce excess sedimentary organic matter associated with shrimp farms (Mathieu-Resuge et al., 2020).

Co-culture with various fish species (*Chanos*; *Trachinotus blochii* and *Lates calcarifer*) was investigated in earthen ponds in the Philippines (Dumalan et al., 2019; Zamora et al., 2018). Dumalan et al. (2019) estimated the viability of growing *H. scabra* to marketable size (>300 g) in a milkfish (*C. chanos*) farm through simultaneous pen grow-out experiments conducted at a fish-farm site and a seagrass site. Small juveniles stocked in the fish farm experienced mass mortality within 15 days, whereas survival was low in the seagrass (12%). In contrast, survival of larger juveniles was high in the fish farm and not different from the seagrass (63%). Also, a system with three tanks was tested, combining *H. scabra*, *C. chanos* and the halophyte sea purslane *Sesuvium portulacastrum* over 8 weeks in Zanzibar, Tanzania (Senff et al., 2020). Changes in dissolved inorganic nitrogen support the notion that the sea cucumber tank was the main site of nitrification, while the hydroponic halophyte tank acted as a net sink of NO_3^-. Watanabe et al. (2015) investigated integrated multi-trophic aquaculture (IMTA) with *C. chanos* and *H. scabra* as the organic extractive species and the seaweed *K. alvarezii* as the inorganic extractive species They concluded that to ensure a complete utilization of particulate nitrogen by the end of milkfish culture period (i.e. zero emission), stocking density should be 805 sea cucumbers m^{-2}, which is 200 times as high as that in existing aquaculture operations in countries such as Viet-Nam and New Caledonia. The purpose of *H. scabra* in IMTA might make more sense in terms of its economic advantages than a footprint mitigation strategy. Sithisak et al. (2013) co-cultured *H. scabra* with red tilapia at a salinity of 30 psu in Thailand. They found that growth and survival of the sea cucumbers was higher in this co-culture than in monoculture, while the growth rates of red tilapia either showed no change or showed an increase. The environmental performances of theoretical production scenarios of red drum (*Sciaenops ocellatus*) sea cage monoculture and an open-water system co-culturing red drum and sea cucumber (*H. scabra*) were assessed in Chary et al. (2020) with mathematical models at local and global scales. The results indicate that co-culture of *H. scabra* in sea cages suspended beneath finfish nets may only slightly decrease particulate waste load (by 0.73%) and benthic impacts.

Besides co-culture with macroalgae, shrimps or fishes, other species are mentioned in the scientific literature, but without detailing the effects on

H. scabra. Co-culture with the sea urchin *Tripneustes gratilla* was tested in recirculation systems in Australia [Guy Carton, unpublished data in Zamora et al. (2018)]. Also, many lobster farmers in Vietnam grow mussels next to the lobster cages and some reportedly integrate *H. scabra* in net enclosures under the lobster cage to reduce the concentrations of organic matter in the water column and in the sediments (Du et al., 2004, 2005; Zamora et al., 2018). In China, *H. scabra* is reportedly farmed in a three-dimensional polyculture with groupers (*Epinephelus* sp.) and pearl oysters (*Pinctada martensii*) (Purcell and Wu, 2017).

Another form of co-culture that could help mitigate the footprint of certain types of aquaculture is bioremediation by *H. scabra*, whereby soil microorganisms, which would be favoured by the presence of sea cucumbers, could naturally depurate the environment. This method could be used to decontaminate discharges from industrial aquaculture (e.g. shrimps, fishes, abalones) including unused food and faeces from animals in culture. In this context, Robinson et al. (2015) investigated the potential for *H. scabra* to provide a sediment remediation service whilst concurrently yielding a high-value secondary product in a land-based recirculating aquaculture system. They observed that the redox regime was the principal factor affecting growth, biochemical composition and behaviour of *H. scabra*, while substrate depth and particle size did not significantly affect growth rate or biomass production. Robinson et al. (2019) reared juveniles of *H. scabra* at high densities solely on particulate organic wastes (in some assays supplemented by starch) from a commercial-scale land-based abalone recirculating aquaculture system. At the start of the experiment juvenile sea cucumbers weighed 4.08 g. At the end of the experiment, after 4 months, sea cucumber juveniles reared solely on effluent had a weight-based density of $711\,g\,m^{-2}$ whereas animals reared on starch-amended effluent (the best performing treatment) had a final density of $1011\,g\,m^{-2}$. Over the course of the trial, the mean growth rate ($0.25\,g\,day^{-1}$) of *H. scabra* reared in the starch treatment was significantly higher than in the treatment with waste alone ($0.16\,g\,day^{-1}$ and $702.12\,g\,m^{-2}$). All of the 64 juveniles used survived the study (Robinson et al., 2019).

Recently, Senff et al. (2022) investigated the performance of laboratory grade cellulose and bagasse agricultural waste products to improve growth and nutrient utilization from the wastes of milkfish aquaculture in *H. scabra*. They found that sea cucumbers were able to assimilate carbon directly from the cellulose and that bagasse supported higher uptake of carbon from aquaculture waste, thus identifying it as a viable resource to improve bioremediation by deposit-feeding sea cucumbers in integrated aquaculture.

21.5 Diseases and predation

With the development of the aquaculture of *H. scabra*, a variety of organisms that have a deleterious effect on sea cucumbers have been observed in rearing systems, including bacteria, protozoans and metazoans (i.e. nematodes, polychaetes, copepods, isopods, decapods and fishes; Table 20). No virus or fungus has been detected to date. The parasites include 13 species of bacteria found in integumentary lesions of juveniles or adults, unidentified Ciliophora present in the gonads of genitors and isopods of the genus *Cymodocea* parasitizing the integument of juveniles in nurseries. Predators include crabs and fishes. Copepods, nematodes and polychaetes are found in larval tanks and their effect on the survival of sea cucumbers is not yet understood, i.e. they could be predators or competitors. Larvae of dipteran flies are also found in nurseries and could have harmful effects on sea cucumber farms.

21.5.1 Diseases

A rapid protocol for the external examination of juveniles of *H. scabra* was developed by Purcell and Eeckhaut (2005) in order to screen for poor health and diseases. Diseases are alterations of the health state of an organism due to multiple causes; they can be innate or acquired, and they can be due to biotic or abiotic factors. Innate diseases are unknown in *H. scabra* and diseases induced by abiotic factors are understudied. It is known that *H. scabra*, in the larval and post-metamorphic stages, only tolerates a certain temperature and salinity range beyond which individuals can weaken and die. The effects of starvation in juveniles were analysed by Watanabe et al. (2012b) in an attempt to establish a method to determine nutritional condition. Although no clear pattern was observed in protein and cholesterol levels in the coelomic fluid during the starvation trial, carbohydrate level increased, as did coelomic fluid density (Watanabe et al., 2012b). These results suggest that coelomic fluid density and carbohydrate level may be used as markers of nutritional condition in *H. scabra* without sacrificing the animal (Watanabe et al., 2012b).

Temperature variations are now well known to affect the health of juveniles of *H. scabra* (Kamyab et al., 2017; Kühnhold et al., 2017, 2019b; Yao et al., 2007) (see previous Section 6.3). Lavitra et al. (2009a) recorded that the salinity near sea pens in South-West Madagascar varied between 17 and 35 psu over a year. Individuals that were submitted to a salinity of 20 psu or lower changed their behaviour: they stayed burrowed into sediment even at night when they are usually surfacing. Their body swelled and showed

Table 20 Pathogens, predators and competitors of *Holothuria scabra* in aquaculture, showing the species involved, the location of the study and the references.

Species	**Location**	**References**
Bacteria		
Arcobacter bivalviorum[a]	Indonesia	Tangestani and Kunzmann (2019)
Pseudoalteromonas citrea[a]	Indonesia	Tangestani and Kunzmann (2019)
Vibrio azureus[a]	Indonesia	Tangestani and Kunzmann (2019)
V. fortis[a]	Indonesia	Tangestani and Kunzmann (2019)
V. owensii[a]	Indonesia	Tangestani and Kunzmann (2019)
V. parahaemolyticus[a]	Indonesia Madagascar	Tangestani and Kunzmann (2019) Delroisse et al. (2020)
V. rotiferianus[a]	Indonesia	Tangestani and Kunzmann (2019)
V. tubiashiand[a]	Indonesia	Tangestani and Kunzmann (2019)
Vibrio sp.[b]	Vietnam Madagascar Indonesia	Pitt and Duy (2004b) Becker et al. (2004), Delroisse et al. (2020) Tangestani and Kunzmann (2019)
V. natriegens[b]	Madagascar	Becker et al. (2004)
V. harveyi[b]	Australia Madagascar Indonesia	Morgan (2001) Becker et al. (2004) Sitoresmi and Pursetyo (2020)
Vibrio alginolyticus[b]	Madagascar Malaysia Indonesia	Becker et al. (2004) Rafidah et al. (2017) Sugama et al. (2019)
Vibrio variabilis	Madagascar	Delroisse et al. (2020)
Bacterium from the CFB group[b]	Madagascar	Becker et al. (2004)
Roseobacter sp.[b]	Madagascar	Becker et al. (2004)
Many taxa evidenced by metagenomic analyses	Madagascar	Delroisse et al. (2020)
Ciliophora		
Unknown species[c]	Madagascar	Eeckhaut (Personal observation)
Nematodes		
Unknown species[d]	Madagascar	Eeckhaut (Personal observation)

Table 20 Pathogens, predators and competitors of *Holothuria scabra* in aquaculture, showing the species involved, the location of the study and the references.—cont'd

Species	Location	References
Polychaetes		
Unknown species[d]	Madagascar	Eeckhaut (Personal observation)
Copepod crustaceans		
Unknown Copepods[d]	India Madagascar Sultanate of Oman Vietnam Philippines	James et al. (1994b) Eeckhaut (Personal observation) Al-Rashdi et al. (2012, 2019) Pitt and Duy (2003) Gamboa et al. (2012)
Isopod *Cymodoce* sp.[e]	Madagascar Philippines	Lavitra et al. (2009b) Cabacaba and Campo (2019)
Decapod *Thalamita crenata*[f]	Madagascar	Eeckhaut et al. (2020), Lavitra et al. (2009b)
Insects		
Larvae of dipteran flies[e]	Madagascar Philippines	Lavitra et al. (2009b) Gamboa et al. (2012)
Fishes		
Fishes in the families Balistidae, Labridae, Lethrinidae and Nemipteridae[g]	Solomon Islands	Dance et al. (2003)

[a]Identified in SKUD of juveniles but no infection assays performed.
[b]In SKUD of juveniles; infection assays were performed in Becker et al. (2004) and indicated that the bacteria was not the inducing agent; infection assays not performed in other studies; considered as opportunistic pathogen after crab attacks in Pitt and Duy (2004b).
[c]Infesting the gonads of broodstock.
[d]In larval rearing tanks.
[e]In ponds with juveniles.
[f]Eating freshly released juveniles and sub-adults in farming sea pens.
[g]Eating juveniles in mangrove/sea grass zones.

blisters (Burel, 2018). When placed in ponds at ambient salinity, diseased specimens recovered within 96h (Burel, 2018). After 9 days at a salinity of 20psu without the possibility to burrow into sediment, individuals became weak, shrank in length, and developed a flabby integument (Lavitra et al., 2009a). After 17 days, the epidermis had disappeared from various parts of the body. Juveniles were found eviscerated after 22 days: they became translucent and died shortly after.

Biotic diseases affecting both larval and post-metamorphic stages of *H. scabra* have been poorly studied. No bacterial, viral or protozoal agent has ever been identified to be the direct cause of disease in the larvae of *H. scabra*. However, larval diseases are present in other sea cucumbers such as *A. japonicus* (Yin-Geng et al., 2005) and *I. fuscus* (Becker et al., 2010). Among metazoans affecting larval development in *H. scabra*, copepods are suspected to be among the most important as they appear in most marine habitats across all oceans (Agudo, 2006; Al-Rashdi et al., 2012; Duy, 2012; Eeckhaut et al., 2012; Mills et al., 2012). However, the identification of the species involved and their effects on sea cucumber larvae have not yet been studied or confirmed. Nonetheless, copepods have been suggested to compete for food with larvae of sea cucumbers in culture tanks (Purcell et al., 2012a). This common problem (often caused by incorporating settlement plates in the culture tanks) can be avoided by conditioning settlement plates in a sterile algal monoculture or by painting plates with a slurry of commercially available *Spirulina* rather than naturally growing diatoms (Agudo, 2006; Pitt and Duy, 2004b; Purcell et al., 2012a). On the other hand, copepods can be treated with brief doses of Trichlorfon/Dipterex at concentrations of about 1–2ppm (Agudo, 2006; Pitt and Duy, 2004b).

At the nursery stage, Lavitra et al. (2008) found that the isopods *Cymodoce* sp. infested individuals of *H. scabra* held in outdoor ponds during the warm season in Madagascar, inducing a high mortality rate. *Cymodoce* sp. has also been observed infesting hapa nets in Guiuan (the Philippines) (Cabacaba and Campo, 2019). The introduction of the carnivorous fish *Terapon jarbua* in ponds in Madagascar proved successful in mitigating this problem. The fish consume isopods and can eliminate them within 10 days in affected ponds. Moreover, outbreaks of the larvae of dipteran flies, commonly called 'bloodworms', can compete with juvenile sea cucumbers for resources in the tanks (Gamboa et al., 2012). Covering tanks with fine mesh was found to solve the issue in some instances, but it often only worked partially.

Skin ulceration diseases (SKUDs) or skin ulceration syndrome (SUS) are a group of diseases where the first symptom is an ulceration of the skin of sea cucumbers, including *H. scabra* (Fig. 31). They are reviewed in Delroisse et al. (2020). The first mention of a skin ulceration was in the work of Zhang and Liu (1998): they reported that juveniles of *A. japonicus* were susceptible to this infection, which tends to occur as a result of elevated temperatures and high stocking densities. The infection is known to spread rapidly from diseased individuals to healthy ones, making it difficult to control. Occasionally, an entire population can be wiped out in a short time,

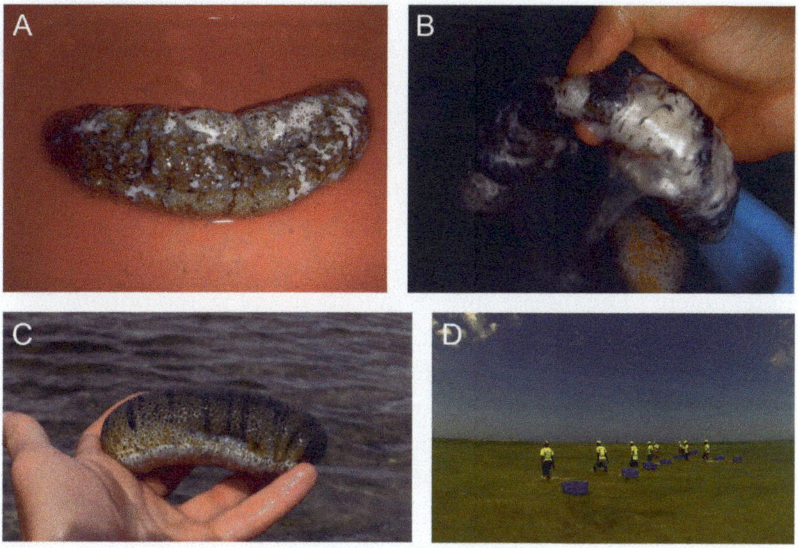

Fig. 31 Skin ulcerations due of an undescribed disease affecting *H. scabra* in Madagascar, showing (A) the early stage and (B) the late stage; (C) skin ulceration due to crab attack; (D) Malagasy workers collecting crab predators in sea pens. *Photos courtesy of I. Eeckhaut.*

once the infection sets in. Bacterial infection was reported to be the main cause of the disease, though the characteristics of the bacteria were not described. Two years after the report of Zhang and Liu (1998), Morgan (2000a) published a short note on a similar disease affecting adults of *H. scabra* at Bribie Island (Australia). The infection involved the loss of epidermal pigmentation associated with the presence of huge amounts of viscous mucus. The lesions first arose around the mouth and/or cloacal opening of individuals and then spread, sometimes to the point of covering the whole body. When infected individuals were not removed from the broodstock, up to 95% of the reared individuals died (Morgan, 2000a). According to this author, *V. harveyi* was the predominant bacterium in the lesions of affected sea cucumbers.

Morgan et al. (2001) observed that brood stock of *H. scabra* collected in October 1997 becamed diseased after a few weeks in captivity. The first visual signs, pigmentation loss and presence of copious amounts of mucous, coincided with a period of weight lost. Although the most common bacteria present on diseased individuals was *V. harveyi*, there were also low numbers of a motile Gram-negative rod bacteria, and the exact nature of the infection was unknown. Surface damage to the epidermis was not a prerequisite for infection, but infection progressed more rapidly from an existing wound. Most infected animals died within 3–7 days (Morgan et al., 2001). A similar condition was noted in juveniles grown in captivity in the Solomon Islands (S. C. Battaglene, Personal communication).

The studies of Becker et al. (2004) and Becker (2007) in Madagascar were the first to describe the histopathology of a SKUD affecting juveniles of *H. scabra* in the hatchery. The SKUD always begins by the appearance of a white lesion near the cloacal opening that extends onto the whole surface of the integument. The disease spreads quickly in the reared populations and destroys the cuticle, the epidermis, the connective tissue and the ossicles of affected individuals. Bacterial infection was studied with classical culture methods but also with denaturing gradient gel electrophoresis (DGGE) and 16S rDNA sequencing. *Bacteroides* and *Vibrio* bacteria occurred in the lesions, but the various transmission assays of cultured bacteria to healthy individuals were not conclusive, suggesting that the disease is not spread by contact. Delroisse et al. (2020) analysed an epidemic SKUD in cultured adults of *H. scabra* in Madagascar. Field observations revealed seasonality of the disease (i.e. wintertime maximum peak). Morphological analyses of integument ulcers showed that sea cucumbers react by forming a collagen fibre plug, and metagenomic analyses revealed a higher proportion of Vibrionaceae (Gammaproteobacteria) in ulcers relative to healthy integuments of the same individuals (Delroisse et al., 2020). Experimental infection assays were performed with crude extracts of ulcer and bacteria isolated from these extracts (e.g. *Vibrio parahaemolyticus*) but they did not significantly induce skin ulceration. The results suggest that the disease is not induced by a pathogen or, at the very least, that the pathogen is not found within the ulcers as the disease is not transmissible by contact (Delroisse et al., 2020). An initial cause of the SKUD in Madagascar might be the repeated and prolonged exposures to cold temperatures. Opportunistic bacteria could settle in the dermis of ulcerated individuals and promote the ulcer extension (Delroisse et al., 2020).

To date, no organism that would be responsible of the induction of a SKUD has been identified in *H. scabra* while two main agents are mentioned in the literature for *A. japonicus*: a viral agent (Deng et al., 2008; Liu et al., 2010) and bacteria, with some studies showing that both types can induce ulcerations in this species. *Vibrio* spp. are the most commonly observed bacterial species in lesions of *H. scabra* (Becker et al., 2004; Rafidah et al., 2017; Sugama et al., 2019; Tangestani and Kunzmann, 2019) and some can induce ulceration in *A. japonicus* (Deng et al., 2008, 2009; Li et al., 2010; Liu et al., 2012; Ma et al., 2006; Morgan et al., 2001; Zhao et al., 2011, 2012). Eeckhaut et al. (2019) demonstrated that sediments enriched with different types of organic matters can also induce skin ulcerations in *H. scabra*. The various types of organic material added to the sediment were: (*i*) crushed integument ulcerations coming from ulcerated *H. scabra*, (*ii*) crushed integument coming from healthy *H. scabra*, (*iii*) crushed skin from healthy fishes and (*iv*) crushed healthy red seaweed. The first three (animal-based) treatment triggered skin ulcerations in <3 days of feeding, sometimes leading to the death of juveniles, whereas the last (non-animal) treatment did not. Even if animal-based feeds induced SKUD in the experiments of Eeckhaut et al. (2019), it was not the case in other studies. For example, experiments conducted by Orozco et al. (2014) showed that nutrients from animal-based feeds are more efficiently digested, suggesting that animal ingredients rich in easily digestible protein could potentially provide an efficiently balanced diet for *H. scabra* fed with diatom. Weerasingha et al. (2015) also suggested that fish waste could be utilized cost effectively as a protein supplement in juvenile sea cucumber feeds.

The most likely hypothesis currently regarding SKUDs is that the induction of ulcerations may be due to multiple factors such as a drop or rise in temperature, incorrect density of individuals, abnormal drop in oxygen, a large variation in salinity or a lesion left by a predator such as a crab. Subsequently, opportunistic bacteria that can become virulent such as some *Vibrio* would proliferate in the ulcerations.

21.5.2 Predation

Predation is one of the most important factors impeding the success of the grow-out phase during the culture of *H. scabra*. In the natural environment, newly released juveniles of *H. scabra* can be injured or ingested by different species of fishes (Hamel et al., 2001; Pitt and Duy, 2004b; Tresnati et al., 2019), crabs (Pitt and Duy, 2004b), shrimps (Pitt and Duy, 2004b) and snails (Tresnati et al., 2019). Francour (1997) reported 76 predators of sea cucumbers and those that co-occur with *H. scabra* in the wild are presumably dangerous for this species. Mortality in juveniles of *H. scabra* that were experimentally released in mangrove-sea grass zones in Solomon Islands was due mainly to predation by fish species in the families Balistidae, Labridae, Lethrinidae and Nemipteridae (Dance et al., 2003). Predation may lead to disappearance of the entire stock in a very short period of time (Mercier et al., 2000a). The crab *Thalamita crenata* was found to be the most important predator in the Toliara region, Madagascar and one of the tasks of the workers in IOT company is to hunt crabs at low tide (Fig. 31D). Unidentified crabs were also observed in floating hapa nets (Cabacaba and Campo, 2019).

Aside from the control of biotic parameters, the choice of adequate sites for building sea pens is one of the key parameters for ensuring the success of sea cucumber farming. In 2007, abnormal mortalities of juveniles of *H. scabra* in sea pens was recorded by Lavitra et al. (2009a) in Madagascar while the adults were not affected. Only juveniles were found dead in the pens during the observation and the crab *T. crenata* (Portunidae) was observed in abundance near and inside the pens. Lavitra et al. (2009a) often found these crabs eating the newly transferred juveniles of *H. scabra*. Experiments in outdoor ponds showed that the crabs did not eat *H. scabra* when they were fed daily. On the other hand, five starved crabs were enough to kill and eat 20 juveniles of an average weight of 17 g within 5 days and 10 individuals of an average weight of 54 g within 10 days when they were kept in outdoor ponds (Lavitra et al., 2009a). Eeckhaut et al. (2020) evaluated the impact of predation by the crab *T. crenata* on the survival of juveniles of *H. scabra* at a sea-farm site in Madagascar where crab predation pressure was high. The juveniles had a wet body weight between 1 and 80 g. Results showed that predation by *T. crenata* is a serious predator of sea cucumber juveniles that can affect the profitability of *H. scabra* farming. The investigators determined that, at a density of 1 crab m^{-2}, all sea cucumber juveniles <15 g were likely to be eaten within 2 weeks, and that juveniles >50 g could also be impacted. The conclusion was that seeding of juveniles at sea should be performed with large individuals weighing at least 30 g to avoid or minimize predation (Eeckhaut et al., 2020). The work also highlighted the importance of covering enclosures against predators, particularly crabs, in farming areas (Eeckhaut et al., 2020). Some success in reducing predation on juvenile sea cucumbers released into the sea has also been obtained with the use of cages to exclude large predators (Dance et al., 2003; Purcell, 2004a; Rougier et al., 2013). On the other hand, experiments carried out by Lavitra et al. (2015) in a sea cucumber farming site with low density of crabs concluded that the use of covered sea pens was not necessary and that juveniles <5 g released in these sites had survival rates between 78% and 84% after 3 months. Similar experiments conducted by Hair et al. (2016) in Papua New Guinea gave comparable results and, contrary to expectations, short-term cage protection did not lead to higher survival.

21.6 Socio-economic considerations

Social and economic factors are key drivers of success in the aquaculture of *H. scabra*, especially in the village farming models (reviewed by Eriksson et al., 2012c; Juinio-Meñez et al., 2012a,b, 2017; Purcell et al., 2012a; Robinson and Pascal, 2009). For example, in Zanzibar (Tanzania), when sea cucumber harvesters were asked if they would like to farm sea cucumbers, about 64% of respondents answered positively (Eriksson, 2012). Their comments highlighted that they perceived farming as an addition, not a replacement, to the fishery, and that they were concerned about the personal risks involved with an investment (Eriksson, 2012). Work on social perception of sea cucumber aquaculture is limited but the socio-economic importance of this fisheries for artisanal harvesters in the Pacific has been analysed by Purcell et al. (2009a, 2010, 2013, 2014a, 2016b). Governance systems must provide adequate property rights and incentives to participants, balancing carefully any negative and exclusionary effects on non-participants (Purcell et al., 2012a). Governance issues associated with the use of existing coastal ponds are straightforward; ownership and leasing arrangements defining clear use rights. Governance of sea ranching is more complex because the strategies must be context specific, and integration with existing governance institutions should provide a robust, locally adapted 'shortcut' to this end (Purcell et al., 2012a). Identifying and engaging with the stakeholders are critical steps during the initial phases of sea ranching or sea farming programs. Consultation needs to include harvesters and managers of the wild fishery and, as highlighted above, the stocking program must be integrated within the fishery management strategy (Fleming, 2012; Hair, 2012; Juinio-Meñez et al., 2012b; Purcell et al., 2012a; Robinson and Pascal, 2012).

Theft (poaching) is certainly one of the most important social parameters to control (Hair et al., 2020b; Juinio-Meñez et al., 2012b; Robinson and Pascal, 2009). In Madagascar and the Philippines, surveillance of sea ranching areas or sea pens may be handled at the community level (Juinio-Meñez et al., 2012b; Robinson and Pascal, 2009). Such surveillance has required villagers to guard released sea cucumber stock at night (e.g. in farms in Madagascar) or around the clock (e.g. in the Philippines), which can incur a cost (Purcell et al., 2012a).

Very little scientific work addresses the economic challenges associated with sea cucumber aquaculture. A first idea of the infrastructure needed and of production was listed by James et al. (1994b) and later by Pitt and Duy (2003). There is no doubt that wild harvests of *H. scabra* provide additional income for coastal villagers engaged in this activity, but it is less obvious that aquaculture of this species is profitable in any region of the world. Aquaculture requires a significant investment, especially with respect to the construction of hatchery and nursery systems. The operating cost is equally important and varies markedly from one region to another, following the cost of labour.

Focusing on the eight countries where aquaculture of *H. scabra* has developed industrially, three economic models can be exemplified, which vary from one country to another (note that in most of these countries, there is only one company practicing aquaculture of *H. scabra*): the 'company farm' model, where employees receive a fixed salary periodically, is found in Saudi Arabia, China, New Caledonia and Australia; the 'village farming' model, where independent coastal harvesters handle the grow-out phase in agreement with a company that delivers juveniles, is found in Vietnam and Philippines; the 'mixed' model where part of a company's production is totally managed internally by employees while the other part is managed by independent coastal workers, is found in the Maldives and Madagascar. The provenance of the sea cucumber adults that will ultimately be marketed is either the open sea or ponds, or both, and processing into beche-de-mer will be carried out by the company or by independent processors.

Purcell et al. (2012a) explored the economics of sea cucumber aquaculture by comparing situations in New Caledonia and Vietnam. They showed clearly that farming of *H. scabra* can only be profitable in some countries and is not cost effective in others. They compared bio-economic model outputs for similar hypothetical 8-ha farms in New Caledonia and the central coast of Vietnam. The model outputs demonstrated a stark contrast: the farm in Vietnam was clearly profitable, while the one in New Caledonia was marginally unprofitable for a harvest size of 500 g (Purcell et al., 2012a). Adult sea cucumbers were unlikely to reach the modelled threshold weight (~590 g) for profitability in 18 months in New Caledonia using published practices. Both grow-out time and component costs were much higher in New Caledonia, with the costs of juvenile purchase, labour and electricity (for pumps) being the major drivers of this difference (Purcell et al., 2012a). Even if mariculture is often most profitable when animals are grown only to the minimum size that markets will accept (Tlusty et al., 2011), that strategy may not hold if the juveniles are expensive to produce and if larger animals command higher prices per unit weight, as is the case for *H. scabra*. Prices for this species tend to rise exponentially with increasing body size and weight (Brown et al., 2010; Kinch et al., 2008b) and Purcell et al. (2012a) wondered whether greater profits could be made by allowing tropical sea cucumbers to grow to large market size (e.g. 1–1.5 kg) instead of harvesting at the minimum size that traders will accept (~300 g). Cost-benefit studies are required on a case-by-case basis to determine if and how harvesting of cultured *H. scabra* is profitable in each of the operating regions. Also in Vietnam, Strehlow (2004) concluded, based on a conservative estimate, that the restocking of *H. scabra* is profitable, although cost-benefit analyses are sensitive to the survival of restocked individuals and their progeny, and the number of boats fishing for them in the release area.

It is difficult to get information on the productivity of private firms; however, in most cases, individuals of *H. scabra* produced in aquaculture are collected when they weigh at least 300–400 g. Processing into beche-de-mer involves a weight loss of up to 95% (i.e. an individual of 400 g produces a beche-de-mer of 20 g; see earlier Section 20.3 for details on processing). Thus 100,000 sea cucumbers collected at the end of the aquaculture process represent 2 tons of beche-de-mer, which are sold on average between 120 and 150 USD per kg to Chinese importers. Most products of *H. scabra* are exported to China or to countries where the Chinese community is important. Purcell et al. (2018c) examined recent changes in retail prices of beche-de-mer in both Hong Kong and mainland China markets and assessed the relationships between product size and price across multiple species (including *H. scabra*) harvested and exported from the tropical Indo-Pacific region. In 2016, average retail prices of beche-de-mer were 1.3–3.8-fold higher in Hong Kong stores than in Guangzhou stores (Purcell et al., 2018c). The tropical species with the highest maximum recorded price was *H. scabra*, for which extra-large, premium quality products were retailing at over 1800 USD per kg at one Hong Kong store (Purcell et al., 2018c). These more expensive products were exceptionally large specimens that would probably have weighed 1.5–2 kg live. The higher market prices for large individuals appears to provide justification for a shift in harvest strategy for longer culture cycles over which animals reach at least 660–1050 g whole body weight. For *H. scabra*, individuals of this size would yield beche-de-mer measuring at least 10 cm in length. It must be acknowledged however, that not all aquaculture operations are conducive to growing sea cucumbers for the several years required to attain large product sizes, either due the culture systems and facilities, or the underlying financial conditions of the operation (Purcell et al., 2018c).

21.7 Perspectives

Since the first review by Hamel et al. (2001), substantial efforts have been deployed to optimize the rearing parameters of *H. scabra* and to expand this aquaculture to new regions. However, many countries are still at the experimental research stage and very few have succeeded in developing a financially profitable holothuriculture. Unlike the wild fishery, the aquaculture of *H. scabra* requires significant investment and production costs, which make it profitable in some countries and not in others. The transition from an experimental to an industrial aquaculture also appears to be challenging because it must involve the appropriation of the technique by private operators who are often cautious to invest substantial sums.

Optimal parameters for the growth and survival of *H. scabra* in aquaculture are now established. In hatchery systems, the conditions for broodstock maintenance, larval development and settlement of juveniles are well known. One must be aware, however, that to reach a break-even point, a hatchery must operate at 100% year-round. As thermal shocks often yield unpredictable spawning results, the combination of methods should become more widespread to obtain fertilizable oocytes. Yet, most hatcheries have an overall survival rate, from eggs to seedling juveniles, of around 2%; improving this figure would increase the final profitability of the operation. One possible way forward is the use of biofloc in holothuriculture as it is the case in other types of aquaculture. This technique is based on in situ microorganism production which plays three major roles: the maintenance of water quality by the uptake of nitrogen compounds generating in situ microbial protein; nutrition, increasing culture feasibility; and competition with pathogens.

In nursery systems, researchers have also identified and optimized the major parameters driving the growth and survival of juveniles, such as temperature, salinity and stocking density. They also found ways to gain cultivation space by developing the floating hapas. Future research should ensure that this optimization is consistent throughout the culture cycles to allow survival rate of 80%.

The grow-out is a critical phase over which aquaculturists have little control, especially in sea farming and sea ranching. Survival is often poor over repeated cycles, often reaching maximum values around 25%. The four main causes are predation, disease, theft and sea cucumbers becoming positively buoyant and escaping the fences. Minimizing their impact should be the focus of future research endeavours. Some of these causes are social, which makes the task arduous and requires case-by-case studies across the regions studied.

Co-culture, if it can be advantageous for species paired with sea cucumbers, does not appear to be very beneficial for sea cucumber farming.

However, rearing with tilapia seems to have a positive impact on *H. scabra* and should thus be investigated further. Likewise, the first attempts at bioremediation seem promising but require additional work.

Current understanding of diseases, parasites, predators and competitors of captive-reared *H. scabra* is worryingly poor. An effort should be made to improve this knowledge both from an identification and an epidemiological point of view with regard to diseases. Understanding the behaviour of predators would prevent losses of sea cucumbers in farming, especially in sea pens.

The up-scaling of holothuriculture poses problems at all levels whether at the hatchery, nursery or grow-out phase. More in-depth analyses should be carried out regarding the limits of this up-scaling. Gregarious behaviour shown by *H. scabra*, for example, is an obstacle to the development of aquaculture over large areas, e.g. mentioning a density of 2 ind. m^{-2} in sea pens of several hectares does not make much sense given that individuals tend to group together.

Today, survival rates of sea cucumber in most farms around the world are preoccupying, i.e. <5% (often 2%) in hatcheries, 60–77% in nurseries and 18.5–77% (often close to 25% when adding multiple cycles) in grow-out. The yield is therefore close to 0.2%, meaning that 1 million fertilized oocytes will yield 2000 individuals of a marketable size. Fig. 32 provides

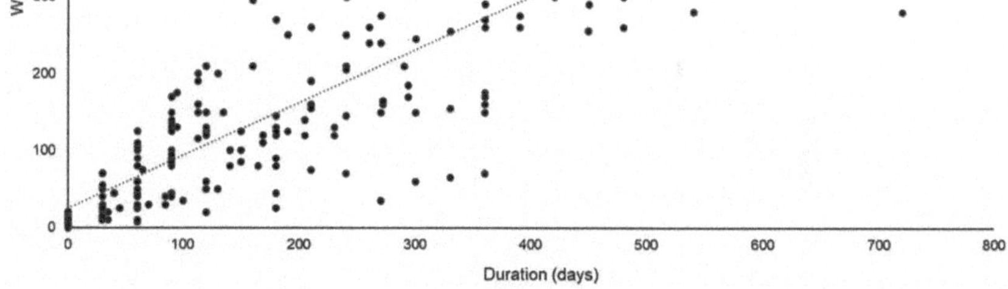

Fig. 32 Growth of *H. scabra* across its distribution range, based on weight gain during the grow-out phase. Data were extracted from studies conducted in Vietnam (Pitt and Duy, 2004a), New Caledonia (Bell et al., 2007; Purcell and Simutoga, 2008), Fiji (Hair, 2012), Maldives (Ahmed et al., 2018), Madagascar (Lavitra et al., 2010a; Robinson and Pascal, 2012) and Sri Lanka (Kumara and Dissanayake, 2017). The dotted line represents a linear regression with least square method ($y=0.6872x+24.962$; $R^2=0.701$).

an indication of the growth rate of *H. scabra* in exploitations around the world. Sea cucumbers of 300 g are obtained on average after 380 days of grow-out, to which must be added the time spent in the hatchery (generally 90 days) and the nursery (generally 60 days). It therefore takes on average 530 days to obtain marketable individuals. As a reminder, sea cucumbers of 300 g yield beche-de-mer of 15 g, i.e. 66 individuals are necessary to make 1 kg of product ready to be exported at a price of around 130–170 USD. Caution must thus be exercised when developing a sea cucumber project. A team composed of biologists and engineers should carry out economic and sociological analyses at the earliest stage to evaluate the profitability of the operations. The development of the aquaculture of *H. scabra* undoubtedly has a bright future, but fundamental knowledge of the driving variables and control over optimal development parameters must first be refined. One major aim for future research is to increase the yields (survival) across the hatchery, nursery and grow-out phases. Values of 5%, 80% and 50%, respectively, would multiply by 10 the overall profitability and would ensure the success of this sea cucumber aquaculture.

22. Conclusion

In view of the present synthesis, it is evident that considerable work has been carried out on *H. scabra* in classic and more innovative fields of research. To keep up with this diversified literature published in several languages was a challenge. Overall, we conclude that future research needs to focus on field monitoring, better understanding of natural predatory pressures, as well as genetic variation and isolation of populations over the distribution range. Detailed study of the reproductive cycle and spawning cues will improve our understanding of the life history of *H. scabra*.

The accumulated knowledge about the biology of the species has paved the way for sustainable management of remaining populations through restocking and stock enhancement, and also offers the potential for increasing production through farming. Further investigations are welcomed given that *H. scabra* is one of the few tropical holothuroids amenable to aquaculture. However, fisheries statistics have to be improved to enable better management of remaining stocks and the use of marine protected areas needs to be examined. Finally, more active collaboration between and within countries of the Indo-Pacific should facilitate the gathering of knowledge on this valuable species.

Acknowledgements

We wish to thank everyone who contributed directly and indirectly to advancing knowledge on *H. scabra* over the years. We also thank Dave Pawson for his contribution to the previous edition of this review. We appreciated the comments of Alessandro Lovatelli (FAO, Rome), Emaline Montgomery (Fisheries and Oceans Canada) and Michel Claeraboudt (Sultan Qaboos University, Oman) on the draft manuscript.

Dedication

We would like to dedicate this book to the memory of Claude Massin, for his tremendous contributions to our understanding of holothuroids.

Appendix

Recent examples of illegal fishing of sea cucumbers

As a growing number of smuggling and illegal fishing cases are reported from various countries (Conand, 2018b), it is important to highlight the existence of this growing problem. As *H. scabra* is now a high value species on the Asian market, cases involving that species are common. Recent examples are presented here.

In India (January 2021), The Indu Journal reported that the Thalamuthu Nagar police seized 300 kg (unprocessed) of banned sea cucumber from a cargo autorickshaw. They seized the vehicle along with the sea cucumbers (https://www.thehindu.com/news/cities/Madurai/300-kg-sea-cucumber-seized/article33503065.ece).

The same journal reported (December 2020) that the Soorankudi police seized a cargo autorickhsaw with 955 kg of sea cucumbers. Inspecting the vehicle, they found a Schedule I species under the Wildlife Protection Act 1972. Police suspect that the sea cucumber was on its way to Sri Lanka (https://www.thehindu.com/news/cities/Madurai/sea-cucumber-seized/article33348176.ece).

Still in India, The Times of India (February 2019) reported that Marine police and forest department personnel seized 152 kg of processed sea cucumber (https://timesofindia.indiatimes.com/city/chennai/10-64-lakh-worth-sea-cucumbers-seized-in-tamil-nadu-man-arrested/articleshow/67797229.cms).

In August 2019, packed in 69 jute bags, more than 800 k of live sea cucumbers that were meant to be smuggled to Sri Lanka were seized by Indian customs officials. The seizure was made when officials were on their usual patrol in the waters of the Gulf of Mannar in Tamil Nadu.

The occupants of the boat jumped into the sea and escaped. The boat, along with the sea cucumbers, was taken into custody. A hunt is on for the smugglers, said The Times of India, quoting Superintendent of Customs Joseph Jayaraj (https://weather.com/en-IN/india/news/news/2019-08-23-sea-cucumber-gulf-of-mannar-illegal-smuggling).

In August 2020 in a coordinated mission that involved air and sea-based assets, the Indian Coast Guard apprehended a fishing boat that was attempting to smuggle sea cucumber, an endangered species, into Sri Lanka. The apprehended fishing boat was carrying 1 ton of sea cucumbers, which are valued at about USD 683,000. The fishing boat along with three crew members have been brought to Tuticorin for further investigation (https://www.wionews.com/india-news/indian-coast-guard-thwarts-smuggling-of-sea-cucumber-into-sri-lanka-323980).

Those reports number by the dozens just for *H. scabra* and by the hundreds if not more when looking at sea cucumbers in general.

https://www.deccanherald.com/national/south/three-tonnes-of-sea-cucumber-worth-rs-1-crore-seized-in-tamil-nadu-910854.html

https://www.thehindu.com/news/cities/Madurai/300-kg-sea-cucumber-seized/article33503065.ece

https://www.thehindu.com/news/cities/Madurai/sea-cucumber-seized/article33348176.ece

https://news.mongabay.com/2020/07/risking-death-and-arrest-madagascar-fishers-chase-dwindling-sea-cucumbers/

References

Abayomi, A., 2020. GFSKLYFamide-like neuropeptide is expressed in the female gonad of the sea cucumber, *Holothuria scabra*. J. Cell Anim. Biol. 14, 1–5.

Abdula, R., 1998. A summary about holothurians in Mozambique. SPC Beche-de-mer Inf. Bull. 10, 34–35.

Abidin, N.A., Muhamad Shaleh, S.R., Ching, F., Othman, R., Manjaji-Matsumoto, M., Senoo, S., Mustafa, S., 2016. Egg and larval development of induced spawned sandfish (*Holothuria scabra*) in hatchery. Int. J. Aquat. Sci. 7, 107–110.

Abidin, N.A., Shaleh, S.R.M., Ching, F.F., Othman, R., Manjaji-Matsumoto, M., Mustafa, S., Senoo, S., 2019. Appropriate diet and stocking density for sea cucumber (*Holothuria scabra*) larvae rearing. J. Phys. Conf. Ser. 1358, 12–15.

Abraham, T.J., Nagarajan, J., Shanmugam, S.A., 2002. Antimicrobial substances of potential biomedical importance from holothurian species. Ind. J. Mar. Sci. 31, 161–164.

Adams, T., 1992. Resource aspects of the Fiji beche-de-mer industry. SPC Beche-de-mer Inf. Bull. 4, 13–16.

Adams, T., Leqata, J., Ramohia, P., Amos, M., Lokani, P., 1992. Pilot survey of the status of trochus and beche-de-mer resources in the Western Province of the Solomon Islands with options for management. South Pacific Commission lnshore Fisheries Research Project assisting the Solomon Islands Government Fisheries Division.

Adithiya, L., 1969. Some notes on the anatomy of *Holothuria*. Loris 11, 385–388.

Afkhami, M., Ehsanpour, M., Khazaali, A., Kamrani, E., Mokhlesi, A., Darvish Bastami, K., 2012. Sea cucumber fisheries of Qeshm Island, Persian Gulf. SPC Beche-de-mer Inf. Bull. 32, 60–61.

Agudo, N., 2006. Sandfish Hatchery Techniques. The WorldFish Center, Secretariat of the Pacific Community and Australian Centre for International Agricultural Research, Noumea, New Caledonia.

Agudo, N.S., 2012. Pond grow-out trials for sandfish (*Holothuria scabra*) in New Caledonia. In: Hair, C.A., Pickering, T.D., Mills, D.J. (Eds.), Asia-Pacific Tropical Sea Cucumber Aquaculture. *Proceedings of an International Symposium Held in Noumea, New Caledonia. ACIAR Proceedings*, pp. 104–112.

Ahmed, M.I., Lawrence, A.J., 2007. The status of commercial sea cucumbers from Egypt's northern Red Sea Coast. SPC Beche-de-mer Inf. Bull. 26, 14–18.

Ahmed, H., Shakeel, H., Naeem, S., Sano, K., 2018. Pilot study on grow-out culture of sandfish (Holothuria scabra) in bottom-set sea cages in lagoon. SPC Beche-de-mer Inf. Bull. 38, 45–50.

Airoldi, L., Connell, S.D., Beck, M.W., 2009. The Loss of Natural Habitats and the Addition of Artificial Substrata. pp. 269–280. Marine Hard Bottom Communities.

Ajayi, A., Withyachumnarnkul, B., 2015. Expression of GFSKLYFamide-like neuropeptide in the digestive system of the sea cucumber *Holothuria scabra* (Echinodermata). Afr. J. Biotechnol. 14, 2124–2129.

Ajayi, A., Amedu, N.O., Withyachumnarnkul, B., 2020. Presence of GFSKLYFamide-like neuropeptide in the tentacles, body wall and tube foot of *Holothuria scabra*: immunohistochemical evidence. Afr. J. Biomed. Res. 23, 25–28.

Akamine, J., 2001. Holothurian exploitation in the Philippines: continuities and discontinuities. Tropics 10, 591–607.

Al Azad, S., Shaleh, S.R.M., Siddiquee, S., 2017. Comparison of fatty acid and proximate composition between *Holothuria edulis* and *Holothuria scabra* collected from coastal water of Sabah, Malaysia. Adv. Biosci. Biotechnol. 8, 91–103.

Alagaraja, K., 1994. Assessment of sea-cucumber resources of India. Bull. Cent. Mar. Fish. Res. Inst. 46, 32–33.

Ali, I.M.Y., Mohamed, S., Gedeiri, A., Babikir, Y., 2015. Some aspects of reproductive biology of the sea cucumbers *Holothuria scabra* and *Holothuria atra* (Echinodermata: Holothuroidea) from the Sudanese Red Sea Coast. Res. J. Fish. Hydrobiol. 10, 1–6.

Al-Rashdi, K.M., Claereboudt, M.R., 2010. Evidence of rapid overfishing of sea cucumbers in the Sultanate of Oman. SPC Beche-de-mer Inf. Bull. 30, 10–13.

Al-Rashdi, K.M., Claereboudt, M.R., 2018. Reproductive biology of the sea cucumber *Holothuria scabra* (Jaeger 1883) in Mahout Bay, Arabian Sea, Oman. Int. J. Fish. Aquat. Stud. 6, 100–108.

Al-Rashdi, K.M., Claereboudt, M.R., Al-Busaidi, S.S., 2007a. Density and size distribution of the sea cucumber, *Holothuria scabra* (Jaeger, 1935), at six exploited sites in Mahout Bay, Sultanate of Oman. J. Agric. Mar. Sci. 12, 43–51.

Al-Rashdi, K.M., Al-Busaidi, S.S., Al-Rassadi, I.H., 2007b. Status of the sea cucumber fishery in the Sultanate of Oman. SPC Beche-de-mer Inf. Bull. 25, 17–21.

Al-Rashdi, K.M., Ibrahim, F.S., Al Habsi, S.H., 2011. Aquaculture Research and Development in the Sultanate of Oman. Ministry of Agriculture and Fisheries Wealth, Muscat, p. 70.

Al-Rashdi, K.M., Eeckhaut, I., Claereboudt, M.R., 2012. A manual on hatchery of sea cucumber *Holothuria scabra* in the Sultanate of Oman. Ministry of Agriculture and Fisheries Wealth, Aquaculture Centre, Muscat, Sultanate of Oman.

Al-Rashdi, K.M., Claereboudt, M.R., Eeckhaut, I., 2019. Preliminary trials on hatchery and larval development of the sea cucumber, *Holothuria scabra* (Jaeger 1883), in the Sultanate of Oman. Asian J. Fish. Aquat. Res. 2 (3), 1–9.

Altamirano, J.P., Noran-Baylon, R.D., 2020. Nursery culture of sandfish *Holothuria scabra* in sea-based floating hapa nets: effects of initial stocking density, size grading and net replacement frequency. Aquaculture 526, 2–11.

Altamirano, J.P., Recente, C.P., Rodriguez Jr., J.C., 2017. Substrate preference for burying and feeding of sandfish *Holothuria scabra* juveniles. Fish. Res. 186, 514–523.

Althunibat, O.Y., Hashim, R.B., Taher, M., Daud, J.M., Ikeda, M.-A., Zali, B., 2009. In vitro antioxidant and antiproliferative activities of three Malaysian sea cucumber species. Eur. J. Sci. Res. 37, 376–387.

Amir, A.A., 1985. Democratic Yemen fisheries: cuttlefish and sea cucumber. Naga 8, 15–16.

Anderson, S.C., Flemming, J.M., Watson, R., Lotze, H.K., 2011. Serial exploitation of global sea cucumber fisheries. Fish Fish. 12, 317–339.

Andriamanamisata, V.L.R., Telesphore, A.F., 2019. The nutritional values of two species of sea cucumbers (*Holothuria scabra* and *Holothuria lessoni*) from Madagascar. Afr. J. Food Sci. 13, 281–286.

Andriawan, S., Hadi, S., Andayani, S., Sanoesi, E., Maftuch, M., 2019. Study of *Holothuria scabra* effect on the immune activity of *Pangasianodon hypophthalmus* against *Aeromonas hydrophila*. AACL Bioflux 12, 2167–2176.

Anjaneyulu, A.S.R., Raju, K.V.S., Rao, G.V., 1995. A new steroid glycoside from the sea cucumber Holothuria scabra off Mandapam coast. Indian J. Chem. Sect. B Med. Chem. 34, 666–668.

Anonymous, 1975. Beche-de-mer: a new export. World Fish. 24, 61.

Anonymous, 1978a. Culture of sea cucumber at Andamans. CMFRI Newsl. *8*, 1–2. April–November 1978.

Anonymous, 1978b. Rupee a slug in Sri Lanka. Fish Newsl. Int. 17, 9.

Anonymous, 1991. Training manual on breeding and culture of scallop and sea cucumber in China. Yellow Sea Fisheries Institute in Qingdao, People's Republic of China, Manual, p. 9.

Anonymous., 1993. Beche-de-mer harvesting in the northern province of New Caledonia. SPC Beche-de-mer Inf. Bull. 5, 7–8.

Anonymous., 1996a. Migration of *H. scabra*. Letter from Aquila Sea Products, Mozambique. SPC Beche-de-mer Inf. Bull. 8, 45.

Anonymous., 1996b. Status and management of inshore fisheries in the Kingdom of Tonga: Beche-de-mer. SPC Beche-de-mer Inf. Bull. 8, 12–19.

Anonymous., 1997. Total catch by species of beche-de-mer for the Queensland east coast fishery, July 1995-June 1996. SPC Beche-de-mer Inf. Bull. 9, 22.

Ansharullah, Tamrin, Patadjai, A.B., Asranuddin, 2020. Powder production of sea cucumber (*Holothuria scabra*): effect of processing methods on the antioxidant activities and physico-chemical characteristics. In: IOP Conference Series: Earth and Environmental Science. IOP Conference Series: Earth and Environmental Science, 443, p. 012024.

Aprianto, R., Amir, N., Tresnati, J., Tuwo, A., 2020. Bycatch sea cucumber *Holothuria scabra* processing and the quality characteristics. In: IOP Conference Series: Earth and Environmental Science, vol. 473. IOP Publishing, p. 012001.

Arakawa, K.Y., 1991. A handbook on the Japanese Sea Cucumber - Its biology, propagation and utilisation. SPC Beche-de-mer Inf. Bull. 3, 8–15.

Aras, T.R., 2013. Uji toksisitas ekstrak teripang *Holothuria scabra* terhadap *Artemia salina*. Thesis, Universitas Hasanuddin, Makassar, Indonesia.

Ardiansyah, A., Rasyid, A., Siahaan, E.A., Pangetistu, R., Murniasih, T., 2020. Nutritional value and heavy metals content of sea cucumber *Holothuria scabra* commercially harvested in Indonesia. Curr. Res. Nutr. Food Sci. J. 8, 765–773.

Arnold, D.C., 1956. A systematic revision of the fishes of the teleost family Carapidae (Percomorphi, Blennioidea), with description of two new species. Bull. Br. Mus. Nat. Hist. 4, 247–307.

Arnull, J., 2018. Investigating the Ecological Role of Farmed Holothurian scabra on Seagrass, Sediment and Blue Carbon Dynamics in Southwest Madagascar. MSc Thesis, University of Edinburgh.

Arnull, J.Z., Wilson, A.M.W., Brayne, K., Dexter, K., Donah, A.G., Gough, C.L.A., Klückow, T., Ngwenya, B., Tudhope, A., 2021. Ecological co-benefits from sea cucumber farming: *Holothuria scabra* increases growth rate of seagrass. Aquac. Environ. Interact. 13, 301–310.

Arsad, N.A., Othman, R., Shaleh, S.R.M., 2017. Reproductive pattern of sea cucumber, *Holothuria scabra* at two different sites in Sabah, Malaysia. J. Nat. Sci. Res. 7, 67–78.

Arsad, N.A., Othman, R., Shaleh, S.R.M., Abdullah, F.C., Matsumoto, M.M., 2020. Effects of physicochemical parameters on the reproductive pattern of sea cucumber *Holothuria scabra* in Sabah. *Songklanakarin*. J. Sci. Technol. 42, 109–116.

Asha, P.S., 2005. Reproductive aspects, larval and juvenile rearing of *Holothuria (Theelothuria) spinifera* Theel. SPC Beche-de-mer Inf. Bull. 22, 55–56.

Asha, P.S., 2007. Unusual colour pattern among *Holothuria scabra*. Mar. Fish. Inf. Serv. Tech. Ext. Ser. 192, 11–12.

Asha, P.S., Diwakar, K., 2013. Effect of stocking density on the hatching rate, larval and early juvenile rearing of edible sea cucumber *Holothuria scabra* (Jaeger, 1883). Indian J. Geo-Mar. Sci. 42, 191–195.

Asha, P.S., Muthiah, P., 2002. Spawning and larval rearing of sea cucumber *Holothuria (Theelothuria) spinifera* Theel. SPC Beche-de-mer Inf. Bull. 16, 11–15.

Asha, P.S., Muthiah, P., 2005. Effects of temperature, salinity and pH on larval growth, survival and development of the sea cucumber *Holothuria spinifera* Theel. Aquaculture 250, 823–829.

Asha, P.S., Muthiah, P., 2006. Effects of single and combined microalgae on larval growth, development and survival of the commercial sea cucumber *Holothuria spinifera* Theel. Aquacult. Res. 37, 113–118.

Asha, P.S., Muthiah, P., 2007. Growth of the hatchery-produced juveniles of commercial sea cucumber *Holothuria (Theelothuria) spinifera* Theel. Aquacult. Res. 38, 1082–1087.

Asha, P.S., Muthiah, P., 2008. Reproductive biology of the commercial sea cucumber *Holothuria spinifera* (Echinodermata: Holothuroidea) from Tuticorin, Tamil Nadu, India. Aquac. Int. 16, 231–242.

Asha, P.S., Rajagopalan, M., Diwakar, K., 2004. Effect of sea weed, sea grass and powdered algae in rearing the hatchery produced juveniles of *Holothuria (Metriatyla) scabra*, jaeger. In: Proceedings of the National Symposium on Recent Trends in Fisheries, pp. 79–85.

Asha, P.S., Rajagopalan, M., Diwakar, K., 2007. Stock enhancement of seacucumbers-a solution for the depletion of natural stocks of *Holothuria scabra* along Gulf of Mannar. Mar. Fish. Inf. Serv. Tech. Ext. Ser. 193, 7–10.

Asha, P.S., Rajagopalan, M., Diwakar, K., 2011. Influence of salinity on hatching rate, larval and early juvenile rearing of sea cucumber *Holothuria scabra* Jaeger. J. Mar. Biol. Assoc. India 53, 218–224.

Asha, P.S., Vinod, K., Ranjith, L., Johnson, B., Vivekanandan, E., 2017. Conservation and sustainable use of sea cucumber resources in India suggestions and way forward. In: Marine Fisheries Policy Series-7. ICAR-Central Marine Fisheries Research Institute.

Assawasuparerk, K., Rawangchue, T., Phonarknguen, R., 2016. Scabraside D derived from sea cucumber induces apoptosis and inhibits metastasis via iNOS and STAT-3 expression in human cholangiocarcinoma xenografts. Asian Pac. J. Cancer Prev. 17, 2151–2157.

Aumeeruddy, R., 2007. Sea cucumbers in Seychelles. In: Conand, C., Muthiga, N. (Eds.), Commercial Sea Cucumbers: A Review for the Western Indian Ocean. WIOMSA Book Series No. 5, pp. 41–52.

Aumeeruddy, R., Conand, C., 2007. Seychelles' sea cucumber fishery: data on processed products and other parameters. SPC Beche-de-mer Inf. Bull. 26, 19–26.

Aumeeruddy, R., Conand, C., 2008. Seychelles: a hotspot of sea cucumber fisheries in Africa and the Indian Ocean region. In: Toral-Granda, V., Lovatelli, A., Vasconcellos, M. (Eds.), Sea Cucumbers. A Global Review of Fisheries and Trade. FAO Fisheries and Aquaculture Technical Paper No. 516, FAO, Rome, pp. 195–209.
Aumeeruddy, R., Payet, R., 2004. Management of the Seychelles sea cucumber fishery: status and prospects. In: Lovatelli, A., Conand, C., Purcell, S., Uthicke, S., Hamel, J.-F., Mercier, A. (Eds.), Advances in Sea Cucumber Aquaculture and Management. FAO Fisheries Technical Paper No. 463, FAO, Rome, pp. 239–246.
Avilov, S.A., Drozdova, O.A., Kalinin, V.I., Kalinovsky, A.I., Stonik, V.A., Gudimova, E.N., Riguera, R., Jimenez, C., 1998. Frondoside C, a new nonholostane triterpene glycoside from the sea cucumber *Cucumaria frondosa*: structure and cytotoxicity of its desulfated derivative. Can. J. Chem. 76, 137–141.
Aydın, M., Sevgili, H., Tufan, B., Emre, Y., Köse, S., 2011. Proximate composition and fatty acid profile of three different fresh and dried commercial sea cucumbers from Turkey. Int. J. Food Sci. Technol. 46, 500–508.
Ayotte, P., 2005. The Relationship Between Holothurians and the Portunid Crab *Lissocarcinus orbicularis. Unpublished Work, University of Hawaii at Hilo*, p. 20.
Azam, K., Singh, J.J., 2013. Development of value added products from dried sea cucumber (*Holothuria scabra*). Oceanography 1, 1–3.
Baine, M.S.P., Choo, P.S., 1999. Sea cucumber fisheries and trade in Malaysia. In: Baine, M. (Ed.), The Conservation of Sea Cucumbers in Malaysia—Proceedings of an International Conference. Heriot-Watt University and the Fisheries Research Institute, Kuala Lumpur, Malaysia, pp. 49–63.
Baird, R.H., Sachithananthan, K., 1974. Beche-de-mer of the South Pacific Islands; A Handbook for fishermen. South Pacific Commission, Noumea, New Caledonia.
Baird, R.H., Sachithananthan, K., 1974. Beche-de-mer of the South Pacific Islands, a Handbook for Fishermen. South Pacific Commission, Noumea, New Caledonia.
Bais, D.J.D., Hiwatig, M.A.R., Jardin, A.R., Lacsamana, M.S.L., Manalo, P.M., Magbojos-Magtibay, C.R., 2018. Antibiofilm activity of coelomic fluid from *Holothuria scabra* (sandfish) against extended spectrum beta-lactamase (ESBL) producing *Escherichia coli*. In: The STETH. vol. 12, pp. 116–131.
Bakus, G.J., 1968. Defensive mechanisms and ecology of some tropical holothurians. Mar. Biol. 2, 23–32.
Bakus, G.J., 1973. The biology and ecology of tropical holothurians. In: Jones, O.A., Endean, R. (Eds.), Biology and Geology of Coral Reefs. Academic Press, New York, pp. 325–367.
Balinski, B.I., 1958. The echinoderms. In: Macnae, W., Kalk, M. (Eds.), A Natural History of Inhaca Island. *Mocambique*. Witwatersrand University Press, Johannesburg, South Africa, pp. 96–107.
Barclay, K., Kinch, J., Fabinyi, M., Waddell, S., Smith, G., Sharma, S., Kichawen, P., Foale, S., Hamilton, R., 2016. Interactive Governance Analysis of the bêche-de-mer 'Fish Chain' From Papua New Guinea to Asian Markets. Report commissioned by the David and Lucile Packard Foundation, University of Technology Sydney, Sydney.
Barclay, K., Fabinyi, M., Kinch, J., 2017. Governance and the Papua New Guinea beche-de-mer value chain. SPC Beche-de-mer Inf. Bull. 37, 3–8.
Barclay, K., Fabinyi, M., Kinch, J., Foale, S., 2019. Governability of high-value fisheries in low-income contexts: a case study of the sea cucumber fishery in Papua New Guinea. Hum. Ecol. 47, 381–396.
Barel, C.D.N., Kramers, P.G.N., 1977. A Survey of the Echinoderm Associates of the North-East Atlantic Area. Brill.
Bartley, D.M., Bell, J.D., 2008. Restocking, stock enhancement, and sea ranching: arenas of progress. Rev. Fish. Sci. 16, 357–365.

Baskar, B.K., 1994. Some observations on the biology of the holothurian *Holothuria (Metriatyla) scabra* (Jaeger). Bull. Cent. Mar. Fish. Res. Inst. 46, 39–43.
Baskar, B.K., James, P.S.B.R., 1989. Size and weight reduction in *Holothuria scabra* processed as beche-de-mer. Mar. Fish. Inf. Serv. Tech. Ext. Ser. 100, 13–16.
Baskar, B.K., James, P.S.B.R., 1995. Studies on the Biology, Ecology and Fishery of the Sea-Cucumber *Holothuria (Metriatyla) scabra* Jaeger From Southeast Coast of India. vol. 61 Central Marine Fisheries Research Institute Special Publication, pp. 13–17.
Battaglene, S.C., 1999a. Culture of tropical sea cucumbers for stock restoration and enhancement. Naga 22, 4–11.
Battaglene, S.C., 1999b. Progress in the production of tropical sea cucumbers *Holothuria scabra* and *Holothuria fuscogilva* for stock enhancement. In: World Aquaculture Society Conference, Sydney, Australia, p. 57.
Battaglene, S.C., 1999c. News from ICLARM Coastal Aquaculture Centre in the Solomon Islands. SPC Beche-de-mer Inf. Bull. 11, 30–31.
Battaglene, S.C., 2000. News from ICLARM Coastal Aquaculture Centre in the Solomon Islands. SPC Beche-de-mer Inf. Bull. 13, 30–31.
Battaglene, S.C., Bell, J.D., 1999. Potential of the tropical Indo-Pacific sea cucumber, *Holothuria scabra*, for stock enhancement. In: Howell, B.R., Moskness, E., Svasand, T. (Eds.), Stock Enhancement and Sea Ranching. Blackwell Science, Oxford, *Oxford, UK*, pp. 478–490.
Battaglene, S.C., Seymour, J.E., 1998. Detachment and grading of the tropical sea cucumber sandfish, *Holothuria scabra*, juveniles from settlement substrates. Aquaculture 159, 263–274.
Battaglene, S.C., Ramofafia, C., Seymour, J.E., 1998. Reproduction, spawning induction, development and larval rearing of the tropical sea cucumber sandfish, *Holothuria scabra*, Jaeger 1833. In: Proceedings of the Third International Larval Biology Meeting, Melbourne, Australia.
Battaglene, S.C., Seymour, J.E., Ramofafia, C., 1999. Survival and growth of cultured juvenile sea cucumbers, *Holothuria scabra*. Aquaculture 178, 293–322.
Battaglene, S.C., Seymour, J.E., Ramofafia, C., Lane, I., 2002. Spawning induction of three tropical sea cucumbers, *Holothuria scabra*, *H. fuscogilva* and *Actinopyga mauritiana*. Aquaculture 207, 29–47.
Becker, P., 2007. Caractérisation et éthiologie des lésions tégumentaires chez deux espèces d'échinodermes d'intérêt économique. PhD Thesis, University of Mons, Belgium.
Becker, P., Gillan, D., Lanterbecq, D., Jangoux, M., Rasolofonirina, R., Rakotovao, J., Eeckhaut, I., 2004. The skin ulceration disease in cultivated juveniles of *Holothuria scabra* (Holothuroidea, Echinodermata). Aquaculture 242, 13–30.
Becker, P., Eeckhaut, I., Ycaza, R.H., Mercier, A., Hamel, J.-F., 2010. Protozoan disease in larval culture of the edible sea cucumber *Isostichopus fuscus*. In: Harris, L.G., Anne Boetger, S., Walker, C.W., Lesser, M.P. (Eds.), Echinoderms. Taylor and Francis, pp. 571–573. Durham.
Bell, F.J., 1887. Report on a collection of Echinodermata from Andaman Islands. Proc. Zool. Soc. London 1887, 139–145.
Bell, J., Nash, W., 2004. When should restocking and stock enhancement be used to manage sea cucumber fisheries? In: Lovatelli, A., Conand, C., Purcell, S., Uthicke, S., Hamel, J.-F., Mercier, A. (Eds.), Advances in Sea Cucumber Aquaculture and Management. FAO Fisheries Technical Paper No. 463, FAO, Rome, pp. 173–179.
Bell, J.D., Rothlisberg, P.C., Munro, J.L., Loneragan, N.R., Nash, W.J., Ward, R.D., Andrew, N.L., 2005. Restocking and stock enhancement of marine invertebrate fisheries. Adv. Mar. Biol. 49, 1–370.
Bell, J.D., Agudo, N.N., Purcell, S.W., Blazer, P., Simutoga, M., Pham, D., Patrona, L.D., 2007. Grow-out of sandfish *Holothuria scabra* in ponds shows that co-culture with shrimp *Litopenaeus stylirostris* is not viable. Aquaculture 273, 509–519.

Bell, J.D., Purcell, S.W., Nash, W.J., 2008. Restoring small-scale fisheries for tropical sea cucumbers. Ocean Coast. Manag. 51, 589–593.

Beltran-Gutierrez, M., Ferse, S.C.A., Kunzmann, A., Stead, S.M., Msuya, F.E., Hoffmeister, T.S., Slater, M.J., 2016. Co-culture of sea cucumber *Holothuria scabra* and red seaweed *Kappaphycus striatum*. Aquacult. Res. 47, 1549–1559.

Bermudes, M., 2018. Did you say pislama, dairo, bislama, kereboki or biche-de-mer? SPC Beche-de-mer Inf. Bull. 38, 81–84.

Boncan, D.A.T., Uy, I.D.C., Tayco, C.C., Lluisma, A.O., 2016. Preliminary discovery of repetitive elements in the genome of the sea cucumber *Holothuria scabra* Jaeger, 1833. Philipp. J. Sci. 145, 339–355.

Boolootian, R.A., 1966. Reproductive physiology. In: Boolootian, R.A. (Ed.), Physiology of Echinodermata. Interscience Publishers, New York, pp. 561–613.

Bordbar, S., Anwar, F., Saari, N., 2011. High-value components and bioactives from sea cucumbers for functional foods—a review. Mar. Drugs 9, 1761–1805.

Bowman, W.M., 2012. Sandfish production and development of sea ranching in northern Australia. In: Asia-Pacific Tropical Sea Cucumber Aquaculture. Proceedings of an international symposium held in Noumea, New Caledonia. ACIAR Proceedings, pp. 75–78.

Brasseur, L., Parmentier, E., Caulier, G., Vanderplanck, M., Michez, D., Flammang, P., Gerbaux, P., Lognay, G., Eeckhaut, I., 2016. Mechanisms involved in pearlfish resistance to holothuroid toxins. Mar. Biol. 163, 14.

Britaev, T.A., Lyskin, S., 2002. Feeding of the symbiotic polychaete *Gastrolepidia clavigera* (Polynoidae) and its interactions with its hosts. In: Doklady Biological Sciences. Consultants Bureau, New York, pp. 352–356. c1965-1992.

Britayev, T.A., Bratova, O., Mekhova, E.S., Chernyshev, A.V., 2019. The first record of symbiosis between a palaeonemertean (Nemertea) and echinoderms (Echinodermata). Symbiosis 78, 135–139.

Brown, E.O., Perez, M.L., Garces, L.R., Ragaza, R.J., Bassig, R.A., Zaragoza, E.C., 2010. Value Chain Analysis for Sea Cucumber in the Philippines. The WorldFish Center, Penang.

Bruce, C., 1983. Sea cucumbers, extraordinary but edible all the same. Infofish Marketing Digest 6, 19–21.

Bruckner, A.W., 2006a. Proceedings of the CITES Workshop on the Conservation of Sea Cucumbers in the families Holothuriidae and Stichopodidae.

Bruckner, A.W., 2006b. Management and conservation strategies and practices for sea cucumbers. In: Bruckner, A.W. (Ed.), *Proceedings of the CITES Workshop on the Conservation and Conservation Strategies and Practices for Sea Cucumbers*. NOAA Technical Memorandum NMFS-OPR-34, USA, pp. 74–103.

Bruckner, A.W., Johnson, K.A., Field, J.D., 2003. Conservation strategies for sea cucumbers: can a CITES Appendix II listing promote sustainable international trade. SPC Beche-de-mer Inf. Bull. 18, 24–33.

Burel, B., 2018. *Holothuria scabra* tegument blister under fresh water influence, south-west of Madagascar. SPC Beche-de-mer Inf. Bull. 38, 68.

Bussarawit, S., Thongtham, N., Baine, M., 1999. Sea cucumber fisheries and trade in Thailand. In: *The Conservation of Sea Cucumbers in Malaysia. Their Taxonomy, Ecology & Trade, Proceedings of an International Conference*. Heriot-Watt University, Stromness, pp. 26–36.

Cabacaba, N.S., Campo, C.J.M., 2019. First-phase juvenile rearing of the sea cucumber *Holothuria scabra* in Eastern Samar, Philippines. Philipp. J. Fish. 26, 72–91.

Campo, C.J.M., Cabacaba, N.S., Cosmiano Jr., D.N., 2019. Growth, development and survival of *Holothuria scabra* larvae in different microalgal regimens and water rearing media. Philipp. J. Fish. 26, 98–110.

Cannon, L.R.G., Silver, H., 1986. Sea Cucumbers of Northern Australia. Queensland Museum, Queensland, Australia.

Cantero, Q., Smith, J., 2016. Effect of stocking density on the growth, survival, and settlement of sandfish sea sucumber (*Holothuria scabra*). J. Health Disparities Res. Pract. 9, 31.

Cariglia, A., Wilson, S.K., Graham, N., Fisher, R., Robinson, J., Aumeeruddy, R., Quatre, R., Polunin, N., 2013. Sea cucumbers in the Seychelles: effects of marine protected areas on high-value species. Aquat. Conserv. Mar. Freshw. Ecosyst. 23, 418–428.

Carter, J., Christopherson, J., Yibarbuk, D., 1997. Trepang studies of the top end. In: *Wetlands Australia* July, 9.

Castro, P., 2015. Symbiotic Brachyura. In: Treatise on Zoology-Anatomy, Taxonomy, Biology. The Crustacea, *vol. 9 Part C (2 vols)*. Brill, pp. 543–581.

Caulier, G., Van Dyck, S., Gerbaux, P., Eeckhaut, I., Flammang, P., 2011. Review of saponin diversity in sea cucumbers belonging to the family Holothuriidae. SPC Beche-de-mer Inf. Bull. 31, 48–54.

Caulier, G., Parmentier, E., Lepoint, G., Van Nedervelde, F., Eeckhaut, I., 2012. Characterization of a population of the Harlequin crab, *Lissocarcinus orbicularis* Dana, 1852, an obligate symbiont of holothuroids, in Toliara bay (Madagascar). Zoosymposia 7, 177–184.

Caulier, G., Flammang, P., Gerbaux, P., Eeckhaut, I., 2013a. When a repellent becomes an attractant: harmful saponins are kairomones attracting the symbiotic Harlequin crab. Sci. Rep. 3, 2639.

Caulier, G., Flammang, P., Rakotorisoa, P., Gerbaux, P., Demeyer, M., Eeckhaut, I., 2013b. Preservation of the bioactive saponins of *Holothuria scabra* through the processing of trepang. Cah. Biol. Mar. 54, 685–690.

Caulier, G., Lepoint, G., Van Nedervelde, F., Eeckhaut, I., 2014. The diet of the Harlequin crab *Lissocarcinus orbicularis*, an obligate symbiont of sea cucumbers (holothuroids) belonging to the genera Thelenota, Bohadschia and Holothuria. Symbiosis 62, 91–99.

Ceccarelli, D.M., Logan, M., Purcell, S.W., 2018. Analysis of optimal habitat for captive release of the sea cucumber Holothuria scabra. Mar. Ecol. Prog. Ser. 588, 85–100.

Chaiyamoon, A., Tinikul, R., Chaichotranunt, S., Poomthong, T., Suphamungmee, W., Sobhon, P., Tinikul, Y., 2018. Distribution and dynamic expression of serotonin and dopamine in the nervous system and ovary of *Holothuria scabra* during ovarian maturation. J. Comp. Physiol. A 204, 391–407.

Chaiyamoon, A., Tinikul, R., Nontunha, N., Chaichotranunt, S., Poomtong, T., Sobhon, P., Tinikul, Y., 2020a. Characterization of TRH/GnRH-like peptides in the sea cucumber, *Holothuria scabra*, and their effects on oocyte maturation. Aquaculture 518, 734814.

Chaiyamoon, A., Tinikul, Y., Chaichotranunt, S., Poomtong, T., Suphamungmee, W., Sobhon, P., Tinikul, R., 2020b. Existence of two mature sequences of cubifrin neuropeptide and their effects on spawning in the sea cucumber, *Holothuria scabra*. Aquaculture 519, 734–753.

Chalorak, P., Jattujan, P., Nobsathian, S., Poomtong, T., Sobhon, P., Meemon, K., 2018. *Holothuria scabra* extracts exhibit anti-Parkinson potential in *C. elegans*: a model for anti-Parkinson testing. Nutr. Neurosci. 21, 427–438.

Chambers, M.R., 1989. Beche-de-mer. In: Done, T., Navin, K.F. (Eds.), The Marine Resources of Vanuatu. Australian Institute of Marine Science, Townsville, Australia, pp. 86–91.

Chary, K., Aubin, J., Sadoul, B., Fiandrino, A., Covès, D., Callier, M.D., 2020. Integrated multi-trophic aquaculture of red drum (*Sciaenops ocellatus*) and sea cucumber (*Holothuria scabra*): assessing bioremediation and life-cycle impacts. Aquaculture 516, 734621.

Chen, C.P., Hsu, H.W., Deng, D.C., 1991. Comparison of larval development and growth of the sea cucumber *Actinopyga echinites*: ovary-induced ova and DTT-induced ova. Mar. Biol. 109, 453–457.

Chen, J., Zheng, A., Xiao, K., Lin, F., 2006. Studies on isolation and partial characterization of acidic mucopolysaccharide from *Holothuria scabra* Jaeger. *Food and Fermentation Industries*, p. 10.

Chen, S.-C., Li, X.-Y., Cao, J.-M., Su, C.-Y., Lu, J.-W., Li, J.-Y., Mo, L.-Z., 2009. Study on the key techniques of artificial propagation for *Holothuria scabra* Jaeger. Guangdong Agric. Sci., 05.

Chen, M., Zhang, L., Yang, X., Jiang, L., Huang, G., Li, Y., 2015. Experimental rearing of sandfish *Holothuria scabra* juveniles in net cages on open sea. Fish. Sci. 34, 571–574.

Cherbonnier, G., 1952. Les holothuries de Quoy et Gaimard. In: Memoires de l'Institut Royal des Sciences Naturelles de Belgique, vol. 44, pp. 1–50.

Cherbonnier, G., 1955. Les holothuries de la Mer Rouge. *Résultats scientifiques des campagnes de la" Calypso". I. Campagne en Mer Rouge (1951–1952).* In: Annales de l'Institut Ocèanographique de Monaco, nouvelle sèrie, vol. 30, pp. 129–183.

Cherbonnier, G., 1980. Holothuries de Nouvelle-Calédonie. In: Bulletin du Museum National d'Historie Naturelles de Belgique, vol. 44, pp. 1–50.

Cherbonnier, G., 1988. Echinodermes: holothurides. Faune Madag. 70, 1–292.

Cherbonnier, G., Féral, J.P., 1984. Les holothuries de Nouvelle-Caledonie. Deuxieme contribution. Premiere partie: Synallactidae et Holothuriidae. Bull. Mus. Natl. Hist. Nat. 6, 659–700.

Chia, D.G., Ng, P.K., 1999. A note on the sea cucumber crab, *Hapalonotus reticulatus* (Crustacea: Brachyura: Eumedonidae). J. South Asian Nat. Hist. 4, 65–70.

Chieu, H.D., Turner, L., Smith, M.K., Wang, T., Nocillado, J., Palma, P., Suwansa-Ard, S., Elizur, A., Cummins, S.F., 2019. Aquaculture breeding enhancement: maturation and spawning in sea cucumbers using a recombinant relaxin-like gonad-stimulating peptide. Front. Genet. 10, 77.

Choo, P.S., 2008a. Population status, fisheries and trade of sea cucumbers in Asia. In: FAO Fisheries and Aquaculture Technical paper, vol. 516, pp. 81–118.

Choo, P.S., 2008b. The Philippines: a hotspot of sea cucumber fisheries in Asia. FAO Fisheries and Aquaculture Technical paper 516, 119–140.

Choo, P.S., 2012. The sea cucumber fishery in Semporna, Sabah, Malaysia. SPC Beche-de-mer Inf. Bull. 32, 43–48.

Choo, P.S., Conand, C., 2015. Sea cucumber products at the China Fisheries Seafood Expo 2014. SPC Beche-de-mer Inf. Bull. 35, 60–63.

Chopra, B., 1931. On some Decapoda Crustacea found in the cloaca of holothurians. In: Further notes on Crustacea Decapoda in the Indian Museum, 2, pp. 303–324.

CITES, 2020. CoP18 Listing of Valuable Teatfish and Cedrela Species in CITES Appendix II Enters into Force. Available from: *https://www.cites.org/eng/teatfish_cedrela_listing_AppendixII_CITES_28082020.*

Claereboudt, E., 2021. Biological Roles and Evolution of Triterpenoids in Holothuroids. PhD Thesis, University of Mons, Belgium.

Claereboudt, M.R., Al-Rashdi, K.M., 2011. Shallow-water sea cucumber inventory in the Sultanate of Oman. SPC Beche-de-mer Inform. Bull. 31, 25–29.

Claereboudt, E.J.S., Eeckhaut, I., Lins, L., Deleu, M., 2018. How different sterols contribute to saponin tolerant plasma membranes in sea cucumbers. Sci. Rep. 8, 1–11.

Clark, A.H., 1921. A monograph of the existing crinoids, Part 2. Bull. US Natl Mus. 82, 1–795.

Clark, H.L., 1931. Echinodermata (other than Asteroidea) of the Great Barrier Reef Expedition. In: Scientific Reports of the Great Barrier Reef Expedition, vol. 4, pp. 197–239.

Clark, H.L., 1938. Echinoderms from Australia: an account of collections made in 1929 and 1932. In: Memoirs of the Museum in Cornparative Zoology. vol. 55. Harvard, pp. 441–558.
Clark, H.L., 1946. The Echinoderm Fauna of Australia: Its Composition and Its Origin. vol. 566 Carnegie Institution, Washington, Publications, pp. 1–567.
Clark, A.M., 1952. The "Manihine" expedition to the Gulf of Aqaba, 1948–1949. VII Echinodermata. Bull. Nat. Hist. Mus. Zool. 1, 203–214.
Clark, A.M., 1971. Monograph of Shallow-Water Indo-West Pacific Echinoderms. vol. 690 British Museum (Natural History) London Publication, pp. 1–238.
Clark, A.M., 1982. Echinoderms of Hong Kong. In: Proceedings of the First International Marine Biological Workshop, Hong Kong, pp. 485–501.
Clark, A.M., 1984. Echinodermata of the Seychelles. In: Stoddart, D.R. (Ed.), Biogeography and Ecology of the Seychelles Islands. Monographiae Biologicae, pp. 83–102.
Clark, A.M., Liao, Y., 1995. The Echinoderms of southern China. Science Press, Beijing.
Collard, M., Eeckhaut, I., Dehairs, F., Dubois, P., 2014. Acid–base physiology response to ocean acidification of two ecologically and economically important holothuroids from contrasting habitats, *Holothuria scabra* and *Holothuria parva*. Environ. Sci. Pollut. Res. 21, 13602–13614.
Conand, C., 1979. Beche-de-mer in New Caledonia: weight loss and shrinkage during processing in three species of holothurians. SPC Fish. Newsl. 19, 14–17.
Conand, C., 1981. Sexual cycle of three commercially important holothurian species (Echinodermata) from the lagoon of New Caledonia. Bull. Mar. Sci. 31, 523–543.
Conand, C., 1983. Methods of studying growth in holothurians (beche-de-mer), and preliminary results from a beche-de-mer tagging experiment in New Caledonia. SPC Fish. Newsl. 26, 31–38.
Conand, C., 1985. Distribution reproductive cycle and morphometric relationships of acanthaster-planci echinodermata asteroidea in new-caledonia western tropical pacific. In: Keegan, B.F., O. C. B. D. S (Eds.), Echinodermata: Fifth International Echinoderm Conference, Galway. A. A. Balkema, Rotterdam, Netherlands, pp. 499–506.
Conand, C., 1986. Les ressources halieutiques des pays insulaires du Pacifique. Deuxième partie: Les Holothuries. FAO, Rome. FAO Documents techniques sur les pêches No. 272.2. 108 p.
Conand, C., 1989. Les holothuries aspidochirotes du lagon de Nouvelle-Calédonie: biologie, écologie et exploitation. PhD Thesis, Université de Bretagne Occidentale, Brest, France.
Conand, C., 1990. The fishery resources of Pacific Island countries. Part 2. Holothurians. FAO, Rome. FAO Fisheries Technical Paper No 272.2. 143 p.
Conand, C., 1991. Long-term movements and mortality of some tropical sea-cucumbers monitored by tagging and recapture. In: Yanagisawa, T., Yasumasu, I., Oguro, C., Suzuki, N., Motokawa, T. (Eds.), Proceedings of the Seventh International Echinoderm Conference. Blakema, Rotterdam, pp. 169–175.
Conand, C., 1993. Reproductive biology of the holothurians from the major communities of the New Caledonian Lagoon. Mar. Biol. 116, 439–450.
Conand, C., 1994. Les holothuries: ressource halieutique des lagons. Rapports Scientifiques et Techniques ORSTOM. Sci. de la mer 65, 1–86.
Conand, C., 1997a. Are holothurian fisheries for export sustainable? In: Lessios, H.A., MacIntyre, I.G. (Eds.), Proceedings of the 8th International Coral Reef Symposium, Panama June 1996. Smithsonian Tropical Research Institute, Balboa, pp. 2021–2026.
Conand, C., 1997b. Mise en œuvre de la gestion durable de la ressource en holothuries. Rapport de l'intervention GREEN/COL.

Conand, C., 1998a. Overexploitation in the present world sea cucumber fisheries and perspectives in mariculture. In: Mooi, R., Telford, M. (Eds.), Echinoderms: San Francisco: Proceedings of the Ninth International Echinoderm Conference, San Francisco, California, USA, 5-9 August. A.A. Balkema, Rotterdam, pp. 449–454.
Conand, C., 1998b. Holothurians. In: Carpente, K.E., Niem, V.H. (Eds.), FAO Species Identification Guide. The Marine Living Resources of the Western Central Pacific. Food and Agriculture Organization of the United Nations, Rome, pp. 1157–1190.
Conand, C., 1999a. Manuel de qualité des holothuries commerciales du sud-ouest de l'Océan Indien. Programme régional environnement. Commission de l'Océan Indien, Secretariat Général.
Conand, C., 1999b. World sea cucumber exploitation and the market for trepang: an overview. In: The Conservation of Sea Cucumbers in Malaysia: Their Taxonomy, Ecology and Trade. Heriot-Watt University, Edinburgh and the Fisheries Research Institute, Kuala Lumpur, Malaysia, pp. 1–10.
Conand, C., 2001. Overview of sea cucumbers fisheries over the last decade-what possibilities for a durable management? Echinoderms 2000, 339–344.
Conand, C., 2004. Present status of world sea cucumber resources and utilisation: an international overview. In: Lovatelli, A., Conand, C., Purcell, S., Uthicke, S., Hamel, J.-F., Mercier, A. (Eds.), Advances in Sea Cucumber Aquaculture and Management. FAO Fisheries Technical Paper No. 463, FAO, Rome, pp. 13–23.
Conand, C., 2005. Commercial Sea Cucumbers and Trepang Markets, in Fisheries and Aquaculture, From Encyclopedia of Life Support Systems (EOLSS), Developed Under the Auspices of the UNESCO. Eolss Publishers, Oxford, UK. http://www.eolss.net.
Conand, C., 2006a. Harvest and trade: utilization of sea cucumbers; sea cucumber fisheries; current international trade; illegal, unreported and unregulated trade; bycatch; socio-economic characteristics of the trade in sea cucumbers. In: Bruckner, A.W. (Ed.), Proceedings of the CITES Workshop on the Conservation of Sea Cucumbers in the Families, pp. 51–73.
Conand, C., 2006b. Sea cucumber biology: taxonomy; distribution; biology; conservation status. In: Bruckner, A. (Ed.), Proceedings of the CITES Workshop on the Conservation of Sea Cucumbers. NOAA Technical Memorandum, Silver Spring, MD, USA, pp. 33–50.
Conand, C., 2008. Population status, fisheries and trade of sea cucumbers in Africa and the Indian Ocean. In: Toral-Granda, V., Lovatelli, A., Vasconcellos, M. (Eds.), Sea Cucumbers. A Global Review of Fisheries and Trade. FAO Fisheries and Aquaculture Technical Paper No. 516, FAO, Rome, pp. 143–193.
Conand, C., 2017. Expansion of global sea cucumber fisheries buoys exports. Rev. Biol. Trop. 65, 1–10.
Conand, C., 2018a. Recent information on worldwide illegal fisheries for sea cucumbers. SPC Beche-de-mer Inf. Bull. 38, 68–71.
Conand, C., 2018b. Tropical sea cucumber fisheries: changes during the last decade. Mar. Pollut. Bull. 133, 590–594.
Conand, C., 2019. Publications related to holothurians published in 2018. SPC Beche-de-mer Inf. Bull. 39, 89.
Conand, C., 2020. Publications related to holothurians published in 2019. SPC Beche-de-mer Inf. Bull. 40, 61.
Conand, C., Byrne, M., 1993. A review of recent developments in the world sea cucumber fisheries. Mar. Fish. Rev. 55, 1–13.
Conand, C., Muthiga, N., 2007. Commercial sea cucumbers: a review for the Western Indian Ocean. In: WIOMSA Book Series No. 5.

Conand, C., Sloan, N.A., 1989. World fisheries for echinoderms. In: Caddy, J.F. (Ed.), Marine Invertebrate Fisheries: Their Assessment and Management. Wiley, Chichester, pp. 647–663.

Conand, C., Tuwo, A., 1996. Commercial holothurians in South Sulawesi, Indonesia: fisheries and mariculture. SPC Beche-de-mer Inf. Bull. 8, 17–20.

Conand, C., De San, M., Refeno, G., Razafintseheno, G., Mara, E., Andriajatovo, S., 1998. Sustainable management of the sea cucumber fishery sector in Madagascar. SPC Beche-de-mer Inf. Bull. 10, 7–9.

Conand, C., Dinhut, V., Quod, J.P., Rolland, R., 2005. Sea cucumber inventory in Mayotte, southwest Indian Ocean. SPC Beche-de-mer Inf. Bull. 22, 19–22.

Conand, C., Muthiga, N., Aumeeruddy, R., de la Torre Castro, M., Frouin, P., Mgaya, Y., Mirault, E., Ochiewo, J., Rasolofonirina, R., 2006. A three years regional project on sea cucumbers in the south-western indian ocean: national and regional analyses to improve management. SPC Beche-de-mer Inf. Bull. 23, 11–15.

Conand, C., Mulochau, T., Chabanet, P., 2013a. The holothurian (Echinodermata) diversity of the glorieuses islands (Eparses islands, France, Mozambique channel). West. Indian Ocean J. Mar. Sci. 12, 71–78.

Conand, C., Eriksson, H., Lovatelli, A., Muthiga, N., Purcell, S., 2013b. Sea cucumber fisheries: an ecosystem approach to management in the Indian Ocean (SCEAM Indian Ocean). SPC Beche-de-mer Inf. Bull. 33, 55–57.

Conand, C., Shea, S., To, A., 2014a. Beche-de-mer trade statistics in Hong Kong in 2012. SPC Beche-de-mer Inf. Bull. 34, 43–46.

Conand, C., Polidoro, B., Mercier, A., Gamboa, R., Hamel, J.-F., Purcell, S., 2014b. The IUCN Red List assessment of aspidochirotid sea cucumbers and its implications. SPC Beche-de-mer Inf. Bull. 34, 3–7.

Conand, C., Mulochau, T., Stöhr, S., Eléaume, M., Chabanet, P., 2016. Inventory of echinoderms in the Iles Eparses (Europa, Glorieuses, Juan de Nova), Mozambique Channel, France. Acta Oecol 72, 53–61.

Conand, C., Ribes-Beaudemoulin, S., Trentin, F., Mulochau, T., Boissin, E., 2018. Marine biodiversity of La Reunion Island: Echinoderms. West. Indian Ocean J. Mar. Sci. 17, 111–124.

Conand, C., Di Simone, M., Horellou, A., 2019. CITES listing of teatfish sea cucumbers. Wiomsa Newsbrief 11, 24–25.

Cowan, M.E., Gomez, E.D., 1982. A preliminary note on the reproductive periodicity of the sea cucumber *Holothuria scabra* [Philippines]. Philipp. J. Biol. 11, 175–178.

Crosnier, A., 1962. Crustacés décapodes. In: Portunidae. Faune de Madagascar, vol. 16, pp. 1–154.

D'Silva, D., 2001. The Torres Strait bêche-de-mer (sea cucumber) fishery. SPC Beche-de-mer Inf. Bull. 15, 2–4.

Dabbagh, A.R., Sedaghat, M.R., 2012. Breeding and rearing of the sea cucumber *Holothuria scabra* in Iran. SPC Beche-de-mer Inf. Bull. 32, 49–52.

Dabbagh, A.R., Sedaghat, M.R., Rameshi, H., Arganji, G.R., Keshavarz, M., 2011. The effect of different substrates on larvae settlement in the sea cucumber, *Holothuria scabra*, Jaeger, 1833. World J. Fish Mar. Sci. 3, 263–264.

Dabbagh, A.R., Keshavarz, M., Mohammadikia, D., Afkhami, M., Nateghi, S.A., 2012. *Holothuria scabra* (Holothuroidea: Aspidochirotida): first record of a highly valued sea cucumber, in the Persian Gulf, Iran. Mar. Biodivers. Rec. 5, E69.

Damaso, M.F., Argente, F.A.T., 2019. Ocean nursery rearing of sea cucumber *Holothuria scabra* early juveniles in hapa nets at different stocking densities. Int. J. Fish. Aquat. Stud. 7 (4), 253–257.

Dance, S.K., Lane, I., Bell, J.D., 2003. Variation in short-term survival of cultured sandfish (*Holothuria scabra*) released in mangrove-seagrass and coral reef flat habitats in Solomon Islands. Aquaculture 220, 495–505.

Dang, N.H., Van Thanh, N., Van Kiem, P., Van Minh, C., Kim, Y.H., 2007. Two new triterpene glycosides from the vietnamese sea cucumber *Holothuria scabra*. Arch. Pharm. Res. 30, 1387–1391.

Darsono, P., Sukarno, Notowinarto, 1995. Siklus reproduksi teripang pasir (*Holothuria scabra*) Jaeger (Holothuroidea, Aspidochirota), di perairan Teluk Lampung. In: Kongres Nasional Biology XI, Depok.

Daud, R., Tangko, A.M., Mansyur, A., Wardoyo, S.E., Sudradjat, A., Cholik, F., 1991. Budidaya rumput laut (*Gracilaria* sp.) dengan teripang (*Holothuria scabra*) dalam sistem polikutur. In: Laporan Penelitian. Balai Penelitian Perikanan Budidaya Pantai Maros, Indonesia.

Daud, R., Tangko, A.M., Mansyur, A., Sudradjat, A., 1993. Polyculture of sea cucumber, *Holothuria scabra* and seaweed, *Eucheuma* sp. in Sopura Bay, Kolaka Regency, Southeast Sulawesi. In: Prosiding Seminar Hasil Penelitian. vol. 11, pp. 95–98.

Davenport, D., 1950. Studies in the physiology of commensalism: 1. The polynoid genus *Arctonoë*. Biol. Bull. 98, 81–93.

De Bary, A., 1879. Die erscheinung der symbiose. Verlag von Karl J Trübner, Strassburg.

De Bruyn, C., David, B., Motreuil, S., Caulier, G., Jossart, Q., Rigaud, T., De Ridder, C., 2016. Should I stay or should I go? Causes and dynamics of host desertion by a parasitic crab living on echinoids. Mar. Ecol. Prog. Ser. 546, 163–171.

de la Torre-de la Cruz, M., Villamor, J.L., Diodoco, R.J.P., 2018. Effect of *Holothuria scabra* (Jaeger) on particle size composition and components of sediments. Environ. Asia 11, 148–160.

Deichmann, E., 1958. The Holothuroidea collected by the Velero 111 and Velero IV during the years 1932 to 1954, Part 11. Aspidochirota. In: Allan Hancock Pacific Expeditions. vol. 11, pp. 239–249.

Delroisse, J., Van Wayneberghe, K., Flammang, P., Gillan, D., Gerbaux, P., Opina, N., Todinanahary, G.G.B., Eeckhaut, I., 2020. Epidemiology of a SKin Ulceration Disease (SKUD) in the sea cucumber *Holothuria scabra* with a review on the SKUDs in Holothuroidea (Echinodermata). Sci. Rep. 10, 1–19.

Delroisse, J., Léonet, A., Alexandre, H., Eeckhaut, I., 2021. Intracellular pathways of holothuroid oocyte maturation induced by the thioredoxin Trx-REES. Antioxidants 10, 1–16.

Demeuldre, M., Eeckhaut, I., 2012. Gonad development in the sea cucumber *Holothuria scabra* Jaeger, 1833. SPC Beche-de-mer Inf. Bull. 2, 15–23.

Deng, H., Zhou, Z.-c., Wang, N.-b., Liu, C., 2008. The syndrome of sea cucumber (*Apostichopus japonicus*) infected by virus and bacteria. Virol. Sin. 23, 63–67.

Deng, H., He, C., Zhou, Z., Liu, C., Tan, K., Wang, N., Jiang, B., Gao, X., Liu, W., 2009. Isolation and pathogenicity of pathogens from skin ulceration disease and viscera ejection syndrome of the sea cucumber *Apostichopus japonicus*. Aquaculture 287, 18–27.

Desmelati, S., Dewita, D., Syafrijal, and P.A. Sari., 2020. Determination of nano-collagen quality from sea cucumber *Holothuria scabra*. In: IOP Conference Series: Earth and Environmental Science. IOP Publishing, p. 012005.

Dewi, K.H., Devi, S., Laili, S., Masturah, M., Hajiral, M., 2010. Ekstraksi teripang pasir (*Holothuria Scabra*) sebagai sumber testosteron Pada Berbagai Kecepatan dan Lama Pengadukan. In: Prosiding Seminar Nasional Teknik Kimia "Kejuangan".

Dewi, K.H., Masturah, M., Wan Daud, W.R., 2011. Parameter optimization in the extraction of sea cucumber (*Holothuria scabra* J) as a source of testosterone. In: Advanced Materials Research. Trans Tech Publ, pp. 1358–1365.

Di Simone, M., Horellou, A., Conand, C., 2020. Follow up on Towards a CITES listing of teatfish. SPC Beche-de-mer Inf. Bull. 40, 3–5.

Din, Z., 1986. Beche-de-mer processing trials along the Egyptian Red Sea coast. FAO Fish. Rep. 329, 286–290.

Ding, W., Jian, C., 2010. Research on nutrient contents of *Holothuria scabra* Jaeger. Food Sci. Technol. 10.

Dissanayake, D.C.T., Stefansson, G., 2012. Present status of the commercial sea cucumber fishery off the north-west and east coasts of Sri Lanka. J. Mar. Biol. Assoc. U. K. 92, 831–841.

Dissanayake, D.C.T., Wijeyaratne, M.J.S., 2007. Studies on the sea cucumber fishery in the North Western coastal region of Sri Lanka. Sri Lanka J. Aquat. Sci. 12, 19–38.

Dissanayake, D.C.T., Athukorala, S., Amarasiri, C., 2010. Present status of the sea cucumber fishery in Sri Lanka. SPC Beche-de-mer Inf. Bull. 30, 14–20.

Dobson, G.T., Duy, N.D.Q., Southgate, P.C., 2020a. First assessment of the potential for coculture of sandfish (*Holothuria scabra*) with Babylon snail (*Babylonia areolata*) in Vietnam. J. World Aquacult. Soc. 51, 527–541.

Dobson, G.T., Duy, N.D.Q., Southgate, P.C., 2020b. Utilisation of organic matter from Babylon snail (*Babylonia areolata*) culture sediments by cultured juvenile sandfish (*Holothuria scabra*). Aquacult. Rep 18, 1–6.

Dobson, G.T., Duy, N.D.Q., Paul, N.A., Southgate, P.C., 2020c. Assessing potential for integrating sea grape (*Caulerpa lentillifera*) culture with sandfish (*Holothuria scabra*) and Babylon snail (*Babylonia areolata*) co-culture. Aquaculture 522, 1–8.

Dobson, G.T., Duy, N.D.Q., Southgate, P.C., 2021. Preliminary assessment of large-scale co-culture of sandfish (*Holothuria scabra*) with the Babylon snail (*Babylonia areolata*) in earthen ponds and in raceways. J. World Aquacult. Soc. 52, 138–154.

Doignon, G., Jangoux, M., Féral, J.-P., Eeckhaut, I., 2001. Seasonal release of the egg capsules of *Anoplodium parasita* Schneider, 1858, intracoelomic turbellarian (Platyhelminthes, Rhabdocoela) symbiotic of the sea cucumber *Holothuria tubulosa* Gmelin, 1788 (Echinodermata, Holothuroida). Echinoderm Res. 200, 261–264.

Dolorosa, R.G., Salazar, C.B., Delfin, M.T.V., Paduga, J.R., Balisco, R.A.T., 2017. Sea cucumber fisheries in Rasa Island Wildlife Sanctuary, Narra, Palawan, Philippines. SPC Beche-de-mer Inf. Bull. 37, 9–20.

Domantay, J., 1936. Philippine edible holothurians. In: Searchlight Manila. vol. 1, pp. 11–18.

Du, P.T., Hoang, D.H., Du, H.T., Thi, V.T.H., 2004. Combined culture of mussel: a tool for providing live feed and improving environmental quality for lobster aquaculture in Vietnam. In: Spiny Lobster Ecology and Exploitation in the South China Sea Region. ACIAR, pp. 57–58.

Du, P.T., Hoang, D.H., Du, H.T., Thi, V.H., 2005. The results of the experimental study on combined culture of mussel *Perna viridis* with lobster at Xuan Tu (Van Ninh, Khanh Hoa). J. Mar. Sci. Technol. 4, 224–235.

Dumalan, R.J.P., Bondoc, K.G.V., Juinio-Meñez, M.A., 2019. Grow-out culture trial of sandfish *Holothuria scabra* in pens near a mariculture-impacted area. Aquaculture 507, 481–492.

Dunn, R.R., Harris, N.C., Colwell, R.K., Koh, L.P., Sodhi, N.S., 2009. The sixth mass coextinction: are most endangered species parasites and mutualists? Proc. R. Soc. B Biol. Sci. 276, 3037–3045.

Durairaj, S., 1982. Evolving quality standards for beche-de-mer. Seafood Exp. J. 14, 19–22.

Durairaj, S., Nainar, M.M., Laine, M.K., Sudhakaran, R., Inbaraj, S., 1984. Study on the quality of beche-de-mer in trade and shrinkage of specimens during processing. Fish. Technol. 21, 19–24.

Duy, N.D.Q., 2005. Improving the Seed Production Technique of Sandfish Holothuria scabra in Nha Trang, Khanh Hoa. Research Institute for Aquaculture No. 3, FSPS Programme. SUMA Project Final Report.

Duy, N.D.Q., 2010. Seed production of sandfish (*Holothuria scabra*) in Vietnam. Aquaculture Department, Southeast Asian Fisheries Development Center.

Duy, N.D.Q., 2012. Large-scale sandfish production from pond culture in Vietnam. In: Hair, C.A., Pickering, T.D., Mills, D.J. (Eds.), *Asia-Pacific Tropical Sea Cucumber Aquaculture. Proceedings of an International Symposium Held in Noumea, New Caledonia. ACIAR Proceedings*. Citeseer, pp. 34–39.

Duy, N.D.Q., 2017. Developing Hatchery Culture of the Tropical Sea Cucumber, *Holothuria scabra*, Using Micro-Algae Concentrates. PhD Thesis, James Cook University.

Duy, N.D.Q., Pirozzi, I., Southgate, P.C., 2015. Ingestion and digestion of live microalgae and microalgae concentrates by sandfish, *Holothuria scabra*, larvae. Aquaculture 448, 256–261.

Duy, N.D.Q., Francis, D.S., Southgate, P.C., 2016a. Development of hyaline spheres in late auriculariae of sandfish, *Holothuria scabra*: is it a reliable indicator of subsequent performance? Aquaculture 465, 144–151.

Duy, N.D.Q., Francis, D.S., Pirozzi, I., Southgate, P.C., 2016b. Use of micro-algae concentrates for hatchery culture of sandfish, *Holothuria scabra*. Aquaculture 464, 145–152.

Duy, N.D.Q., Francis, D.S., Southgate, P.C., 2017. The nutritional value of live and concentrated micro-algae for early juveniles of sandfish, *Holothuria scabra*. Aquaculture 473, 97–104.

Edullantes, C.M., Juinio-Meñez, M.A., 2009. The effects of seawater nutrient concentration on the growth and survival of juvenile sandfish *Holothuria scabra*. In: 10th National Symposium in Marine Science, 22–24 October 2009, Davao City, Philippines.

Eeckhaut, I., 2003. Symbiosis. In: Grzimeck's Animal Life Encyclopedia. Gale, Detroit, pp. 31–34.

Eeckhaut, I., 2021. From fundamental research to its applications: the story of Indian Ocean Trepang. SPC Beche-de-mer Inf. Bull. 41, 40–47.

Eeckhaut, I., Puozzo, N., 2018. MetaMorphos: An Educational Tool Illustrating Three-Dimensional Body Plans. https://www.bio-mar.com/metamorphos-application. (last accessed 14 September 2021).

Eeckhaut, I., Parmentier, E., Becker, P., Gomez da Silva, S., Jangoux, M., 2004. Parasites and biotic diseases in field and cultivated sea cucumbers. In: Lovatelli, A., Conand, C., Purcell, S., Uthicke, S., Hamel, J.-F., Mercier, A. (Eds.), Advances in Sea Cucumber Aquaculture and Management. FAO Fisheries Technical Paper No. 463, FAO, Rome, pp. 311–325.

Eeckhaut, I., Lavitra, T., Rasoforinina, R., Rabenevanana, M.W., Gildas, P., Jangoux, M., 2008. Madagascar Holothurie (MH SA): the first trade company based on sea cucumber aquaculture in Madagascar. SPC Beche-de-mer Inf. Bull. 28, 22–23.

Eeckhaut, I., T. Lavitra, A. Léonet, M. Jangoux, and R. Rasolofonirina. 2012. In-vitro fertilisation: a simple, efficient method for obtaining sea cucumber larvae year round. Pages 40-49 *in* C. A. Hair, T. D. Pickering, and D. J. Mills, editors. *Asia-Pacific Tropical Sea Cucumber Aquaculture. Proceedings of an International Symposium Held in Noumea, New Caledonia. ACIAR Proceedings.*

Eeckhaut, I., Caulier, G., Brasseur, L., Flammang, P., Gerbaux, P., Parmentier, E., 2015. Effects of Holothuroid Ichtyotoxic Saponins on the Gills of Free-Living Fishes and Symbiotic Pearlfishes. Biol. Bull. 228, 253–265.

Eeckhaut, I., Van Wayenberghe, K., Nicolas, F., Delroisse, J., 2019. Skin ulcerations in *Holothuria scabra* can be induced by various types of food. SPC Beche-de-mer Inf. Bull. 39, 31–35.

Eeckhaut, I., Février, J., Todinanahary, G., Delroisse, J., 2020. Impact of *Thalamita crenata* (Decapoda; Portunidae) predation on *Holothuria scabra* juvenile survival in sea farming pens. SPC Beche-de-mer Inf. Bull. 40, 11–16.

Elanganayagam, P., Ganesalingam, V.K., Sachithananthan, K., 1980. Nutritional composition of some raw Sri Lankan holothurians. In: Proceedings of the Sri Lanka Association for the Advancement of Science. vol. 36, pp. 62–63.

Elanganayagam, P., Mahendran, M., Krishnarajah, S.R., Ganesalingam, V.K., Sachithananthan, K., 1982. Estimation of saponin content of some Sri Lankan holothurians. In: Proceedings of the Institute of Chemistry of Ceylon. vol. 11, p. 5.

Elfath, N.A., Putri, R.M.S., Apriandi, A., 2020. MINUMAN FUNGSIONAL TERIPANG PASIR (*Holothuria scabra*) DAN TERIPANG HITAM (*Holothuria atra*) - Functional Drinks of Sand Sea Cucumber (*Holothuria scabra*) and Black Sea Cucumber (*Holothuria atra*). Marinade 3, 47–58.

Endean, R., 1953. Queensland faunistic records part III.~ Echinodeflnata (excluding Crinoidea). In: Papers From the Department of Zoology of the University of Queensland. vol. 1, pp. 51–60.

Endean, R., 1956. Queensland faunistic records. Part IV. Further records of echinodermata (excluding Crinoidea). In: Papers from the Department of Zoology of the University of Queensland. vol. 1, pp. 121–140.

Endean, R., 1957. The biogeography of Queensland's shallow-water echinoderm fauna (excluding Crinoidea), with a rearrangement of the faunistic provinces of tropical Australia. Mar. Freshw. Res. 8, 233–273.

Eriksson, H., 2012. Sandfish (*Holothuria scabra*) farming in a social–ecological context: conclusions from Zanzibar. In: Hair, C.A., Pickering, T.D., Mills, D.J. (Eds.), Asia-Pacific Tropical Sea Cucumber Aquaculture. *Proceedings of an International Symposium Held in Noumea, New Caledonia. ACIAR Proceedings*, pp. 196–202.

Eriksson, H., Byrne, M., 2015. The sea cucumber fishery in Australia's Great Barrier Reef Marine Park follows global patterns of serial exploitation. Fish Fish. 16, 329–341.

Eriksson, H., Clarke, S., 2015. Chinese market responses to overexploitation of sharks and sea cucumber. Biol. Conserv. 184, 163–173.

Eriksson, H., de la Torre-Castro, M., Eklöf, J., Jiddawi, N., 2010. Resource degradation of the sea cucumber fishery in Zanzibar, Tanzania: a need for management reform. Aquat. Living Resour. 23, 387–398.

Eriksson, H., De La Torre-castro, M., Olsson, P., 2012a. Mobility, expansion and management of a multi-species scuba diving fishery in East Africa. PLoS One 7, e35504.

Eriksson, H., Byrne, M., De la Torre-Castro, M., 2012b. Sea cucumber (Aspidochirotida) community, distribution and habitat utilization on the reefs of Mayotte, Western Indian Ocean. Mar. Ecol. Prog. Ser. 452, 159–170.

Eriksson, H., Robinson, G., Slater, M.J., Troell, M., 2012c. Sea cucumber aquaculture in the Western Indian Ocean: challenges for sustainable livelihood and stock improvement. Ambio 41, 109–121.

Eriksson, H., Conand, C., Lovatelli, A., Muthiga, N., Purcell, S., 2012d. Sea cucumber fisheries: an ecosystem approach to management in the Indian Ocean (SCEAM Indian Ocean). In: Workshop Conducted at Mazizini, Zanzibar in the United Republic of Tanzania, November, pp. 12–16.

Eriksson, H., de la Torre-Castro, M., Purcell, S.W., Olsson, P., 2015a. Lessons for resource conservation from two contrasting small-scale fisheries. Ambio 44, 204–213.

Eriksson, H., Conand, C., Lovatelli, A., Muthiga, N.A., Purcell, S.W., 2015b. Governance structures and sustainability in Indian Ocean sea cucumber fisheries. Mar. Policy 56, 16–22.

Eriksson, H., Österblom, H., Crona, B., Troell, M., Andrew, N., Wilen, J., Folke, C., 2015c. Contagious exploitation of marine resources. Front. Ecol. Environ. 13, 435–440.

Eriksson, H., Adhuri, D.S., Adrianto, L., Andrew, N.L., Apriliani, T., Daw, T., Evans, L., Garces, L., Kamanyi, E., Mwaipopo, R., 2016. An ecosystem approach to small-scale fisheries through participatory diagnosis in four tropical countries. Glob. Environ. Chang. 36, 56–66.

FAO, 2012. Report on the FAO workshop on sea cucumber fisheries: an ecosystem approach to management in the pacific (SCEAM Pacific). In: FAO Fisheries and Aquaculture Report. No. 1003, FAO, Rome.

FAO, 2013. Report on the FAO workshop on sea cucumber fisheries: an ecosystem approach to management in the Indian Ocean (SCEAM Indian Ocean). In: FAO Fisheries and Aquaculture Report. No. 1038, FAO, Rome.

FAO, 2019. Report of the sixth FAO expert advisory panel for the assessment of proposals to amend Appendices I and II of CITES concerning commercially exploited aquatic species, Rome, 21–25 January. In: FAO Fisheries and Aquaculture Report No. 1255.

FAO, 2020. FishStatJ—Software for fishery and aquaculture statistical time series 2020. FAO Fisheries and Aquaculture Department [online], Rome. Updated 2021 www.fao.org/fishery/statistics/software/fishstatj/en.

Féral, J.-P., Cherbonnier, G., 1986. Les holothuries. In: Guide des Étoiles de Mer, Oursins et autres Échinodermes du Lagon de Nouvelle-Calédonie. ORSTOM, Paris, pp. 1–238.

Firdaus, M., Indriana, L.F., 2019. Nursery performance of sandfish *Holothuria scabra* juveniles in tidal earthen pond using different types of cage. In: IOP Conference Series: Earth and Environmental Science. IOP Publishing, pp. 1–7.

Fitch, A.J., Leeworthy, G., Li, X., Bowman, W., Turner, L., Gardner, M.G., 2013. Isolation and characterisation of eighteen microsatellite markers from the sea cucumber *Holothuria scabra* (Echinodermata: Holothuriidae). Aust. J. Zool. 60, 368–371.

Fleming, A.E., 2012. Sea ranching of sandfish in an Indigenous community within a well-regulated fishery (Northern Territory, Australia). In: Hair, C.A., Pickering, T.D., Mills, D.J. (Eds.), *Asia-Pacific Tropical Sea Cucumber Aquaculture. Proceedings of an International Symposium Held in Noumea, New Caledonia. ACIAR Proceedings*, pp. 156–160.

Forbes, B., Ilias, Z., 1999. The taxonomy and ecology of sea cucumbers in Malaysia. In: The Conservation of Sea Cucumbers in Malaysia—Proceedings of an International Conference. Heriot-Watt University and the Fisheries Research Institute, Kuala Lumpur, Malaysia, pp. 42–48.

Fourgon, D., Jangoux, M., Eeckhaut, I., 2007. Biology of a "babysitting" symbiosis in brittle stars: analysis of the interactions between *Ophiomastix venosa* and *Ophiocoma scolopendrina.* Invertebr. Biol. 126, 385–395.

Francour, P., 1997. Predation on holothurians: a literature review. Invertebr. Biol. 116, 52–60.

Fransen, C.H., Hoeksema, B.W., 2014. Going for the stars: extending the host record for the reef-dwelling Emperor shrimp, *Periclimenes imperator* (Pontoniinae). Mar Biodivers. 44, 465–466.

Friedman, K., Purcell, S., Bell, J., Hair, C., 2008. Sea Cucumber Fisheries: A Manager's Toolbox. ACIAR Monograph No. 135. Australian Centre for International Agricultural Research, Canberra, Australia. 32 p.

Friedman, K., Eriksson, H., Tardy, E., Pakoa, K., 2011. Management of sea cucumber stocks: patterns of vulnerability and recovery of sea cucumber stocks impacted by fishing. Fish Fish. 12, 75–93.

Froese, R., 1998. Length-weight relationships for 18 less-studied fish species. J. Appl. Ichthyol. 14, 117–118.

Fujiwara, A., Yamano, K., Ohno, K., Yoshikuni, M., 2010. Spawning induced by cubifrin in the Japanese common sea cucumber *Apostichopus japonicus.* Fisheries Science 76, 795–801.

Furqon, A.D.C., Maulana, F., Prihantari, E.T., Prabowo, R.E., 2020. The induction of asexual reproduction on *Holothuria scabra* and *Bohadschia marmorata*: The conservation effort in Tanimbar Archipelago, Maluku. IOP Conference Series: Earth and Environmental Science, vol. 550 IOP Publishing, p. 012025.

Gamboa, R., Juinio-Meñez, M.A., 2003. Pilot-testing the laboratory production of the sea cucumber, *Holothuria scabra* (Jaeger), in the Philippines. In: The Philippine Scientist, vol. 40, pp. 111–121.

Gamboa, R., Gomez, A.L., N.M. F., 2004. The status of sea cucumber fisheries and mariculture in the Philippines. In: Lovatelli, A., Conand, C., Purcell, S., Uthicke, S., Hamel, J.-F., Mercier, A. (Eds.), Advances in Sea Cucumber Aquaculture and Management. FAO Fisheries Technical Paper No. 463, FAO, Rome, pp. 69–78.

Gamboa, R.U., Aurelio, R.A., Ganad, D.A., Concepcion, L.B., Abreo, N.A.S., 2012. Small-scale hatcheries and simple technologies for sandfish (*Holothuria scabra*) production. In: Hair, C.A., Pickering, T.D., Mills, D.J. (Eds.), *Asia-Pacific Tropical Sea Cucumber Aquaculture. Proceedings of an International Symposium Held in Noumea, New Caledonia. ACIAR Proceedings*, pp. 63–74.

Gao, F., Xu, Q., Yang, H., 2011. Seasonal biochemical changes in composition of body wall tissues of sea cucumber *Apostichopus japonicus*. Chinese J. Oceanol. Limnol. 29, 252–260.

Gentle, M.T., 1985. People's Democratic Republic of Yemen. Commercial sea cucumber resources. A Report Prepared for the Beche-de-Mer Fishery Development (Phase II) Project. Report.

Gianasi, B.L., Verkaik, K., Hamel, J.-F., Mercier, A., 2015. Novel use of PIT tags in sea cucumbers: promising results with the commercial species *Cucumaria frondosa*. PLoS One 10, e0127884.

Gibbs, P.E., Clark, A.M., Clark, C.M., 1976. Echinoderms from the northern region of the Great Barrier Reef, Australia. Bull Br. Mus. Nat. Hist. Zool. 30, 102–144.

Gilliland, P.M., 1993. The skeletal morphology, systematics and evolutionary history of holothurians. Spec. Pap Palaeontol. 47, 1–147.

Giraspy, D.A.B., Ivy, G., 2005. Australia's first commercial sea cucumber culture and sea ranching project in Hervey Bay, Queensland, Australia. SPC Beche-de-mer Inf. Bull. 21, 29–31.

Giraspy, D.A.B., Walsalam, I.G., 2010. Aquaculture potential of the tropical sea cucumbers *Holothuria scabra* and *H. lessoni* in the Indo-Pacific region. SPC Beche-de-mer Inf. Bull. 30, 29–32.

Giri, I.N.A., Sembiring, S.B.M., Marzuqi, M., Andamari, R., 2018. Formulasi dan Aplikasi pakan buatan berbasis rumput laut untuk pendederan benih teripang pasir (*Holothuria scabra*). J. Ris. Akuakultur 12, 263–273.

Gonzales, D., 1975. Beche-de-mer: a new export. World Fish. 24, 61.

González-Wangüemert, M., Domínguez-Godino, J.A., Cánovas, F., 2018. The fast development of sea cucumber fisheries in the Mediterranean and NE Atlantic waters: from a new marine resource to its over-exploitation. Ocean Coast. Manag. 151, 165–177.

Goodwill, R.H., Fautin, D.G., Furey, J., Daly, M., 2009. A sea anemone symbiotic with gastropods of eight species in the Mariana Islands. Micronesica 41, 117–130.

Gorospe, J.C., Altamirano, J.P., Juinio-Meñez, M.A., 2017. Viability of a bottom-set tray ocean nursery system for *Holothuria scabra* Jaeger 1833. Aquacult. Res. 48, 5984–5992.

Gorospe, J.R.C., Juinio-Meñez, M.A., Southgate, P.C., 2019. Effects of shading on periphyton characteristics and performance of sandfish, *Holothuria scabra* Jaeger 1833, juveniles. Aquaculture 512, 734307.

Govan, H., 2017a. A review of sea cucumber fisheries and management in Melanesia. SPC Fish. Newsl. 154, 31–42.

Govan, H., 2017b. Melanesian Spearhead Group of countries addresses sea cucumber fisheries management. SPC Fish. Newsl. 154, 19–21.

Govan, H., 2019. 2018 Market price update for beche-de-mer in Melanesian countries. Fish. Newsl. 154, 16–20.

Gowda, N.M., Goswami, U., Khan, M.I., 2008a. Purification and characterization of a T-antigen specific lectin from the coelomic fluid of a marine invertebrate, sea cucumber (*Holothuria scabra*). Fish Shellfish Immunol. 24, 450–458.

Gowda, N.M., Goswami, U., Khan, M.I., 2008b. T-antigen binding lectin with antibacterial activity from marine invertebrate, sea cucumber (*Holothuria scabra*): possible involvement in differential recognition of bacteria. J. Invertebr. Pathol. 99, 141–145.

Gowda, N.M., Gaikwad, S.M., Khan, M.I., 2013. Kinetics and thermodynamics of glycans and glycoproteins binding to *Holothuria scabra* lectin: a fluorescence and surface plasmon resonance spectroscopic study. J. Fluoresc. 23, 1147–1155.

Gravely, F.H., 1927. The littoral fauna of Krusadai Island in the Gulf of Mannar: Echinodermata. In: Bulletin of Madras Governemnt Museum (Natural History), vol. 1, pp. 163–173.

Gurumani, O.N., Krishnamurthy, S., 1994. Some aspects of processing and quality control of beche-de-mer for export. CMFRI Bull. 46, 81–84.

Hacker, C.J.O., 2017. Small-Scale Farming of Sandfish (*Holothuria scabra*) at Unguja Ukuu, Zanzibar. Independant Study Project Collection 2615. https://digitalcollections.sit.edu/isp_collection/2615.

Hair, C., 2012. Sandfish (*Holothuria scabra*) production and sea-ranching trial in Fiji. In: Hair, C.A., Pickering, T.D., Mills, D.J. (Eds.), *Asia-Pacific Tropical Sea Cucumber Aquaculture. Proceedings of an International Symposium Held in Noumea, New Caledonia. ACIAR Proceedings*, pp. 129–141.

Hair, C., Pickering, T., Meo, S., Vereivalu, T., Hunter, J., Cavakiqali, L., Hamilton, R., Lokani, P., Tsiresy, G., Pascal, B., 2011. Sandfish culture in Fiji Islands. SPC Beche-de-mer Inf. Bull. 31, 3–11.

Hair, C., T.D. Pickering, and D.J. Mills. 2012. Asia-Pacific Tropical Sea Cucumber Aquaculture: Proceedings of an International Symposium Held in Noumea, New Caledonia, *15-17 February, 2011*. Australian Centre for International Agricultural Research.

Hair, C., Mills, D.J., McIntyre, R., Southgate, P.C., 2016. Optimising methods for community-based sea cucumber ranching: experimental releases of cultured juvenile *Holothuria scabra* into seagrass meadows in Papua New Guinea. Aquacult. Rep 3, 198–208.

Hair, C., Ram, R., Southgate, P.C., 2018a. Is there a difference between bêche-de-mer processed from ocean-cultured and wild-harvested sandfish (*Holothuria scabra*)? Aquaculture 483, 63–68.

Hair, C., Kinch, J., Galiurea, T., Kanawi, P., Mwapweya, M., Noiney, J., 2018b. Re-opening of the sea cucumber fishery in Papua New Guinea: a case study from the Tigak Islands in the New Ireland Province. SPC Beche-de-mer Inf. Bull. 38, 3–10.

Hair, C., Foale, S., Kinch, J., Frijlink, S., Lindsay, D., Southgate, P.C., 2019. Socioeconomic impacts of a sea cucumber fishery in Papua New Guinea: is there an opportunity for mariculture? Ocean Coast. Manag. 179, 104826.

Hair, C., Militz, T., Daniels, N., Southgate, P.C., 2020a. Comparison of survival, growth and burying behavior of cultured and wild sandfish (*Holothuria scabra*) juveniles: implications for ocean mariculture. Aquaculture 526, 1–11.

Hair, C., Foale, S., Daniels, N., Minimulu, P., Aini, J., Southgate, P.C., 2020b. Social and economic challenges to community-based sea cucumber mariculture development in new Ireland province, Papua New Guinea. Mar. Policy 117, 103940.

Hair, C., Militz, T.A., Daniels, N., Southgate, P.C., 2021. Performance of a trial sea ranch for the commercial sea cucumber, *Holothuria scabra*, in Papua New Guinea. Aquaculture 547, 737500.

Halstead, B.W., 1988. Poisonous and Venomous Marine Animals of the World. Darwin Press, Princeton, NJ, USA.

Hamad, M.I., Mwandya, A.W., Munubi, R.S., Chenyambuga, S., Lamtane, H.A., 2019. Comparative growth and survival performance of sea cucumber (*Holothuria scabra*) in co-cultured pen system with commercial macroalgae. Afr. J. Biol. Sci. 1, 32–41.

Hamel, J.-F., Mercier, A., 1999. Mucus as a mediator of gametogenic synchrony in the sea cucumber *Cucumaria frondosa* (Holothuroidea: Echinodermata). J. Mar. Biol. Assoc. U. K. 79, 121–129.

Hamel, J.-F., Mercier, A., 2008. Population status, fisheries and trade of sea cucumbers in temperate areas of the northern hemisphere. In: Toral-Granda, V., Lovatelli, A., Vasconcellos, M. (Eds.), Sea Cucumbers. A Global Review of Fisheries and Trade. FAO Fisheries and Aquaculture Technical Paper 516, FAO, Rome, pp. 257–292.
Hamel, J.-F., Himmelman, J.H., Dufresne, L., 1993. Gametogenesis and spawning of the sea cucumber *Psolus fabricii* (Duben and Koren). Biol. Bull. 184, 125–143.
Hamel, J.-F., Ng, P.K.L., Mercier, A., 1999. Life cycle of the pea crab *Pinnotheres halingi* sp. nov., an obligate symbiont of the sea cucumber *Holothuria scabra* Jaeger. Ophelia 50, 149–175.
Hamel, J.-F., Conand, C., Pawson, D.L., Mercier, A., 2001. The sea cucumber *Holothuria scabra* (Holothuroidea: Echinodermata): its biology and exploitation as beche-de-mer. Adv. Mar. Biol. 41, 129–223.
Hamel, J.-F., Mercier, A., Conand, C., Purcell, S., Toral-Granda, T.G., Gamboa, R., 2013. *Holothuria scabra*. The IUCN Red List of Threatened Species. 2013, https://doi.org/10.2305/IUCN.UK.2013-1.RLTS.T180257A1606648.en.
Hamel, J.-F., Ng, P.K., McCurry, S., Mercier, A., 2019a. Note on the pea crab *Holotheres semperi* (Bürger, 1895) parasitising the sea cucumber *Holothuria scabra* in Rempang, Indonesia. SPC Beche-de-mer Inf. Bull. 39, 36–38.
Hamel, J.-F., Ng, P.K.L., McCurry, S., Mercier, A., 2019b. Note on the pea crab *Holotheres semperi* (Burger, 1895) parasiting the sea cucumber *Holothuria scabra* in Rempang (Indonesia). SPC Beche-de-mer Inf. Bull. 39, 36–38.
Hamel, J.-F., Sun, J., Gianasi, B.L., Montgomery, E.M., Kenchington, E.L., Burel, B., Rowe, S., Winger, P.D., Mercier, A., 2019c. Active buoyancy adjustment increases dispersal potential in benthic marine animals. J. Anim. Ecol. 88, 820–832.
Hammond, L.S., 1982a. Analysis of grain-size selection by deposit-feeding holothurians and echinoids (Echinodermata) from a shallow reef lagoon. Discovery Bay, Jamaica. Mar. Ecol. Prog. Ser. 8, 25–36.
Hammond, L.S., 1982b. Patterns of feeding and activity in deposit-feeding holothurians and echinoids (Echinodermata) from a shallow back-reef lagoon, Discovery Bay, Jamaica. Bull. Mar. Sci. 32, 549–571.
Han, H., Yi, Y., La, M., Yan, L., Liu, B., 2008. Studies on antifungal and antitumor activities of scabraside A, B from *Holothuria scabra* Jaeger. Zhongguo Yaolixue Tongbao 24, 1111–1112.
Han, H., Yi, Y., Xu, Q., La, M., Zhang, H., 2009a. Two new cytotoxic triterpene glycosides from the sea cucumber *Holothuria scabra*. Planta Med. 75, 1608–1612.
Han, H., Yi, Y., Li, L., Liu, B., La, M., Zhang, H., 2009b. Antifungal active triterpene glycosides from sea cucumber *Holothuria scabra*. Acta Pharm. Sin. 44, 620–624.
Han, H., Yi, Y., Li, L., Pan, M., Sun, P., 2010. Cytotoxicitic triterpene glycosides from sea cucumber *Holothuria scabra* Jaeger. Chin. J. Med. Chem. 20, 290–297.
Haran, N., Athauda, A., Mallikarachchi, U., 2012. Selection of Feasible Location for Pen Culture of *Holuthuria scabra* in Jaffana Peninsula. NARA.
Harriott, V.J., 1980. The Ecology of Holothurian Fauna of Heron Reef and Moreton Bay. MSc Thesis, University of Queensland, Australia.
Hartati, R., Widianingsih, D.A., Djunaedi, A., 2016. Ultrastruktur alimentary canal teripang *Holothuria scabra* dan *Holothuria atra* (Echinodermata: Holothuroidea). Buletin Oseanografi Marina 5, 86–96.
Hasan, M.H., 2005. Destruction of a *Holothuria scabra* population by overfishing at Abu Rhamada Island in the Red Sea. Mar. Environ. Res. 60, 489–511.
Hasan, M.H., 2008. Fisheries status and management plan for Saudi Arabian sea cucumbers. SPC Beche-de-mer Inf. Bull. 28, 14–21.
Hasan, M.H., 2019. Destruction of sea cucumber populations due to overfishing at Abu Ghosoun area, Red Sea. J. Basic Appl. Zool. 80, 5.

Hashim, R., Mohamed, N., Yasin, Z., Sasekumar, A., Awang, N., Thongtham, N., Baine, M., Conand, C., Battaglene, S.C., Mustafa, S., 1999. Towards a holothurians conservation and management strategy for Malaysia. In: The Conservation of sea cucumbers in Malaysia—Proceedings of an International Conference. Heriot-Watt University and the Fisheries Research Institute, Kuala Lumpur, Malaysia, pp. 64–77.

Hiruy, K., Murphy, R., Lewis, T., White, W.T., Purcell, S.W., 2019. Measuring scientific impact of fisheries and aquaculture research-for-development projects in South East Asia and the Pacific. Res. Eval. 28, 313–325.

Holland, A., 1994a. The Status of Global beche-de-mer Fisheries With Special Reference to the Solomon Islands and the Potentials of Holothurian Culture. Unpublished M. Sc, Thesis, University of Newcastle, United Kingdom.

Holland, A., 1994b. The beche-de-mer industry in the Solomon Islands: recent trends and suggestions for management. SPC Beche-de-mer Inf. Bull. 6, 2–9.

Hornell, J., 1917. The Indian beche-de-mer Industry: Its History and Recent Revival. Government Press.

Houck, M.A., 1994. Adaptation and transition into parasitism from commensalism: a phoretic model. In: Mites. Springer, pp. 252–281.

Hsueh, P.-W., 2001. Intertidal distribution, symbiotic association and reproduction of *Pinnotheres bidentatus* (Brachyura: Pinnotheridae) from Taiwan. J. Nat. Hist. 35, 1681–1692.

Humes, A.G., 1967. A new species of *Scambicornus* (Copepoda, Cyclopoida, Lichomolgidae) associated with a holothurian in Madagascar, with notes on several previously described species. Beaufortia 14, 135–155.

Humes, A.G., 1980. A review of the copepods associated with holothurians, including new species from the Indo-Pacific. Beaufortia 30, 31–123.

Humes, A.G., Ho, J.-S., 1969. Cyclopoid copepods parasitic in holothurians in Madagascar. J. Parasitol. 1969, 877–894.

Humphreys, W.F., 1981. The Echinoderms of Kenya's Marine Parks and Adjacent Regions. vol. 19 Koninklijk Museum Voor Midden-Afrika, pp. 1–39.

Ibrahim, H.A.H., 2012. Antibacterial carotenoids of three *Holothuria* species in Hurghada, Egypt. Egypt. J. Aquat. Res. 38, 185–194.

Ibrahim, M.Y., Elamin, S.M., Gideiri, Y.B.A., Ali, S.M., 2015. The proximate composition and the nutritional value of some sea cucumber species inhabiting the Sudanese Red Sea. Food Sci. Qual. Manag. 41, 11–16.

Indap, M., Thakur, N., Gaikwad, S., 1996. Acute toxicity of *Holothuria scabra* (Jaeger) extracts on fish *Tilapia mossambica* (Peters). Environ. Ecol. 14, 917–919.

Indriana, L.F., Firdaus, M., 2020. Growth performance of sea cucumber *Holothuria scabra* juvenile in different initial size of pond culture in Lombok, Indonesia. In: E3S Web of Conferences. EDP Sciences, p. 01003.

Indriana, L.F., Kunzmann, A., 2014. The correlation between wet and dry weight of intestine and body without intestine of sea cucumber (*Holothuria scabra*). Indones. Aquac. J. 9, 51.

Indriana, L.F., Afrianti, Y., Hilyana, S., Firdaus, M.F., 2017. Preferensi Penempelan, Pertumbuhan, Dan Sintasan Larva Teripang Pasir, *Holothuria scabra* Pada Substrat Lamun Yang Berbeda. J. Ris. Akuakultur 11, 249–258.

Indriana, L.F., A'an, J.W., Kunzmann, A., 2018. Assimilation dynamics of different diet sources by the sea sucumber *Holothuria scabra*, with evidence from stable isotope signature. Annu. Res. Rev. Biol. 28, 1–10.

Intes, A.G., Menou, J.L., 1979. Quelques holothuries (Echinodermes) des environs de Nouméa et leur répartition. In: Rapport Scientifique et Technique ORSTOM, vol. 3, pp. 1–23.

Ito, M., 2014. A hatchery operations manual for rearing sandfish, *Holothuria scabra*. Secretariat of the Pacific Community, Noumea, Tarawa, Republic of Kiribati.
Ivy, G., Giraspy, D.A.B., 2006. Development of large-scale hatchery production techniques for the commercially important sea cucumber *Holothuria scabra veriscolor* (Conand, 1986) in Queensland, Australia. Beche-de-Mer Inf. Bull. 24, 28–34.
Jaeger, G.F., 1833. De Holothuriis. In: Dissertracio, Turici, pp. 1–40.
Jagadees, K., Sasidharan, S., 1990. Beche-de-mer handling and processing for export. In: Marine Products Export Development Authority, Extension Section, India.
James, D.B., 1973. Beche-de-mer resources of India. In: Proceedings of a Symposium on Living Resources of the Seas around India. Central Marine Fisheries Research Institute Special Publication, pp. 706–711.
James, D.B., 1976. Studies on Indian Echinoderms 6. Redescription of two little known holothurians with a note on an early juvenile of *Holothuria scabra* Jaeger from the Indian Seas. J. Mar. Biol. Assoc. India 18, 55–61.
James, D.B., 1978. Studies on the systematics of some shallow water Asteroidea, Ophiuroidea and Holothuroidea of the Indian seas. PhD Thesis, Andhra University.
James, D.B., 1982. Ecology of intertidal echinoderms of the Indian seas. J. Mar. Biol. Assoc. India 24, 124–129.
James, D.B., 1983. Sea cucumber and sea urchin resources and beche-de-mer industry. Bull. Cent. Mar. Fish. Res. Inst. 34, 85–93.
James, D.B., 1986a. Zoogeography of shallow-water echinoderms of Indian seas. In: James, P.S.B.R. (Ed.), Recent Advances in Marine Biology. Today and Tomorrow Printers and Publishers, New Delhi, India, pp. 569–591.
James, D.B., 1986b. The holothurian resources. In: R&D Series for Marine Fishery Resources Management. Central Marine Fisheries Research Institute, Cochin, India, p. 4.
James, D.B., 1986c. Quality improvement in beche-de-mer. Seafood export journal 18, 5–10.
James, D.B., 1987a. Research, Conservation and Management of Edible Holothurians and Their Impact on the beche-de-mer Industry. vol. 44 Central Marine Fisheries Research Institute, Special Publication, pp. 648–660.
James, D.B., 1987b. Prospects and problems of beche-de-mer industry in Andaman and Nicobar Islands. In: Proceedings of the Symposium on management of Coastal Aquaculture and Oceanic Resources of Andamans, pp. 110–118.
James, D.B., 1989a. Beche-de-mer: Its resources, fishery and industry. In: Marine Fisheries Information Service, Technical and Extension Series, vol. 92, pp. 2–30.
James, D.B., 1989b. A Hand Book on Beche-de-mer. Issued on the occasion of the Workshop on Beche-de-mer, at Mandapam Camp, Central Marine Fisheries Research Institute, Cochin.
James, P.S.B.R., 1990. Progress of research on sea ranching at the CMFRI. In: Marine Fisheries Information Service, Technical and Extension Series, vol. 105, pp. 1–5.
James, D.B., 1991. Echinoderms of the Marine National Park, South Andamans. J. Andaman Sci. Assoc. 7, 19–25.
James, D.B., 1994a. Systematics, biology, ecology and zoogeography of holothurians: ecology of commercially important holothurians of India. Bull. Cent. Mar. Fish. Res. Inst. 46, 37–38.
James, D.B., 1994b. Holothurian resources from India and their exploitation. In: Proceedings of the National Workshop on Beche-de-mer, vol. 46, pp. 27–31.
James, D.B., 1994c. Zoogeography and systematics of holothunans used for beche-de-mer in India. In: Proceedings of the National Workshop on Beche-de-mer, vol. 46, pp. 34–36.
James, D.B., 1994d. Culture of holothurians: hatchery and production techniques. A review of the hatchery and culture practices in Japan and China with special reference to possibilities of culturing holothurians in India. Bull. Cent. Mar. Fish. Res. Inst. 46, 63–65.

James, D.B., 1994e. An annotated bibliography on sea-cucumbers. Bull. Cent. Mar. Fish. Res. Inst. 58, 1–92.
James, D.B., 1994f. Processing, quality control and utilization of beche-de-mer: improved methods of processing holothurians for beche-de-mer. Bull. Cent. Mar. Fish. Res. Inst. vol. 46, 71–75.
James, D.B., 1994g. A review of the hatchery and culture practiced in Japan and China with special reference to possibilities of culturing holothurians in India. In: Proceedings of the National Workshop on Beche-de-mer. Bulletin of the Central Marine Fisheries Research Institute. Central Marine Fisheries Research Institute, Indian Council of Agricultural Research, Cochin, India, pp. 63–65.
James, D.B., 1994h. Seed production in sea cucumbers. Aquac. Int. 1, 15–26.
James, D.B., 1995a. Taxonomic studies of the species of *Holothuria* (Linnaeus, 1767) from the seas around India. Part 1. J. Bombay Nat. Hist. Soc. 92, 43–62.
James, D.B., 1995b. Animal associations in echinoderms. J. Mar. Biol. Assoc. India 37, 272–276.
James, D.B., 1996a. Culture of sea-cucumber. Bull. Cent. Mar. Fish. Res. Inst. 48, 120–126.
James, P.S.B.R., 1996b. Technologies and potential for seafarming in India, Part I. Aquac. Mag. 22, 50–60.
James, D.B., 1998. On the occurrence of the gastropod parasite *Prostilifer* sp. on the holothurian *Holothuria scabra* Jaeger at Tuticorin. In: Marine Fisheries Information Service, Technical and Extension Series, 157, p. 26.
James, D.B., 1999. Hatchery and culture technology for the sea cucumber, *Holothuria scabra* Jaeger, in India. Naga 22, 12–16.
James, C.M., 2012. Sea cucumbers in the Maldives. Aquac. Asia Pac. Mag (September/October), 43–44.
James, D.B., Ali Manikfan, M., 1994. Some remarks on the present status of beche-de-mer industry of Maldives and its lesson for the Lakshadweep. CMFRI Bull. 46, 101–105.
James, D.B., Baskar, B.K., 1994. Present status of the beche-de-mer industry in the Palk Bay and the Gulf of Mannar. In: Proceedings of the National Workshop on Beche-de-mer, Central Marine Fisheries Research Institute, Indian Council of Agricultural Research, Cochin, India, pp. 85–90.
James, P.S.B.R., James, D.B., 1993. Ecology, breeding, seed production and prospects for farming of sea cucumbers from the seas around India. Fish. Chimes 13, 24–34.
James, D.B., James, P.S.B.R., 1994a. Hand-book on Indian sea-cucumbers. Bull. Cent. Mar. Fish. Res. Inst. 59, 1–47.
James, P.S.B.R., James, D.B., 1994b. Management of the beche-de-mer industry in India. Bull. Cent. Mar. Fish. Res. Inst. 46, 17–22.
James, D.B., Rajapandian, M.E., Baskar, B.K., Gopinathan, C.P., 1988. Successful induced spawning and rearing of the holothurian *Holothuria* (*Metriatyla*) *scabra* Jaeger at Tuticorin. In: Marine Fisheries Information Service. Technical and Extension Series, vol. 87, pp. 30–33.
James, D.B., Rajapandian, M.E., Baskar, B.K., Gopinathan, C.P., 1989. In: Breakthrough in the Hatchery of the Holothurian *Holothuria* (*Metriatyla*) *scabra*. National Workshop on beche-de-mer at Mandapam Camp. Central Marine Fisheries Research Institute, Cochin, India.
James, D.B., Rajapandian, M.E., Gopinathan, C.P., Baskar, B.K., 1994a. Breakthrough in induced breeding and rearing of the larvae and juveniles of *Holothuria* (*Metriatyla*) *scabra* Jaeger at Tuticorin. Bull. Cent. Mar. Fish. Res. Inst. 46, 66–70.
James, D.B., Gandhi, A.D., Palaniswamy, N., Rodrigo, J.X., 1994b. Hatchery Techniques and Culture of the Sea-Cucumber *Holothuria scabra*. Cochin, India.
James, D.B., Lordson, A.J., Ivy, W.G., Gandhi, A.D., 1996. Experiments on rearing of the juveniles of *Holothuria scabra* Jaeger produced in the hatchery. In: Paulraj, S. (Ed.), Proceeding of Symposium on Aquaculture for 2000 AD. Palani Paramount Publications, pp. 207–214.

Jangoux, M., 1984. Diseases of echinoderms. Helgoländer Meeresun. 37, 207.

Jangoux, M., 1987. Diseases of Echinodermata. 111. Agents metazoans (Annelida to Pisces). In: Diseases of Aquatic Organisms, vol. 3, pp. 59–83.

Jaquemet, S., Conand, C., 1999. The beche-de-mer trade in 1995/1996 and an assessment of exchanges between the main world markets. SPC Beche-de-mer Inf. Bull. 12, 11–14.

Jasmadi, A.K.T.T., 2018. A preliminary study of Difur coast, Tual City, Kei Islands as site for sandfish *Holothuria scabra* culture. In: Proceedings of The 1st International Conference on Applied Marine Science. Tual State Fisheries Polytechnic, p. 133.

Jattujan, P., Chalorak, P., Siangcham, T., Sangpairoj, K., Nobsathian, S., Poomtong, T., Sobhon, P., Meemon, K., 2018. *Holothuria scabra* extracts possess anti-oxidant activity and promote stress resistance and lifespan extension in *Caenorhabditis elegans*. Exp. Gerontol. 110, 158–171.

Jiang, S., Wang, J., Fan, S., Wen, W., Huang, G., Liu, B., Zhang, B., Yu, D., 2014. Influence of photosynthetic bacteria on breeding of *Holothuria scabra* larvae. South China Fish. Sci. 10, 78–82.

Jiang, S., Zhou, F., Mo, X., Huang, J., Yang, Q., 2017. Polyculture of sea cucumber *Holothuria scabra* with Pacific white shrimp *Litopenaeus vannamei*. Isr. J. Aquac. 69, 1–8.

Jimmy, R.A., Pickering, T.D., Hair, C.A., 2012. Overview of sea cucumber aquaculture and stocking research in the Western Pacific region. In: Hair, C.A., Pickering, T.D., Mills, D.J. (Eds.), *Asia-Pacific Tropical Sea Cucumber Aquaculture. Proceedings of an International Symposium Held in Noumea, New Caledonia. ACIAR Proceedings*, pp. 12–21.

Jones, S., Mahadevan, S., 1965. Notes on animal associations. 5. The pea crab *Pinnotheres deccanensis* Chopra inside the respiratory tree of the sea cucumber, *Holothuria scabra* Jager. J. Mar. Biol. Assoc. India 7, 377–380.

Juinio-Meñez, M.A., de Peralta, G.M., Dumalan, R.J.P., Edullantes, C.M., Catbagan, T.O., 2012a. Ocean nursery systems for scaling up juvenile sandfish (*Holothuria scabra*) production: ensuring opportunities for small fishers. In: Hair, C.A., Pickering, T.D., Mills, D.J. (Eds.), *Asia-Pacific Tropical Sea Cucumber Aquaculture. Proceedings of an International Symposium Held in Noumea, New Caledonia. ACIAR Proceedings*, pp. 57–62.

Juinio-Meñez, M.A., Paña, M.A., de Peralta, G.M., Catbagan, T.O., Olavides, R.D.D., Edullantes, C.M.A., Rodriguez, B.D.D., 2012b. Establishment and Management of Communal Sandfish (*Holothuria scabra*) Sea Ranching in the Philippines. Citeseer, pp. 121–127.

Juinio-Meñez, M.A., Evangelio, J.C., Olavides, R.D., Paña, M.A.S., De Peralta, G.M., Edullantes, C.M.A., Rodriguez, B.D.R., Casilagan, I.L.N., 2013. Population dynamics of cultured *Holothuria scabra* in a sea ranch: implications for stock restoration. Rev. Fish. Sci. 21, 424–432.

Juinio-Meñez, M.A., Evangelio, J.C., Miralao, S.J.A., 2014. Trial grow-out culture of sea cucumber *Holothuria scabra* in sea cages and pens. Aquacult. Res. 45, 1332–1340.

Juinio-Meñez, M.A., Tech, E.D., Ticao, I.P., Gorospe, J.R.C., Edullantes, C.M.A., Rioja, R.A.V., 2017. Adaptive and integrated culture production systems for the tropical sea cucumber *Holothuria scabra*. Fish. Res. 186, 502–513.

Junus, S., Kwong, P.J., Khoo, G., 2018. A review on the recent advances in the biology and aquaculture technology of *Holothuria scabra*. Surv. Fish. Sci. 4, 5–25.

Kamarudin, K.R., Rehan, M.M., Bahaman, N.A., 2017. Morphological and molecular identifcation of sea cucumber species *Holothuria scabra*, *Stichopus horrens* and *Stichopus ocellatus* from Kudat, Sabah, Malaysia. Pertanika J. Trop. Agric. Sci. 40, 161–172.

Kamarudin, K.R., Rehan, M.M., Rehan, A.M., 2018. Species Identification and molecular phylogenetics of processed sea cucumbers from Malaysian market based on 12S mitochondrial rRNA gene. Pertanika J. Trop. Agric. Sci. 41, 1833–1851.

Kamarudin, K.R., Rehan, M.M., Rehan, A.M., 2019. Species identification of processed sea cucumbers from Malaysian market based on concatenated gene sequences of mitochondrial rRNA genes. Marit. Tech. Res. 1, 28–44.

Kamyab, E., Kühnhold, H., Novais, S.C., Alves, L.M.F., Indriana, L., Kunzmann, A., Slater, M., Lemos, M.F.L., 2017. Effects of thermal stress on the immune and oxidative stress responses of juvenile sea cucumber *Holothuria scabra*. J. Comp. Physiol. B 187, 51–61.

Karnila, R., Astawan, M., Sukarno, S., Wresdiyati, T., 2011. Karakteristik konsentrat protein teripang pasir (*Holothuria scabra* J.) dengan bahan pengekstrak aseton. JPK 16, 90–102.

Kato, S., Tsurumaru, S., Taga, M., Yamane, T., Shibata, Y., Ohno, K., Fujiwara, A., Yamano, K., Yoshikuni, M., 2009. Neuronal peptides induce oocyte maturation and gamete spawning of sea cucumber, *Apostichopus japonicus*. Dev. Biol. 326, 169–176.

Katow, H., Katow, T., Moriyama, A., 2004. Gonad-stimulating substance-like molecule from the radial nerve of the sea cucumber. Int. J. Dev. Biol. 53, 483–491.

Kerr, A.M., 1994. Shallow-water holothuroids (Echinodermata) of Kosrae, Eastern Caroline Islands. Pac. Sci. 48, 161–174.

Kharisma, H., 2012. Pengaruh ekstrak air teripang pasir (*Holothuria scabra*) terhadap kolesterol total pada tikus hiperlipidemia. PhD Thesis, Universitas Muhammadiyah Surakarta, Indonesia.

Kim, S.K., Himaya, S., 2012. Triterpene glycosides from sea cucumbers and their biological activities. In: Advances in Food and Nutrition Research. Elsevier, pp. 297–319.

Kinch, J., 2002. Overview of the beche-de-mer fishery in Milne Bay Province, Papua New Guinea. SPC Beche-de-mer Bull. 17, 2–16.

Kinch, J., Purcell, S., Uthicke, S., Friedman, K., 2008a. Papua New Guinea: a hotspot of sea cucumber fisheries in the Western Central Pacific. In: Toral-Granda, V., Lovatelli, A., Vasconcellos, M. (Eds.), Sea Cucumbers. A Global Review of Fisheries and Trade. FAO Fisheries and Aquaculture Technical Paper 516, FAO, Rome, pp. 57–77.

Kinch, J., Purcell, S., Uthicke, S., Friedman, K., 2008b. Population status, fisheries and trade of sea cucumbers in the Western Central Pacific. In: Toral-Granda, V., Lovatelli, A., Vasconcellos, M. (Eds.), Sea Cucumbers. A Global Review of Fisheries and Trade. FAO Fisheries and Aquaculture Technical Paper 516, FAO, Rome, pp. 7–55.

Kithakeni, T., Ndaro, S.G.M., 2002. Some aspects of sea cucumber, *Holothuria scabra* (Jaeger, 1935), along the coast of Dar es Salaam. West. Indian Ocean J. Mar. Sci. 1, 163–168.

Knauer, J., 2011. Growth and survival of larval sandfish, *Holothuria scabra* (Echinodermata: Holothuroidea), fed different microalgae. J. World Aquacult. Soc. 42, 880–887.

Kobayashi, M., Hori, M., Kan, K., Yasuzawa, T., Matsui, M., Suzuki, S., Kitagawa, I., 1991. Marine natural products. 27. Distribution of lanostane-type triterpene oligoglycosides in ten kinds of Okinawan sea cucumbers. Chem. Pharm. Bull. 39, 2282–2287.

Kodama, M., Sumbing, J.G., Lebata-Ramos, M.J.H., Watanabe, S., 2015. Metabolic rate characteristics and sediment cleaning potential of the tropical sea cucumber *Holothuria scabra*. Jpn. Agric. Res. Q. 49, 79–84.

Krishnan, S., 1967. Biochemical and cytochemical observations of the nucleic acids in the gonads of *Holothuria scabra* Jäger. Acta Biol. Med. (Gdánsk) 11, 307–313.

Krishnan, S., 1968. Histochemical studies on reproductive and nutritional cycles of the holothurian, *Holothuria scabra*. Mar. Biol. 2, 54–65.

Krishnan, S., 1970. Studies on Reproductive and Nutritional Cycles of the *Holothuria scabra* Jaeger. PhD Thesis, University of Madras, India.

Krishnan, S., 1971. Autoradiographic studies on sugar transport in the sea cucumber *Holothuria scabra*. Mar. Biol. 10, 189–191.

Krishnan, S., Krishnaswamy, S., 1970. Studies on the transport of sugars in the holothurian *Holothuria scabra*. Mar. Biol. 5, 303–306.

Krishnan, S., Krishnaswamy, S., 1989. Studies on reproductive and nutritional cycles of the holothurian *Holothuria* (*Metriatyla*) *scabra* Jaeger. In: C. M. F. R. Institute (Ed.), National Workshop on beche-de-mer at Mandapam Camp, Cochin, India.

Krishnaswamy, S., Krishnan, S., 1967. A report on the reproductive cycle of the holothurian *Holothuria scabra* Jaeger. Curr. Sci. 36, 155–156.

Kroll, A., Jangoux, M., 1989. Les grégarines (Sporozoa) et les umagillides (Turbellaria) parasites du coelome et du système hémal de l'holothurie *Holothuria tubulosa*, Gmelin (Echinodermata). In: Vie marine. Hors série, 10, pp. 193–204.

Kühnhold, H., 2017. Temperature Tolerance of the Sea Cucumber *Holothuria scabra*: Towards a Systematic Understanding of Multi-Level Temperature Effects. PhD Thesis, Universität Bremen, Germany.

Kühnhold, H., Kamyab, E., Novais, S., Indriana, L., Kunzmann, A., Slater, M., Lemos, M., 2017. Thermal stress effects on energy resource allocation and oxygen consumption rate in the juvenile sea cucumber, *Holothuria scabra* (Jaeger, 1833). Aquaculture 467, 109–117.

Kühnhold, H., Steinmann, N., Huang, Y.-H., Indriana, L., Meyer, A., Kunzmann, A., 2019a. Temperature-induced aerobic scope and Hsp70 expression in the sea cucumber *Holothuria scabra*. PLoS One 14, e0214373.

Kühnhold, H., Novais, S.C., Alves, L.M.F., Kamyab, E., Lemos, M.F.L., Slater, M.J., Kunzmann, A., 2019b. Acclimation capability inferred by metabolic performance in two sea cucumber species from different latitudes. J. Therm. Biol. 84, 407–413.

Kumara, A., Dissanayake, C., 2017. Preliminary study on broodstock rearing, induced breeding and grow-out culture of the sea cucumber *Holothuria scabra* in Sri Lanka. Aquacult. Res. 48, 1058–1069.

Kumara, P., Cumaranathunga, P.R.T., Linden, O., 2005. Present status of the sea cucumber fishery in southern Sri Lanka: a resource depleted industry. SPC Beche-de-mer Inf. Bull. 22, 24–29.

Kumara, P.A., Jayanatha, J., Pushpakumara, J., Bandara, W., Dissanayake, D.C.T., 2013. Artificial breeding and larval rearing of three tropical sea cucumber species–*Holothuria scabra*, *Pseudocolochirus violaceus* and *Colochirus quadrangularis*–in Sri Lanka. SPC Beche-de-mer Inf. Bull. 33, 30–37.

Lampe-Ramdoo, K., Pillay, R.M., Conand, C., 2014. An assessment of holothurian diversity, abundance and distribution in the shallow lagoons of Mauritius. SPC Beche-de-mer Inf. Bull. 34, 17–24.

Lanchester, W., 1900. On a collection of crustaceans made at Singapore and Malacca. Part 1. Crustacea Brachyura. In: Proceedings of the Zoological Society of London, pp. 44–47.

Lantang, D., Paulangan, Y.P., 2015. Potensi Pengembangan Komoditas Teripang Pasir (*Holothuria scabra* Jaenger) secara Berkelanjutan dengan Model Integrasi Sea Farming di Kepulauan Padaido Kabupaten Biak Numfor. Jurnal Biologi Papua 7, 37–46.

Lavitra, T., Rasolofonirina, R., Jangoux, M., Eeckhaut, I., 2008. Effect of food quality and settling density on growth and survival of epibiotic juveniles of the sea cucumber *Holothuria scabra*. SPC Beche-de-mer Inf. Bull. 27, 49.

Lavitra, T., Rasolofonirina, R., Jangoux, M., Eeckhaut, I., 2009a. Problems related to the farming of *Holothuria scabra* (Jaeger, 1833). SPC Beche-de-mer Inf. Bull. 29, 20–30.

Lavitra, T., Rasolofonirina, R., Jangoux, M., Grosjean, P., Eeckhaut, I., 2009b. Effect of food quality and settling density on growth and survival of epibiotic juveniles of the sea cucumber *Holothuria scabra*. West. Indian Ocean J. Mar. Sci. 8, 1–10.

Lavitra, T., Rasolofonirina, R., Eeckhaut, I., 2010a. The effect of sediment quality and stocking density on the survival and growth of the sea cucumber *Holothuria scabra* reared in nursery ponds and in sea pens. West. Indian Ocean J. Mar. Sci. 9, 152–165.

Lavitra, T., Fohy, N., Gestin, P.-G., Rasolofonirina, R., Eeckhaut, I., 2010b. Effect of water temperature on the survival and growth of endobenthic *Holothuria scabra* (Echinodermata: Holothuroidea) juveniles reared in outdoor ponds. SPC Beche-de-mer Inf. Bull. 30, 25–28.

Lavitra, T., Tsiresy, G., Rasolofonirina, R., Eeckhaut, I., 2015. Effect of nurseries and size of released *Holothuria scabra* juveniles on their survival and growth. SPC Beche-de-mer Inf. Bull. 35, 37–41.

Lawrence, A.J., Ahmed, M., Hanafy, M., Gabr, H., Ibrahim, A., Gab-Alla, A., 2004. Status of the sea cucumber fishery in the Red Sea-the Egyptian experience. In: Lovatelli, A., Conand, C., Purcell, S.W., Uthicke, S., Hamel, J.-F., Mercier, A. (Eds.), Advances in Sea Cucumber Aquaculture and Management. FAO Fisheries Technical Paper No. 463, FAO, Rome, pp. 79–90.

Lawrence, A.J., Ahmed, M., Hanafy, M., Gabr, H., Ibrahim, A., Gab-Alla, A., 2006. Status of the sea cucumber fishery along the Egyptian Red Sea Coast. In: Bruckner, A.W. (Ed.), *Proceedings of the CITES Workshop on the Conservation of Sea Cucumbers in the Families Holothuridae and Stichopodidae*. Technical memorandum 44, National Oceanographic and Atmospheric Administration, National Marine Fisheries Service Office of Protected Resources, Silver Springs, Maryland, USA, pp. 205–206.

Laxminarayana, A., 2005. Induced spawning and larval rearing of the sea cucumbers, Bohadschia marmorata and Holothuria atra in Mauritius. SPC Beche-de-mer Inf. Bull. 22, 48–52.

Lebata-Ramos, M.J.H., Solis, E.F.D., Sibonga, R.C., Watanabe, S., 2012. Co-culture trials of sandfish *Holothuria scabra* and black tiger shrimp *Penaeus monodon* in mangroves. JIRCAS Working Report, pp. 87–95.

Lee, S., 2016. Sedimentary Response to Sea Cucumber (*Holothuria scabra*) Removal: Insights From Experimental Manipulations on a Fijian Reef Flat. MSc thesis, University of Bremen.

Lee, S., Ford, A.K., Mangubhai, S., Wild, C., Ferse, S.C.A., 2018a. Length-weight relationship, movement rates, and in situ spawning observations of *Holothuria scabra* (sandfish) in Fiji. SPC Beche-de-mer Inf. Bull. 38, 11–14.

Lee, S., Ford, A.K., Mangubhai, S., Wild, C., Ferse, S.C.A., 2018b. Effects of sandfish (*Holothuria scabra*) removal on shallow-water sediments in Fiji. PeerJ 6, e4773.

Leonardo, L.R., Cowan, M.E., 1984. Shallow-Water Holothurians of the Calatagan, Batangas. University of the Phillipines, Marine Sciences Center, Philippines.

Léonet, A., 2010. Caractérisation d'un agent induisant la maturation ovocytaire chez *Holothuria scabra* (Holothuroidea), espèce à haute valeur commerciale. PhD Thesis, University of Mons, Belgium.

Léonet, A., Rasolofonirina, R., Wattiez, R., Jangoux, M., Eeckhaut, I., 2009. A new method to induce oocyte maturation in holothuroids (Echinodermata). Invertebr. Reprod. Dev. 53, 13–21.

Léonet, A., Delroisse, J., Schuddinck, C., Wattiez, R., Jangoux, M., Eeckhaut, I., 2019. Thioredoxins induce oocyte maturation in holothuroids (Echinodermata). Aquaculture 510, 293–301.

Léopold, M., Govinden, R., 2018. SEACUSEY: co-management of the sea cucumber fishery in the Seychelles (2017–2018). SPC Beche-de-mer Inf. Bull. 38, 85–87.

Léopold, M., Cornuet, N., Andréfouët, S., Moenteapo, Z., Duvauchelle, C., Raubani, J., Ham, J., Dumas, P., 2013a. Comanaging small-scale sea cucumber fisheries in New Caledonia and Vanuatu using stock biomass estimates to set spatial catch quotas. Environ. Conserv. 40, 367–379.

Léopold, M., Ham, J., Kaku, R., Kaltavara, J., Raubani, J., Gereva, S., Moenteapo, Z., Andréfouët, S., Dumas, P., 2013b. Towards a new management strategy for Pacific island sea cucumber fisheries. SPC Fish. Newsl. 140, 43–48.

Léopold, M., Ham, J., Kaku, R., Gereva, S., Raubani, J., Moenteapo, Z., 2015. Spatial sea cucumber management in Vanuatu and New Caledonia. SPC Beche-de-mer Inf. Bull. 35, 3–18.

Levin, V.S., Ho, D.T., 1989. Holothurians of the upper sublittoral zone of the coastal waters of Phukhan Province (southern Vietnam). In: Zhirmunsky, A.V., Phan, L.T. (Eds.), Biology of the Coastal Waters of Vietnam. Benthic Invertebrates of Southern Vietnam. The Academy of Sciences of the USSR, Vladivostok, pp. 54–60.
Li, H., Qiao, G., Gu, J.-Q., Zhou, W., Li, Q., Woo, S.-H., Xu, D.-H., Park, S.-I., 2010. Phenotypic and genetic characterization of bacteria isolated from diseased cultured sea cucumber *Apostichopus japonicus* in northeastern China. Dis. Aquat. Organ. 91, 223–235.
Levin, V.S., Kalinin, V.I., Stonik, V.A., 1984. Chemical characters and taxonomic revision of holothurian *Bohadschia graeffei* (Semper) as refer to elucidation of a new genus. Biol. Morya 3, 33–38.
Li, Z.B., Dai, G., Shangguan, J.B., Ning, Y.F., Li, Y.Y., Chen, R.B., Huang, Y.S., Yuan, Y., 2015. Isolation and characterization of polymorphic microsatellite loci in the sea cucumber *Holothuria scabra*. Genet. Mol. Res. 14, 6529–6532.
Li, Y., Li, M., Xu, B., Li, Z., Qi, Y., Song, Z., Zhao, Q., Du, B., Yang, Y., 2021. The current status and future perspective in combination of the processing technologies of sulfated polysaccharides from sea cucumbers: a comprehensive review. J. Funct. Foods 87, 104744.
Liao, Y., 1980. The aspidochirote holothurians of China with erection of a new genus in echinoderms: present and past. In: Lawrence, J.M., Jangoux, M. (Eds.), Echinoderms: Present and Past, Proceedings of the European Colloquium on Echinoderms. AA Balkema, Rotterdam, pp. 115–120.
Liao, Y., 1984. The aspidochirote holothurians of China. Stud. Mar. Sin. 23, 221–247.
Liao, T.F., 1997. Fauna Sinica: Phylum Echinodermata: Class Holothuroidea. Science Press, Beijing.
Lin, D., 2015. Holothurian Host Size and Its Effects on Symbiont Colonization. Unpublished Work-Moorea UCB 2015-student papers, p. 7.
Liu, H.-H., Ko, W.-C., Hu, M.-L., 2002. Hypolipidemic effect of glycosaminoglycans from the sea cucumber *Metriatyla scabra* in rats fed a cholesterol-supplemented diet. J. Agric. Food Chem. 50, 3602–3606.
Liu, H., Zheng, F., Sun, X., Hong, X., Dong, S., Wang, B., Tang, X., Wang, Y., 2010. Identification of the pathogens associated with skin ulceration and peristome tumescence in cultured sea cucumbers *Apostichopus japonicus* (Selenka). J. Invertebr. Pathol. 105, 236–242.
Liu, Z.-M., Ma, Y.-X., Yang, Z.-P., Li, M., Liu, J., Bao, P.-Y., 2012. Immune responses and disease resistance of the juvenile sea cucumber *Apostichopus japonicus* induced by *Metschnikowia* sp. C14. Aquaculture 368, 10–18.
Lokani, P., 1990. Beche-de-mer research and development in Papua New Guinea. SPC Beche-de-mer Inf. Bull. 1, 8–11.
Lokani, P., 1995. Fisheries Dynamics, Ecology and Management of Beche-de-mer at the Warrior Reef, Torres Strait, Papua New Guinea. MSc Thesis, James Cook University, Australia.
Lokani, P., 1996. Illegal fishing for sea-cucumber (beche-de-mer) by Papua New Guinea artisanal fishermen in the Torres Strait protected zone. SPC Beche-de-mer Inf. Bull. 8, 2–6.
Lokani, P., Dalzell, P., Adams, T.J.H., 1995a. Illegal fishing for sea cucumber (beche-de-mer) by Papua New Guinea artisanal fishermen in the Torres Strait protected zone. In: South Pacific Commission and Forum Fisheries Agency Workshop on the Management of South Pacific Inshore Fisheries, Noumea, New Caledonia, pp. 279–289.
Lokani, P., Polon, P., Lari, R., 1995b. Fisheries and management of bêche-de-mer fisheries in Western Province of Papua New Guinea. In: SPC Integrated Coastal Fisheries Management Project Technical Document. vol. 11, pp. 267–275.

Lokani, P., Polon, P., Lari, R., Dalzell, P., Adams, T.J.H., 1995c. Fisheries and Management of beche-de-mer Fisheries in Western Province of Papua New Guinea. In: South Pacific Commission.

Lokani, P., Polon, P., Lari, R., 1996. Management of beche-de-mer fisheries in the Western Province of Papua New Guinea. SPC Beche-de-mer Inf. Bull. 8, 7–11.

Long, B., Skewes, T., 1997. Distribution and abundance of beche-de-mer on Torres Strait reefs. SPC Beche-de-mer Inf. Bull. 9, 17–22.

Long, B., Skewes, T., Dennis, D., Poiner, I., Pitcher, C., Taranto, T., Manson, F., Polon, F., Karre, B., Evans, C., 1996. Distribution and abundance of beche-de-mer on Torres Strait reefs. Final Report to the Queensland Fisheries Management Authority.

Lovatelli, A., Conand, C., Purcell, S., Uthicke, S., Hamel, J.-F., Mercier, A. (Eds.), 2004. Advances in sea cucumber aquaculture and management. In: FAO Fisheries Technical Paper 463. FAO, Rome.

Ma, Y., Xu, G., Chang, Y., Zhang, E.P., Zhou, W., Song, L.S., 2006. Bacterial pathogens of skin ulceration disease in cultured sea cucumber *Apostichopus japonicus* (Selenka) juveniles. J. Dalian Fish. Univ. 21, 13–18.

Macnae, W., Kalk, M., 1962. The Fauna and Flora of Sand Flats at Inhaca Island, Moçambique. J. Anim. Ecol. 31, 93–128.

Madeali, M.I., Tangko, A.M., Pantai, D.E.R., Maros, B.P.P.B., 1993. Polyculture of sea cucumber, *Holothuria scabra* and seaweed, *Eucheuma cottoni* in Battoa waters, Polmas Regency, South Sulawesi. In: Prosiding Seminar Hasil Penelitian, vol. 11, pp. 105–109.

Mageswaran, R., Balakrishnan, V., 1985. Chemical constituents of some species of holothurians found in Sri Lanka. J. Natl. Sci. Found. 13, 115–130.

Magno-Tan, M.J., Coronado, A.S., Garcia, R.R., Maulani, M.D., 2016. Weight measurement of *Holothuria scabra* Jaeger, 1833 utilizing the surface area of digitized image captured under water. In: Proceedings of the Fifth International Conference on Network, Communication and Computing, pp. 1–5.

Mahendran, M., Abraham, T.W., Krishnarajah, S.R., Elanganayagam, P., 1983. A comparative study of the glycoside fractions of some holothurians found in Sri Lankan waters. J. Natl. Sci. Counc. Sri Lanka 11, 185–190.

Mahishalini, V., Athula, J.M., Nirooparaj, B., 2019. Growth Performance of *Holothuria scabra* (sand fish) with different stocking density in open sea pen culture in Jaffna, Sri Lanka. In: Uva Wellassa University of Sri Lanka, International Research Conference 2019.

Maier, M.S., 2008. Biological activities of sulfated glycosides from echinoderms. In: Studies in Natural Products Chemistry. Elsevier, pp. 311–354.

Makatutu, D., Yunus, A.T., Rusdi, I., 1993. Study on the larval growth and survival of the sea cucumber *Holothuria scabra* fed different natural feeds. Journal Penelitian Budidaya Patain 9, 97–102.

Mallawaarachchi, M., Arulnanthan, K., 2014. A Preliminary Study on Potentiality of Captive Breeding of *Holothuria scabra*. National Aquatic Resources Research and Development Agency, Colombo.

Mangubhai, S., Nand, Y., Ram, R., Fox, M., Tabunakawai-Vakalalabure, M., Vodivodi, T., 2017. Value-chain analysis of the wild-caught sea cucumber fishery. In: Fiji's Sea Cucumber Fishery: Advances in Science for Improved Management. Wildlife Conservation Society, *Suva, Fiji*, pp. 23–29.

Manikandan, K.P., 2000. From KP Manikandan. SPC Beche-de-mer Inf. Bull. 13, 33–34.

Mantel, N., 1967. The detection of disease clustering and a generalized regression approach. Cancer Res. 27, 209–220.

Marizal, D., Jaya, Y.V., Irawan, H., 2012. Aplikasi SIG Untuk Kesesuaian Kawasan Budidaya Teripang *Holothuria Scabra* Dengan Metode Penculture Di Pulau Mantang, Kecamatan Mantang, Kabupaten Bintan. In: Repository UMRAH, pp. 1–8. https://jurnal.umrah.ac.id/wp-content/uploads/2013/08/Dendi-M-070210450051.pdf.

Markle, D.F., Olney, J.E., 1990. Systematics of the pearlfishes (Pisces: Carapidae). Bull. Mar. Sci. 47, 269–410.
Marsh, L.M., 1994. Echinoderms of the Cocos (Keeling) Islands. Atoll Res. Bull. 411, 1–12.
Marsh, L., Vail, L., Hoggett, A., Rowe, F., 1993. Echinoderms of Ashmore Reef and Cartier Island. *Marine faunal surveys of Ashmore Reef and Cartier Island, North-western Australia. Records of the Western.* Aust. Mus. Suppl. 44, 53–65.
Martens, E., De Clerck, G., 1994. Interstitial and parasitic Platyhelminthes from the coast of the Seychelles. In: Oceanic Reefs of the Seychelles, Netherlands Indian Ocean programme, Cruise Reports 2, pp. 97–106.
Martin, D., Britayev, T., 1998. Symbiotic polychaetes: review of known species. Oceanogr. Mar. Biol. Annu. Rev. 36, 217–340.
Martin, J.W., Crandall, K.A., Felder, D.L., 2016. Decapod Crustacean Phylogenetics. CRC Press.
Mary Bai, M., 1971a. Regeneration in the holothurian, *Holothuria scabra* Jager. Indian J. Exp. Biol. 9, 467–471.
Mary Bai, M., 1971b. Studies on Regeneration in the Holothurian *Holothuria (Metriatyla) scabra* Jaeger. PhD Thesis, University of Madurai, India.
Mary Bai, M., 1978. The anatomy and histology of the digestive system of *Holothuria scabra* Jaeger. J. Mar. Biol. Ass. India 20, 22–31.
Mary Bai, M., 1980. Monograph on *Holothuria (Metriatyla) scabra* Jaeger. In: Memoirs of the Zoological Survey of India, vol. 16, pp. 1–75.
Mary Bai, M., 1994. Studies on regeneration in the holothurian *Holothuria (Metriatyla) scabra* Jaeger. Bull. Cent. Mar. Fish. Res. Inst. 46, 44–50.
Mashjoor, S., Yousefzadi, M., 2017. Holothurians antifungal and antibacterial activity to human pathogens in the Persian Gulf. J. Mycol. Méd. 27, 46–56.
Mashjoor, S., Yousefzadi, M., 2019. Cytotoxic effects of three Persian Gulf species of Holothurians. Iran. J. Vet. Res. 20, 19.
Mashjoor, S., Yousefzadi, M., Pishevarzad, F., 2018. *In vitro* biological activities of holothurians edible sea cucumbers in the Persian Gulf. Indian J. Geo-Mar. Sci. 47, 1518–1526.
Mashjoor, S., Yousefzadi, M., Pishevarzad, F., 2019. Assessment of anticancer potential of selected *Holothuria* species. Indian J. Tradit. Knowl. 18, 272–280.
Massin, C., 1982. Effects of feeding on the environment: Holothuroidea. In: Jangoux, M., Lawrence, J.M. (Eds.), Echinoderm Nutrition. AA Balkema, Rotterdam, pp. 493–497.
Massin, C., 1996. Results of the Rumphius Biohistorical Expedition to Ambon (1990) Part. 4. The Holothurioidea (Echinodermata) collected at Ambon during the Rumphius Biohistorical Expedition. Zool. Verh. 307, 1–53.
Massin, C., 1999. Reef-dwelling Holothuroidea (Echinodermata) of the Spermonde archipelago (South-West Sulawesi, Indonesia). Zool. Verh. 329, 1–144.
Massin, C., Mercier, A., Hamel, J.-F., 2000. Ossicle change in *Holothuria scabra* with a discussion of ossicle evolution within the Holothuriidae (Echinodermata). Acta Zool. 81, 77–91.
Massin, C., Uthicke, S., Purcell, S.W., Rowe, F.W., Samyn, Y., 2009. Taxonomy of the heavily exploited Indo-Pacific sandfish complex (Echinodermata: Holothuriidae). Zool. J. Linn. Soc. 155, 40–59.
Mathieu-Resuge, M., Le Grand, F., Schaal, G., Kraffe, E., Lorrain, A., Letourneur, Y., Lemonnier, H., Benoît, J., Hochard, S., 2020. Assimilation of shrimp farm sediment by *Holothuria scabra*: a coupled fatty acid and stable isotope approach. Aquat. Living Resour. 33, 3.
Matthes, H., 1983. People's Democratic Republic of Yemen. Beche-de-mer resources of PDRY. A report prepared for the Beche-de-mer Fishery Development Project. Report.
Mazlan, N., Hashim, R., 2015. Spawning induction and larval rearing of the sea cucumber *Holothuria scabra* in Malaysia. SPC Beche-de-mer Inf. Bull. 35, 32–36.

McClanahan, T.R., 1988. Seasonality in East Africa's coastal waters. Mar. Ecol. Prog. Ser. 44, 191–199.
McElroy, S., 1990. Beche-de-mer species of commercial value-an update. SPC Beche-de-mer Inf. Bull. 2, 2–7.
McVean, A.R., Hemery, G., Walker, R.C.J., Ralisaona, B.L.R., Fanning, E., 2005. Traditional sea cucumber fisheries in southwest Madagascar: a case-study of two villages in 2002. SPC Beche-de-mer Inf. Bull. 21, 15–18.
Medagedara, C.B., Fahim, M.S.M., Pushpakumara, J., Jayatissa, A.J., Pad, A.K., 2019. Survival and Growth of the Sea Cucumber *Holothuria scabra* Farmed in Two Selected Sites in Puttalam Lagoon, Sri Lanka. National Aquatic Resources Research and Development Agency (NARA), Scientific Sessions.
Meirinawati, H., Prayitno, H.B., Indriana, L.F., Firdaus, M., 2020. Water quality assessment and monitoring of closed rearing system of the sea cucumber *Holothuria scabra*. ASEAN Journal on Science and Technology for Development 37, 73–80.
Mercier, A., Hamel, J.-F., 1997. Finding solutions for beche-de-mer industry. Island Business Magazine (September), 44–47.
Mercier, A., Hamel, J.-F., 2009. Endogenous and exogenous control of gametogenesis and spawning in echinoderms. Adv. Mar. Biol. 55, 1–302.
Mercier, A., Hamel, J.-F., 2013. Sea cucumber aquaculture: hatchery production, juvenile growth and industry challenges. In: Allan, G., Burnell, G. (Eds.), Advances in Aquaculture Hatchery Technology. Woodhead Publishing Ltd, Cambridge, pp. 431–454.
Mercier, A., Battaglene, S.C., Hamel, J.-F., 1999. Daily burrowing cycle and feeding activity of juvenile sea cucumbers *Holothuria scabra* in response to environmental factors. J. Exp. Mar. Biol. Ecol. 239, 125–156.
Mercier, A., Battaglene, S.C., Hamel, J.-F., 2000a. Settlement preferences and early migration of the tropical sea cucumber *Holothuria scabra*. J. Exp. Mar. Biol. Ecol. 249, 89–110.
Mercier, A., Battaglene, S.C., Hamel, J.-F., 2000b. Periodic movement, recruitment and size-related distribution of the sea cucumber *Holothuria scabra* in Solomon Islands. Hydrobiologia 440, 81–100.
Mercier, A., Ycaza, R.H., Hamel, J.-F., 2007. Long-term study of gamete release in a broadcast-spawning holothurian: predictable lunar and diel periodicities. Mar. Ecol. Prog. Ser. 329, 179–189.
Mgaya, Y., Mmbaga, T., 2007. Sea cucumbers in Tanzania. In: Conand, C., Muthiga, N. (Eds.), Commercial Sea Cucumbers: A Review for the Western Indian Ocean. WIOMSA Book Series No. 5, WIOMSA, pp. 53–59.
Militz, T.A., Leini, E., Duy, N.D.Q., Southgate, P.C., 2018. Successful large-scale hatchery culture of sandfish (*Holothuria scabra*) using micro-algae concentrates as a larval food source. Aquacult. Rep 9, 25–30.
Mills, D., Duy, N.D.Q., Johnson, W., 2008. Review of sandfish pond-culture progress in Vietnam. FIS/2007/117, ACIAR Final Project Report.
Mills, D.J., Duy, N.D.Q., Juinio-Meñez, M.A., Raison, C.M., Zarate, J.M., 2012. Overview of sea cucumber aquaculture and sea-ranching research in the South-East Asian region. In: Hair, C.A., Pickering, T.D., Mills, D.J. (Eds.), Asia-Pacific Tropical Sea Cucumber Aquaculture. *Proceedings of an International Symposium Held in Noumea, New Caledonia. ACIAR Proceedings.*
Minami, Y., Tamaki, S., 2013. Rearing experiments in land-based nursery tank, feeding artificial food to cultured juvenilesasha *Holothuria nobilis* and *H. scabra*. Aquacult. Sci. 61, 171–176.
Minami, Y., Yoshikuni, M., 2014. The reproductive cycle of the sea cucumber *Holothuria scabra* around Shioya Fishing Port, Okinawa Prefecture. Aquacult. Sci. 62, 235–242.

Mitu, S.A., Bose, U., Suwansa-ard, S., Turner, L.H., Zhao, M., Elizur, A., Ogbourne, S.M., Shaw, P.N., Cummins, S.F., 2017. Evidence for a saponin biosynthesis pathway in the body wall of the commercially significant sea cucumber *Holothuria scabra*. Mar. Drugs 15, 349.

Miyazaki, S., Ichiba, T., Reimer, J.D., Tanaka, J., 2014. Chemoattraction of the pearlfish *Encheliophis vermicularis* to the sea cucumber *Holothuria leucospilota*. Chemoecology 24, 121–126.

Mmbaga, T.K., 2002. Some Aspects of the Fishery, Ecology and the Reproductive Biology of *Holothuria Scabra* (Jaeger, 1935) Along the Coast of Dar es Salaam. PhD Thesis, University of Dar es Salaam.

Mmbaga, T.K., 2013. The effect of fishing on the ecology of sea cucumber (Holothuroidea: Echinodermata) *Holothuria scabra* and *Holothuria nobilis* in Tanzanian sea water. Int. J. Dev. Sustain. 2, 1099–1126.

Mmbaga, T.K., 2015. Sea cucumber fishery characteristics in Tanzania. J. Dev. Sustain. 4, 661–701.

Mmbaga, T., Mgaya, Y., 2004. Sea cucumber fishery in in Tanzania: identifying the gaps in resource inventory and management. In: Lovatelli, A., Conand, C., Purcell, S.W., Uthicke, S., Hamel, J.-F., Mercier, A. (Eds.), Advances in Sea Cucumber Aquaculture and Management. FAO Fisheries Technical Paper No. 463, FAO, Rome, pp. 193–204.

Mohammadizadeh, F., Ehsanpor, M., Afkhami, M., Mokhlesi, A., Khazaali, A., Montazeri, S., 2013. Evaluation of antibacterial, antifungal and cytotoxic effects of *Holothuria scabra* from the north coast of the Persian Gulf. J. Mycol. Méd. 23, 225–229.

Mohammadizadeh, M., Bastami, K.D., Ehsanpour, M., Afkhami, M., Mohammadizadeh, F., Esmaeilzadeh, M., 2016. Heavy metal accumulation in tissues of two sea cucumbers, *Holothuria leucospilota* and *Holothuria scabra* in the northern part of Qeshm Island, Persian Gulf. Mar. Pollut. Bull. 103, 354–359.

Moiyadeen, N.M., 1994. The biannual reproductive activity in *Holothuria (Metriatyla) scabra* (Jaeger, 1833)—the most abundant commercial holothuroid of the north-western coastal waters. In: *Proceedings of the First Annual Scientific Session NARA, Colombo, Sri Lanka, 2 November 1993*, pp. 123–129.

Mojica, E.-R.E., Merca, F.E., 2005a. Biological properties of lectin from sea cucumber (*Holothuria scabra* Jäger). J. Biol. Sci. 5, 472–477.

Mojica, E.-R.E., Merca, F.E., 2005b. Anodic stripping voltammetric determination of mercury (II) using lectin-modified carbon paste electrode. J. Appl. Sci. 5, 1461–1465.

Mondal, K., Kunzmann, A., 2018. Inferior assimilation of algae-based diets by sea cucumber *Holothuria scabra* under laboratory condition expressed by stable isotope mixing model. Annu. Res. Rev. Biol. 28, 1–8.

Moore, A., 1998. Preliminary notes on the exploitation of holothurians in the new Wakatobi Marine National Park, Sulawesi, Indonesia. SPC Beche-de-mer Inf. Bull. 10, 31–33.

Morgan, A.D., 1996. Sea cucumber, potential aquaculture in Queensland. Queensland Aquacult. News, 7.

Morgan, A.D., 1999. Husbandry and Spawning of the Sea Cucumber *Holothuria scabra* (Echinodermata: Holothuroidea). MSc Thesis, University of Queensland, Australia.

Morgan, A.D., 2000a. Aspects of the reproductive cycle of the sea cucumber *Holothuria scabra* (Echinodermata: Holothuroidea). Bull. Mar. Sci. 66, 47–57.

Morgan, A.D., 2000b. Induction of spawning in the sea cucumber *Holothuria scabra* (Echinodermata: Holothuroidea). J. World Aquacult. Soc. 31, 186–194.

Morgan, A.D., 2000c. Spawning: Holothurian reproductive behaviour and egg quality in culture. SPC Beche-de-mer Inf. Bull. 13, 36.

Morgan, A.D., 2001. The effect of food availability on early growth, development and survival of the sea cucumber *Holothuria scabra* (Echinodermata: Holothuroidea). SPC Beche-de-mer Inf. Bull. 14, 6–12.

Morgan, A.D., Schoppe, S., Samyn, Y., Lambeth, L., Levin, V.S., Gudimova, E.N., Battaglene, S., 2001. Aspects of sea cucumber broodstock management (Echinodermata: Holothuroidea). SPC Beche-de-mer Inf. Bull. 13, 2–8.

Mortensen, T., 1937. Contributions to the study of the development and larval forms of echinoderms III. D. Kgl. Danske Vidensk. Selsk. Skrifter, Naturv. og Math. Afd., 9. Raekke 7, pp. 1–65.

Mortensen, T., 1954. Echinoderms of Hong Kong. In: Hong Kong Naturalist Supplement, vol. 3, pp. 1–14.

Mosher, C., 1980. Distribution of *Holothuria arenicola* Semper in the Bahamas with observations on habitat, behaviour, and feeding activity (Echinodermata: Holothuroidea). Bull. Mar. Sci. 30, 1–12.

MPEDA, 1989. Sea Weed, Sea Urchin and Sea Cucumber: Handbook on Aquafarming. Marine Products Export Development Authority, Kochi, India.

Mukai, H., Koike, I., Nishihira, M., Nojima, S., 1990. Oxygen consumption and ammonium excretion of mega-sized benthic invertebrates in a tropical seagrass bed. J. Exp. Mar. Biol. Ecol. 134, 101–115.

Mukhopadhyay, S.K., 1988. On some holothurians from the Gulf of Mannar, India. Rec. Zool. Surv. India 85, 1–17.

Mulian, M., 1993. Effect of different supplemental feeds and stocking densities on the growth rate and survival of sea cucumber *Holothuria scabra* in Tallo river mouth, South Sulawest. Journal Penelitian Budidaya Pantai 9, 15–22.

Mulochau, T., Conand, C., Stöhr, S., Eléaume, M., Chabanet, P., 2014. First inventory of the echinoderms from Juan de Nova (Iles Eparses, France) in the Mozambique Channel, South Western Indian Ocean. West. Indian Ocean J. Mar. Sci. 13, 23–30.

Mulyani, L.F., Marsoedi, M., Guntur, G., 2017. An application of geographic information system to identify the suitability of sea cucumbers (*Holothuria scabra*) in West Lombok waters. J. Indones. Tour. Dev. Stud. 5, 155–160.

Mulyani, L.F., Marsoedi, M., Guntur, G., 2019. Identifying of land suitability sea cucumber (*Holothuria scabra*) using geographic information systems and eemote sensing based on physics parameters in West Lombok. J. Aquac. Dev. Environ. 2, 98–102.

Murdy, E., Cowan, M., 1980. Observations on the behavior and symbiotic relationship of the pearlfish *Encheliophis vermicularis* (Osteichthys: Carapidae). Kalikasan 9, 309–312.

Murniasih, T., Putra, M.Y., Pangestuti, R., 2015. Antioxidant capacities of *Holothuria* sea cucumbers. Ann. Bogor 19, 21–26.

Muthiga, N., Conand, C., 2014. Sea cucumbers, a poorly understood but important coastal resource: a regional analysis to improve management. In: WIOMSA Book Series, vol. 14. WIOMSA.

Muthiga, N.A., Kawaka, J.A., Ndirangu, S., 2009. The timing and reproductive output of the commercial sea cucumber *Holothuria scabra* on the Kenyan coast. Estuar. Coast. Shelf Sci. 84, 353–360.

Muthiga, N., Ochiewo, J., Kawaka, J., Al-Rashdi, K.M., Claereboudt, M.R., Dissanayake, D.C.T., Athukorala, S., Amarasiri, C., Ram, R., Friedman, K., 2010. Strengthening capacity to sustainably manage sea cucumber fisheries in the western Indian Ocean. SPC Beche-de-mer Inf. Bull. 30, 3–9.

Nair, M.R., Iyer, T.S.G., Gopakumar, K., 1994. Processing and quality requirements of beche-de-mer. In: Rengarajan, K., James, D.B. (Eds.), *Proceedings of the National Workshop on Beche-de-mer*. Central Marine Fisheries Research Institute, Indian Council of Agricultural Research, Cochin, India, pp. 76–78.

Namukose, M., Msuya, F.E., Ferse, S.C.A., Slater, M.J., Kunzmann, A., 2016. Growth performance of the sea cucumber *Holothuria scabra* and the seaweed *Eucheuma denticulatum*: integrated mariculture and effects on sediment organic characteristics. Aquac. Environ. Interact. 8, 179–189.

Nash, W., Ramofafia, C., 2006. Recent developments with the sea cucumber fishery in Solomon Islands. SPC Beche-de-mer Inf. Bull. 23, 3–4.

Natan, Y., Uneputty, P.A., Lewerissa, Y.A., Pattikawa, J.A., 2015. Species and size composition of sea cucumber in coastal waters of Un Bay, Southeast Maluku, Indonesia. Int. J. Fish. Aquat. Stud. 3, 251–256.

Navarro, P.G., García-Sanz, S., Conand, C., 2014. An Indonesian sea cucumber fishing village: the case of Pulau Misa. SPC Beche-de-mer Inf. Bull. 34, 14–16.

Ng, P., Manning, R., 2003. On two new genera of pea crabs parasitic in holothurians (Crustacea: Decapoda: Brachyura: Pinnotheridae) from the Indo-West Pacific, with notes on allied genera. Proc. Biol. Soc. Wash. 116 (4), 901–919.

Nievales, M.F., Sibonga, R., Figurado, H., 2015. Post-larval rearing strategies in sandish (*Holothuria scabra*) culture. In: Resource Enhancement and Sustainable Aquaculture Practices in Southeast Asia: Challenges in Responsible Production of Aquatic Species: Proceedings of the International Workshop on Resource Enhancement and Sustainable Aquaculture Practices in Southeast Asia 2014 (RESA). Aquaculture Department, Southeast Asian Fisheries Development Center, p. 309.

Nimah, S., Ma'ruf, W.F., Trianto, A., 2012. Uji bioaktivitas ekstrak teripang pasir (*Holothuria scabra*) terhadap bakteri *Pseudomonas aeruginosa* dan Bacillus cereus. JPBHP 1, 9–17.

Ninwichian, P., Klinbunga, S., 2020. Population genetics of sandfish (*Holothuria scabra*) in the Andaman Sea, Thailand inferred from 12S rDNA and microsatellite polymorphism. Reg. Stud. Mar. Sci. 35, 101189.

Nishanthan, G., de Croos, M., Prasada, D., Dissanayake, D., 2018. Effects of processing on proximate and fatty acid compositions of six commercial sea cucumber species of Sri Lanka. J. Food Sci. Technol. 55, 1933–1941.

Nishanthan, G., Kumara, A., Prasada, P., Dissanayake, C., 2019. Sea cucumber fishing pattern and the socio-economic characteristics of fisher communities in Sri Lanka. Aquat. Living Resour. 32, 12.

Nobsathian, S., Tuchinda, P., Sobhon, P., Tinikul, Y., Poljaroen, J., Tinikul, R., Sroyraya, M., Poomton, T., Chaichotranunt, S., 2017. An antioxidant activity of the whole body of *Holothuria scabra*. Chem. Biol. Technol. Agric. 4, 4.

Nontunha, N., Chaichotranunt, S., Poomtong, T., Tinikul, Y., Sobhon, P., Suphamungmee, W., 2020. Oocyte development and maturation in the sea cucumber, *Holothuria scabra*. Agric. Nat. Resour. 54, 491–498.

Northern Province Fisheries Division, 1993. Beche-de-mer harvesting in the Northern Province of New Caledonia. SPC Beche-de-mer Bull. 5, 7–8.

Notowinarto, Putro, D.H., 1992. Artificial spawning on sea cucumber (*Holothuria scabra*) by using environmental manipulation. Buletin Budidaya Laut (Indonesia) 4, 15–25.

Notowinarto, Putro, D.H., 1993. Artificial spawning on sea cucumber (*Holothuria scabra*) by using environmental manipulation. AGRIS 4, 1–8.

Nowland, S.J., Southgate, P.C., Basiita, R.K., Jerry, D.R., 2017. Elucidation of fine-scale genetic structure of sandfish (*Holothuria scabra*) populations in Papua New Guinea and northern Australia. Mar. Freshw. Res. 68, 1901–1911.

Oedjoe, M.D.R., 2017. Composition of nutritional content of sea cucumbers (Holothuroidea) in mania waters, Sabu Raijua Regency, East Nusa Tenggara. J. Aquac. Res. Dev. 8, 1–3.

Olavides, R.D.D., Edullantes, C.M.A., Menez, M.A.J., 2010. Assessment of the sea cucumber resource and fishery in the Bolinao-Anda reef system. Sci. Diliman 22, 1–12.

Olavides, R.D.D., Rodriguez, B.D.R., Juinio-Meñez, M.A., 2011. Simultaneous mass spawning of *Holothuria scabra* in sea ranching sites in Bolinao and Anda municipalities, Philippines. SPC Beche-de-mer Inf. Bull. 31, 23–24.

Omran, N.E.-S.E.-S., 2013. Nutritional value of some Egyptian sea cucumbers. Afr. J. Biotechnol. 12, 5466–5472.

Ong Che, R.G., Gomez, E.D., 1985. Reproductive periodicity of *Holothuria scabra* Jaeger at Calatagan, Batangas, Philippines. Asian Mar. Biol. 2, 21–30.

Ongkers, O.T., Pattinasarany, M., Mamesah, J., Uneputty, P.A., Pattikawa, J., 2018. Size distribution and growth pattern of *Holothuria atra* and *Holothuria scabra* in the coastal waters of Morella, Central Maluku Indonesia. Int. J. Fish. Aquat. Stud. 6, 301–305.

Ongkers, O.T.S., Pattinasarany, M., Mamesah, J.A.B., Uneputty, P., Pattikawa, J.A., Natan, Y., Tetelepta, J.M.S., 2020. Asexual reproduction experiment of sandfish (*Holothuria scabra*) on natural condition of Morella, central maluku Indonesia. Dent. Surv. 100, 100.

Ordoñez, J.F.F., Galindez, G.G.S.T., Gulay, K.T., Ravago-Gotanco, R., 2021. Transcriptome analysis of growth variation in early juvenile stage sandfish *Holothuria scabra*. Comp. Biochem. Physiol. Part D Genomics Proteomics 40, 100904.

Orozco, Z.G.A., Sumbing, J.G., Lebata-Ramos, M.J.H., Watanabe, S., 2014. Apparent digestibility coefficient of nutrients from shrimp, mussel, diatom and seaweed by juvenile *Holothuria scabra* Jaeger. Aquacult. Res. 45, 1153–1163.

Otero-Villanueva, M., Ngoc Ut, V., 2007. Sea cucumber fisheries around Phu Quoc Archipelago: a cross-border issue between South Vietnam and Cambodia. SPC Beche-de-mer Inf. Bull. 25, 32–36.

Özer, N.P., Mol, S., Varlık, C., 2004. Effect of the handling procedures on the chemical composition of sea cucumber. Turk. J. Fish. Aquat. Sci. 4, 71–74.

Padang, A., Lukman, E., Sangadji, M., Subiyanto, R., 2016. Pemeliharaan teripang pasir (*Holothuria scabra*) di kurungan tancap. Agrikan: Jurnal Agribisnis Perikanan 9, 11–18.

Padang, A., Sangadji, M., Lukman, E., Subiyanto, R., 2017. Pertumbuhan Dan Kelulusan Hidup Teripang Pasir (*Holothuria Scabra*) Yang Dipelihara Di Keramba Jaring Apung. TRITON: Jurnal Manajemen Sumberdaya Perairan 13, 115–124.

Pakoa, K., 2013. Standardising sea cucumber resource assessments in the Pacific Islands. SPC Fish. Newsl. 142, 14–17.

Pakoa, K., Lasi, F., Tardy, E., Friedman, K., 2009. The Status of Sea Cucumbers Exploited by Palau's Subsistence Fishery. Secretariat of the Pacific Community, Noumea, New Caledonia.

Palacios-Theil, E., Cuesta, J.A., Campos, E., Felder, D.L., 2009. Molecular genetic re-examination of subfamilies and polyphyly in the family Pinnotheridae (Crustacea: Decapoda). Decapod Crustacean Phylogenetics 18, 457–474.

Pangan, R.S., Yaptenco, K.F., Pardua, S.N., Duque, J.A.C., 2017. Development of an alternative technology package for processing sea cucumber. Philipp. J. Agric. Biosyst. Eng. 10, 16–29.

Panning, A., 1929. Die Gattung *Holothuria* (1. Teil). In: Mitteilungen aus dem Zoologischen Staut-sinstitut und Zoologischen Museum in Hamburg, vol. 44, pp. 92–138.

Panning, A., 1934a. Die Gattung *Holothuria* (2. Teil). In: Mitteilungen Aus dem Zoologischen Staut-sinstitut und Zoologischen Museum in Hamburg, vol. 45, pp. 24–50.

Panning, A., 1934b. Die Gattung *Holothuria* (3. Teil). In: Mitteilungen Aus dem Zoologischen Staut-sinstitut und Zoologischen Museum in Hamburg, vol. 45, pp. 65–84.

Panning, A., 1935. Die Gattung *Holothuria* (4. Teil). In: Mitteilungen Aus dem Zoologischen Staut-sinstitut und Zoologischen Museum in Hamburg, vol. 45, pp. 85–107.

Panning, A., 1944. Die Trepang Fischerei. In: Mitteilungen Aus dem Zoologischen Staut-sinstitut und Zoologischen Museum in Hamburg, vol. 49, pp. 1–76.

Parmentier, E., 2004. *Encheliophis chardewalli*: a new species of Carapidae (Ophidiiformes) from French Polynesia, with a redescription of *Encheliophis vermicularis*. Copeia 2004, 62–67.

Parmentier, E., Michel, L., 2013. Boundary lines in symbiosis forms. Symbiosis 60, 1–5.

Parmentier, E., Vandewalle, P., 2003. Morphological adaptations of pearlfish (Carapidae) to their various habitats. In: Fish Adaptations, pp. 261–276.

Parmentier, E., Gennotte, V., Focant, B., Goffinet, G., Vandewalle, P., 2003. Characterization of the primary sonic muscles in *Carapus acus* (Carapidae): a multidisciplinary approach. Proc. R. Soc. London, Ser. B 270, 2301–2308.

Parmentier, E., Lanterbecq, D., Eeckhaut, I., 2016. From commensalism to parasitism in Carapidae (Ophidiiformes): heterochronic modes of development? PeerJ 4, e1786.

Parulekar, A.H., 1981. Marine fauna of Malvan, central west coast of India. Mahasagar 14, 33–44.

Pascal, B., Robinson, G., 2011. Handbook for Sandfish Farming. Regional Coastal Management Programme of the Indian Ocean Countries (ReCoMaP), Indian Ocean Commission, Mauritius. 66 pp.

Pattarapongpan, S., Puchakarn, S., Darbanandana, T., 2016. Population biology of white sea cucumber (*Holothuria scabra* Jeager, 1833) at Libong Island, Trang province. In: *Agricultural Innovation for Global Value Chain: Proceedings of 54th Kasetsart University Annual Conference*. Plants, Animals, Veterinary Medicine, Fisheries, Agricultural Extension and Home Economics, *vol. 1*. Kasetsart University, Thailand, pp. 797–805.

Pattinasarany, M.M., Pattiasina, B.J., 2020. Application of asexual reproduction techniques and grow ability of the sandfish *Holothuria scabra*. In: IOP Conference Series: Earth and Environmental Science, vol 584. IOP Publishing, p. 012061.

Pearson, J., 1903. Report on the Holothuroidea. In: Herdman, W.A. (Ed.), Report to the Government of Ceylon on the Pearl Oyster Fisheries of the Gulf of Mannar. Royal Society, London, pp. 181–208.

Pearson, J., 1910. Littoral Marine Fauna: Kerimba Archipelago, Portuguese East Africa. Collected by James J. Simpson, MA, B.Sc., University of Aberdeen, September 1907-May 1908: Holothurioidea. Proc. Zool. Soc. London 80, 167–182.

Peranginangin, R., Setiabudi, E., Murniayati, and Suparno., 1994. Improvement of quality of dried-smoked sea cucumber by enzymic treatments. In: Research Contributions Presented at the Ninth Session of the Working Party on Fish Technology and Marketing: Cochin*, India, 7-9 March 1994.*

Phillips, D.W., 1977. Avoidance and escape responses of the gastropod mollusc *Olivella biplicata* (Sowerby) to predatory asteroids. J. Exp. Mar. Biol. Ecol. 28, 77–86.

Pierrat, J., Bédier, A., Eeckhaut, I., Magalon, H., Frouin, P., 2022. Sophistication in a seemingly simple creature: a review of wild holothurian nutrition in marine ecosystems. Biol. Rev. 97 (1), 273–298.

Pinault, M., Conand, C., 2007. Sea Cucumber Fisheries in the WIO: Masma Project Seychelles. Report from the visit in May 2007, 19 pp., available online www.wiomsa.org.

Pinn, W., Yasin, Z., Hwai, T.S., 2014. First record of *Lissocarcinus orbicularis* associated with sea cucumber, *Thelenota ananas*, in the north east of Celebes Sea. Malayan Nat. J. 66, 368–375.

Pitogo, K.M.E., Sumin, J.P., Ortiz, A.T., 2016. Effect of lunar phases in the size distribution of *Holothuria scabra* on intertidal areas in Sarangani Bay, Philippines. SPC Beche-de-mer Inf. Bull. 36, 48–53.

Pitt, R., 2001. Review of sandfish breeding and rearing methods. SPC Beche-de-mer Inf. Bull. 14, 14–21.

Pitt, R., Duy, N.D.Q., 2003. How to produce 100 tonnes of sandfish. SPC Beche-de-mer Inf. Bull. 18, 15–17.

Pitt, R., Duy, N.D.Q., 2004a. Length–weight relationship for sandfish, *Holothuria scabra*. SPC Beche-de-mer Inf. Bull. 19, 39–40.
Pitt, R., Duy, N.D.Q., 2004b. Breeding and rearing of the sea cucumber *Holothuria scabra* in VietNam. In: Lovatelli, A., Conand, C., Purcell, S., Uthicke, S., Hamel, J.-F., Mercier, A. (Eds.), Advances in Sea Cucumber Aquaculture and Management. FAO Fisheries Technical Paper No 463, FAO, Rome, pp. 333–346.
Pitt, R., Thu, N.T.X., Minh, M.D., Phuc, H.N., 2001. Preliminary sandfish growth trials in tanks, ponds and pens in Vietnam. SPC Beche-de-mer Inf. Bull. 15, 17–27.
Pitt, R., Duy, N.D.Q., Duy, T.V., Long, H.T.C., 2004. Sandfish (*Holothuria scabra*) with shrimp (*Penaeus monodon*) co-culture tank trials. SPC Beche-de-mer Inf. Bull. 20, 12–22.
Plotieau, T., 2012. Analyse de certains éléments nutritionnels essentiels à *Holothuria scabra* (Echinodermata Holothuroidea): influence de la qualité du sédiment sur le développement des holothuries en aquaculture et importance des bactéries. PhD Thesis, University of Mons, Belgium.
Plotieau, T., Baele, J.-M., Vaucher, R., Hasler, C.-A., Koudad, D., Eeckhaut, I., 2013. Analysis of the impact of *Holothuria scabra* intensive farming on sediment. Cahiers De Biologie Marine 54, 703–711.
Plotieau, T., Lepoint, G., Lavitra, T., Eeckhaut, I., 2014a. Isotopic tracing of sediment components that are assimilated by epibiontic juveniles of *Holothuria scabra* (Holothuroidea). J. Mar. Biol. Assoc. U. K. 94, 1485–1490.
Plotieau, T., Lepoint, G., Baele, J.-M., Tsiresy, G., Rasolofonirina, R., Lavitra, T., Eeckhaut, I., 2014b. Caractéristiques minérales et organiques des sédiments des enclos d'élevage d'*Holothuria scabra* (Holothuroidea, Echinodermata). SPC Beche-de-mer Inf. Bull. 34, 29–33.
Polon, P., 2004. The Papua New Guinea national beche-de-mer fishery management plan. In: Lovatelli, A., Conand, C., Purcell, S.W., Uthicke, S., Hamel, J.-F., Mercier, A. (Eds.), Advances in Sea Cucumber Aquaculture and Management. FAO Fisheries Technical Paper No. 463, FAO, Rome, pp. 205–220.
Pouget, A., 2005. Abundance and distribution of holothurians on the fringing reef flats of Grande Terre, Mayotte, Indian Ocean. SPC Beche-de-mer Inf. Bull. 21, 22–26.
Pranoto, E.N., Ma'ruf, W.F., Pringgenies, D., 2012. Kajian aktivitas bioaktif ekstrak teripang pasir (*Holothuria scabra*) terhadap jamur *Candida albicans*. JPBHP 1, 1–8.
Pranweerapaiboon, K., Apisawetakan, S., Nobsathian, S., Itharat, A., Sobhon, P., Chaithirayanon, K., 2019. An ethyl-acetate fraction of *Holothuria scabra* modulates inflammation in vitro through inhibiting the production of nitric oxide and pro-inflammatory cytokines via NF-κB and JNK pathways. Inflammopharmacology 28 (4), 1–11.
Preston, G.L., 1990. Mass beche-de-mer production in Fiji. SPC Beche-de-mer Inf. Bull. 1, 4–5.
Preston, G.L., 1993. Beche-de-mer. In: Nearshore Marine Resources of the South Pacific: Information for Fisheries Development and management. Forum Fisheries Agency, Honiara, Solomon Islands, pp. 371–401.
Price, A.R.G., 1981. Echinoderm fauna of the western arabian gulf. J. Nat. Hist. 15, 1–16.
Price, A.R.G., Harris, A., McGowan, A., Venkatachalam, A.J., Sheppard, C.R.C., 2010. Chagos feels the pinch: assessment of holothurian (sea cucumber) abundance, illegal harvesting and conservation prospects in British Indian Ocean Territory. Aquat. Conserv.: Mar. Freshw. Ecosyst. 20, 117–126.
Price, A.R.G., Evans, L.E., Rowlands, l.N., Hawkins, J.P., 2013. Negligible recovery in Chagos holothurians (sea cucumbers). Aquat. Conserv.: Mar. Freshw. Ecosyst. 23, 811–819.
Prompoon, Y., Weerachatyanukul, W., Withyachumnarnkul, B., Vanichviriyakit, R., Wongprasert, K., Asuvapongpatana, S., 2015. Lectin-based profiling of coelomocytes in *Holothuria scabra* and expression of superoxide dismutase in purified coelomocytes. Zoolog. Sci. 32, 345–351.

Purcell, S.W., 2004a. Criteria for release strategies and evaluating the restocking of sea cucumbers. In: Advances in Sea Cucumber Aquaculture and Management. FAO Fisheries Technical Paper 463, FAO, pp. 181–191.
Purcell, S.W., 2004b. Rapid growth and bioturbation activity of the sea cucumber *Holothuria scabra* in earthen ponds. In: Proceedings of Australasian Aquaculture, p. 244.
Purcell, S.W., 2010. Diel burying by the tropical sea cucumber *Holothuria scabra*: effects of environmental stimuli, handling and ontogeny. Mar. Biol. 157, 663–671.
Purcell, S.W., 2012. Principles and science of stocking marine areas with sea cucumbers. In: Hair, C.A., Pickering, T.D., Mills, D.J. (Eds.), *Asia-Pacific Tropical Sea Cucumber Aquaculture. Proceedings of an International Symposium Held in Noumea, New Caledonia. ACIAR Proceedings*. Citeseer, pp. 92–103.
Purcell, S.W., 2014a. Value, market preferences and trade of beche-de-mer from Pacific Island sea cucumbers. PLoS One 9, e95075.
Purcell, S.W., 2014b. Processing Sea Cucumbers Into beche-de-mer: A Manual for Pacific Island Fishers. SPC/SCU.
Purcell, S.W., Agudo, N.S., 2013. Optimisation of mesh enclosures for nursery rearing of juvenile sea cucumbers. PLoS One 8, e64103.
Purcell, S.W., Blockmans, B.F., 2009. Effective fluorochrome marking of juvenile sea cucumbers for sea ranching and restocking. Aquaculture 296, 263–270.
Purcell, S.W., Eeckhaut, I., 2005. An external check for disease and health of hatchery-produced sea cucumbers. SPC Beche-de-mer Inf. Bull. 22, 34.
Purcell, S.W., Eriksson, H., 2015. Echinoderms piggybacking on sea cucumbers: benign effects on sediment turnover and movement of hosts. Mar. Biol. Res. 11, 666–670.
Purcell, S.W., Eriksson, H.B., 2016. *How to Tackle the Rising Tide of Poaching in Australia's Tropical Seas*, The Conversation. 23 December. Accessed January 2021 https://theconversation.com/how-to-tackle-the-rising-tide-of-poaching-in-australias-tropical-seas-70640.
Purcell, S.W., Kirby, D.S., 2006. Restocking the sea cucumber *Holothuria scabra*: sizing no-take zones through individual-based movement modelling. Fish. Res. 80, 53–61.
Purcell, S.W., Lalavanua, W., 2018. Boat fuel consumption by sea cucumber fishers: new study raises concern. SPC Fish. Newsl. 155, 23–24.
Purcell, S.W., Simutoga, M., 2008. Spatio-temporal and size-dependent variation in the success of releasing cultured sea cucumbers in the wild. Rev. Fish. Sci. 16, 204–214.
Purcell, S.W., Wu, M., 2017. Large-scale sandfish (*Holothuria scabra*) aquaculture in multitrophic polyculture ponds in southern China. SPC Beche-de-mer Inf. Bull. 37, 51–52.
Purcell, S.W., Blockmans, B.F., Agudo, N.N.S., 2006a. Transportation methods for restocking of juvenile sea cucumber, *Holothuria scabra*. Aquaculture 251, 238–244.
Purcell, S.W., Patrois, J., Fraisse, N., 2006b. Experimental evaluation of co culture of juvenile sea cucumbers, *Holothuria scabra* (Jaeger), with juvenile blue shrimp, *Litopenaeus stylirostris* (Stimpson). Aquacult. Res. 37, 515–522.
Purcell, S.W., Blockmans, B.F., Nash, W.J., 2006c. Efficacy of chemical markers and physical tags for large-scale release of an exploited holothurian. J. Exp. Mar. Biol. Ecol. 334, 283–293.
Purcell, S.W., Gossuin, H., Agudo, N.S., 2009a. Status and Management of the Sea Cucumber Fishery of La Grande Terre, New Caledonia. WorldFish Center Studies and Review 1901, Penang.
Purcell, S.W., Gossiun, H., Agudo, N.S., 2009b. Changes in Weight and Length of Sea Cucumbers During Conversion to Processed beche-de-mer: Filling Gaps for Some Exploited Tropical Species. Secretariat of the Pacific Community.
Purcell, S.W., Lovatelli, A., Vasconcellos, M., Ye, Y., 2010. Managing sea cucumber fisheries with an ecosystem approach. In: Fisheries and Aquaculture Technical Paper No. 520, Food and Agriculture Organization of the United Nations, Rome.

Purcell, S.W., Hair, C.A., Mills, D.J., 2012a. Sea cucumber culture, farming and sea ranching in the tropics: progress, problems and opportunities. Aquaculture 368–369, 68–81.

Purcell, S.W., Samyn, Y., Conand, C., 2012b. Commercially important sea cucumbers of the world. In: FAO Species Catalogue for Fishery Purposes No 6, FAO, Rome.

Purcell, S.W., Mercier, A., Conand, C., Hamel, J.-F., Toral-Granda, V., Lovatelli, A., Uthicke, S., 2013. Sea cucumber fisheries: global review of stock status, management measures and drivers of overfishing. Fish Fish. 14, 34–59.

Purcell, S.W., Lovatelli, A., Pakoa, K., 2014a. Constraints and solutions for managing Pacific Island sea cucumber fisheries with an ecosystem approach. Mar. Policy 45, 240–250.

Purcell, S.W., Choo, P.S., Akamine, J., Fabinyi, M., 2014b. Alternative product forms, consumer packaging and extracted derivatives of tropical sea cucumbers. SPC Beche-de-mer Inf. Bull. 34, 47–52.

Purcell, S.W., Polidoro, B.A., Hamel, J.-F., Gamboa, R.U., Mercier, A., 2014c. The cost of being valuable: predictors of extinction risk in marine invertebrates exploited as luxury seafood. Proc. R. Soc. B Biol. Sci. 281, 20133296.

Purcell, S.W., Eriksson, H., Byrne, M., 2016a. Rotational Zoning Systems in Multi-Species Sea Cucumber Fisheries. South Pacific Commission, Fisheries, Aquaculture, and Marine Ecosystems Division.

Purcell, S.W., Conand, C., Uthicke, S., Byrne, M., 2016b. Ecological roles of exploited sea cucumbers. Oceanogr. Mar. Biol. Annu. Rev. 54, 367–386.

Purcell, S.W., Piddocke, T.P., Dalton, S.J., Wang, Y.-G., 2016c. Movement and growth of the coral reef holothuroids *Bohadschia argus* and *Thelenota ananas*. Mar. Ecol. Prog. Ser. 551, 201–214.

Purcell, S.W., Williamson, D.H., Ngaluafe, P., 2018a. Chinese market prices of beche-de-mer: implications for fisheries and aquaculture. Mar. Policy 91, 58–65.

Purcell, S.W., Lalavanua, W., Cullis, B.R., Cocks, N., 2018b. Small-scale fishing income and fuel consumption: Fiji's artisanal sea cucumber fishery. ICES J. Mar. Sci. 75, 1758–1767.

Purcell, S.W., Ngaluafe, P., Lalavanua, W., Ceccarelli, D.M., 2018c. Market price trends of Latin American and Caribbean sea cucumbers inform fisheries management. Reg. Stud. Mar. Sci. 17, 127–132.

Purwati, P., 2006a. Souvenir from west Lombok, eastern Indonesia. SPC Beche-de-mer Inf. Bull. 23, 26–29.

Purwati, P., 2006b. Reproductive patterns of *Holothuria scabra* (Echinodermata: Holothuroidea) in Indonesian waters. Mar. Res. Indones. 30, 47–55.

Queiroz, V., de Souza, L.S., Pimenta, A.D., Cunha, C.M., 2013. New host records to *Melanella* (Caenogastropoda: Eulimidae) from the Brazilian coast. Mar. Biodivers. Rec. 6, e67.

Rachmansyah, M.I., Madeali, A.M., Tangko, S.T., Ismail, D.A., 1992. Polyculture of sea cucumber, *Holothuria scabra* and seaweed, *Eucheuma* sp. in pen culture at Parepare Bay, South Sulawesi. Journal Penelitian Budidaya Pantai 8, 63–70.

Radhakrishnan, N., 1994. The role of fisherwomen in the beche-de-mer industry. CMFRI Bull. 46, 99–100.

Rafidah, O., Firdaus-Nawi, M., Raehanah, S., Ina-Salwany, M.Y., Ching, F.F., Abidin, N.A., Zamri-Saad, M., 2017. An outbreak of *Vibrio alginolyticus* infection in juvenile sea cucumbers *Holothuria scabra* in Sabah, Malaysia. Pertanika J. Trop. Agric. Sci. 40, 691–696.

Rahael, K.P., Rahantoknam, S.P.T., Hamid, S.K., 2019. The amino acid of sandfish sea cucumber (*Holothuria scabra*): dry method with various feeding enzyme. In: Journal of Physics: Conference Series, IOP Publishing, p. 012005.

Rahantoknam, S.P.T., 2017. Maturity gonad sea cucumber Holothuria scabra under the month cycle. In: IOP Conference Series: Earth and Environmental Science, vol. 89. IOP Publishing, p. 012015.

Rahman, M.A., Chowdhury, S.H., Hasan, M.J., Rahman, M.H., Yeasmin, S.M., Farjana, N., Molla, M.H.R., Parvez, M.S., 2020. Status, prospects and market potentials of the sea cucumber fisheries with special reference on their proper utilization and trade. Annu. Res. Rev. Biol. 35 (7), 84–101.

Raison, C.M., 2008. Advances in sea cucumber aquaculture and prospects for commercial culture of Holothuria scabra. In: CAB Reviews: Perspectives in Agriculture, Veterinary Science, Nutrition and Natural Resources, vol. 3, pp. 1–15.

Rakotonjanahary, F., Tsiresy, G., Rasolofonirina, R., Eeckhaut, I., Lavitr, T., 2015. Test of liquid injection and elastomer implant for tagging edible sea cucumber *Holthuria scabra*. In: Presentation at the 8th WIOMSA Scientific Symposium, Durban, South Africa.

Rakotonjanahary, F., Lavitra, T., Fohy, N., Eeckhaut, I., 2016. Essais d'optimisation de la croissance des juvéniles d'*Holothuria scabra* pendant la phase de pré-grossissement. SPC Beche-de-mer Inf. Bull. 36, 76–81.

Ram, R., 2017. Influence of Processing Techniques on Quality and Nutritional Composition of the Tropical Sea Cucumber *Holothuria scabra*. PhD Thesis, James Cook University, Australia.

Ram, R., Chand, R.V., Southgate, P.C., 2014. Effects of processing methods on the value of bêche-de-mer from the Fiji Islands. J. Mar. Sci. Res. Dev. 4, 1.

Ram, R., Chand, R.V., Zeng, C., Southgate, P.C., 2016. Recovery rates for eight commercial sea cucumber species from the Fiji Islands. Reg. Stud. Mar. Sci. 8, 59–64.

Ram, R., Francis, D.S., Chand, R.V., Southgate, P.C., 2017a. Nutritional value of the sea cucumber *Holothuria scabra* from Fiji Islands. SPC Fish. Newsl. 152, 29–31.

Ram, R., Chand, R.V., Forrest, A., Southgate, P.C., 2017b. Effect of processing method on quality, texture, collagen and amino acid composition of sandfish (*Holothuria scabra*). LWT- Food Sci. Technol. 86, 261–269.

Ramofafia, C., Battaglene, S.C., Bell, J.D., Byrne, M., 2000. Reproductive biology of the commercial sea cucumber *Holothuria fuscogilva* in the Solomon Islands. Mar. Biol. 136, 1045–1056.

Ramofafia, C., Byrne, M., Battaglene, S.C., 2001. Reproductive biology of the intertidal sea cucumber *Actinopyga mauritiana* in the Solomon Islands. J. Mar. Biol. Assoc. U. K. 81, 523–531.

Ramofafia, C., Byrne, M., Battaglene, C.S., 2003a. Reproduction of the commercial sea cucumber *Holothuria scabra* (Echinodermata: Holothuroidea) in the Solomon Islands. Mar. Biol. 142, 281–288.

Ramofafia, C., Byrne, M., Battaglene, S.C., 2003b. Development of three commercial sea cucumbers, *Holothuria scabra, H fuscogilva* and *Actinopyga mauritiana*: larval structure and growth. Mar. Freshw. Res. 54, 657–667.

Ramofafia, C., Lane, I., Oengpepa, C., 2004. Customary marine tenure in Solomon islands: a shifting paradigm for management of sea cucumber in artisanal fisheries. In: Lovatelli, A., Conand, C., Purcell, S.W., Uthicke, S., Hamel, J.-F., Mercier, A. (Eds.), Advances in Sea Cucumber Aquaculture and Management. FAO Fisheries Technical Paper No. 463, FAO, Rome, pp. 259–260.

Rao, D.S., James, D.B., Girijavallabhan, K.G., Muthusamy, S., Najmuddin, M., 1985a. Bioactivity in echinoderms. Mar. Fish. Inf. Serv. Tech. Ext. Ser. 63, 10–12.

Rao, D.S., James, D.B., Girijavallabhan, K.G., Muthusamy, S., Najumddin, M., 1985b. Biotoxicity in echinoderms. J. Mar. Biol. Ass. India 27, 88–96.

Rasolofonirina, R., 2004. Reproduction et développement de l'holothurie comestible *Holothuria scabra* (Jaëger, 1833) (Holothroidea: Echinodermata). PhD Thesis, Université libre de Bruxelles.

Rasolofonirina, R., 2007. Sea cucumbers in Madagascar. In: Conand, C., Muthiga, N. (Eds.), Commercial Sea Cucumbers: A Review for the Western Indian Ocean. WIOMSA Book Series No. 5, WIOMSA, pp. 31–40.

Rasolofonirina, R., Jangoux, M., 2005. Appearance and development of skeletal structures in *Holothuria scabra* larvae and epibiont juveniles. SPC Beche-de-mer Inf. Bull. 22, 6–10.
Rasolofonirina, R., Vaïtilingon, D., Eeckhaut, I., Jangoux, M., 2005. Reproductive cycle of edible echinoderms from the southwestern Indian Ocean. II The sandfish *Holothuria scabra* (Jaeger, 1833). West. Indian Ocean J. Mar. Sci. 4, 61–75.
Rasyid, A., Murniasih, T., Putra, M.Y., Pangestuti, R., Harahap, I.A., Untari, F., Sembiring, S.B.M., 2020. Evaluation of nutritional value of sea cucumber *Holothuria scabra* cultured in Bali, Indonesia. Aquac. Aquar. Conserv. Legis. 13, 2083–2093.
Ravago-Gotanco, R., Kim, K.M., 2019. Regional genetic structure of sandfish *Holothuria* (*Metriatyla*) *scabra* populations across the Philippine archipelago. Fish. Res. 209, 143–155.
Reyes-Leonardo, L.D., 1984. A taxonomic report of shallow-water holothurians of Calatagan, Batangas. Philipp. J. Sci. 113, 137–172.
Reyes-Leonardo, L.D., Monzon, R.B., Navarro, V.C., 1985. A taxonomic account of shallow water holothurians of Bolinao, Pangasinan. Nat. Appl. Sci. Bull. 37, 261–284.
Riani, E., Maheswari, H., Dzikrifishofa, M., Kusumorini, N., 2016. Sandifish (*Holothuria scabra*) ameliorates aging in menopausal women by increasing estradiol hormones. Physiol. Pharmacol. 20, 206–214.
Richards, A.H., Bell, L.J., Bell, J.D., 1994. Inshore fisheries resources of Solomon Islands. Mar. Pollut. Bull. 29, 90–98.
Ricker, W.E., 1981. Changes in the average size and average age of Pacific salmon. Can. J. Fish. Aquat. Sci. 38, 1636–1656.
Ridwanudin, A., Indriana, L.F., Kunzmann, A., 2018. No difference in nutritional profiles of wild and cultured juvenile sandfish, *Holothuria scabra*. Annu. Res. Rev. Biol. 26, 1–11.
Ridzwan, B.H., Kaswandi, M.A., Azman, Y., Fuad, M., 1995. Screening for antibacterial agents in three species of sea cucumbers from coastal areas of Sabah. Gen. Pharmacol. 26, 1539–1543.
Ridzwan, B.H., Hanita, M.H., Nurzafirah, M., Norshuhadaa, M.P.S., Hanis, Z.F., 2014. Free fatty acids composition in lipid extracts of several sea cucumbers species from Malaysia. Int. J. Biosci. Biochem. Bioinforma. 4, 204.
Rizal, A., Muskita, W.H., 2018. Pengaruh Kepadatan yang Berbeda terhadap Pertumbuhan Teripang Pasir (Holothuria scabra). Jurnal Media Akuatika 3, 760–768.
Rizqi, M.P., 2019. Site suitability analysis for community-based tropical sea cucumber *Holothuria scabra* grow-out in North Sulawesi waters. Mar. Res. Indones. 44, 1–10.
Robinson, G., 2011. Community based sea cucumber aquaculture in Madagascar. In: Mariculture in the WIO Region-Challenges and Prospects. WIOMSA Book Series No. 11. WIOMSA, pp. 47–53.
Robinson, G., 2013. A bright future for sandfish aquaculture. World Aquac. Soc. Mag. 2013, 18–24.
Robinson, G., Pascal, B., 2009. From hatchery to community–Madagascar's first village-based holothurian mariculture programme. SPC Beche-de-mer Inf. Bull. 29, 38–43.
Robinson, G., Pascal, B., 2012. Se cucumber farming experiences in south-western Madagascar. In: Hair, C.A., Pickering, T.D., Mills, D.J. (Eds.), *ACIAR Proceedings*, Asia-Pacific Tropical Sea Cucumber Aquaculture. Proceedings of an International Symposium Held in Noumea, New Caledonia. ACIAR Proceedings, pp. 142–155.
Robinson, G., Slater, M.J., Jones, C.L.W., Stead, S.M., 2013. Role of sand as substrate and dietary component for juvenile sea cucumber *Holothuria scabra*. Aquaculture 392, 23–25.
Robinson, G., Caldwell, G.S., Jones, C.L.W., Slater, M.J., Stead, S.M., 2015. Redox stratification drives enhanced growth in a deposit-feeding invertebrate: implications for aquaculture bioremediation. Aquac. Environ. Interact. 8, 1–13.
Robinson, G., Caldwell, G.S., Jones, C.L.W., Stead, S.M., 2019. The effect of resource quality on the growth of *Holothuria scabra* during aquaculture waste bioremediation. Aquaculture 499, 101–108.

Rochette, R., Maltais, M.-J., Dill, L.M., Himmelman, J.H., 1999. Interpopulation and context-related differences in responses of a marine gastropod to predation risk. Anim. Behav. 57, 977–987.

Rougier, A., Ateweberhan, M., Harris, A., 2013. Strategies for improving survivorship of hatchery-reared juvenile *Holothuria scabra* in community-managed sea cucumber farms. SPC Beche-de-mer Inf. Bull. 33, 14–22.

Rowe, F.W.E., 1969. A review of the family Holothuriidae (Holothuroidea: Aspidochirotida). Bull. Br. Mus. Nat. Hist. Zool. 18, 119–170.

Rowe, F.W.E., Gates, J., 1995. Echinodermata. In: Zoological Catalogue of Australia, vol. 33, pp. 294–295.

Rowe, F.W.E., Richmond, M.D., 1997. Echinodermata. In: Richmond, M.D. (Ed.), A Guide to the Seashores of Eastern Africa and the Western Indian Ocean Islands. The Sea Trust, Zanzibar, pp. 290–321.

Sabilu, K., Supriyono, E., Nirmala, K., Jusadi, D., Widanarni, W., 2020. Production performance and physiological responses of sea cucumber (*Holothuria scabra*) reared using Penaeus vannamei pond sediment as a source of nutrients. Aquac. Aquar. Conserv. Legis. 13, 3507–3519.

Sachithananthan, K., 1972. South Pacific Islands bêche-de-mer fishery. A report prepared for the South Pacific Islands Fisheries Development Agency, Food and Agriculture Organization of the United Nations, Rome.

Sachithananthan, K., 1975. De-scummer for beche-de-mer processing. In: Bulletin of the Fisheries Research Station of Sri Lanka, vols. 1–2, pp. 11–15.

Sachithananthan, K., 1979. Beche-de-mer of the South Pacific Islands: A Handbook for Fishermen. South Pacific Commission, Noumea, New Caledonia.

Sachithananthan, K., 1994a. A small-scale unit to process sand-fish *Holothuria* (*Metriatyla*) *scabra*. CMFRI Bull. 46, 79–80.

Sachithananthan, K., 1994b. Beche-de-mer trade: global perspectives. CMFRI Bull. 46, 106–109.

Sachithananthan, K., Natesan, P., Algaratnam, C., Thevathasan, A., Philip, L.B., 1986. Artisanal handling and processing of sea cucumber (sand fish). In: Infofish Marketing Digest, vol. 2, pp. 35–36.

Saengsuwan, M.J., 2017. The Effects of Sea Cucumber Extracts (*Holothuria scabra*) on Human Placenta Mesenchymal Stem Cells. Master Thammasat University.

Safitri, Y., Mardhia, D., Syafikri, D., Kautsari, N., 2020. Pengaruh Jenis Plankton yang Berbeda Terhadap Kelulushidupan dan Laju Pertumbuhan Larva Teripang Pasir (*Holothuria Scabra*) pada Fase Planktonik. Indones. J. Appl. Sci. Technol. 1, 1–9.

Samyn, Y., 2000. Conservation of aspidochirotid holothurians in the littoral waters of Kenya. SPC Beche-de-mer Inf. Bull. 13, 12–17.

Sangpairoj, K., Chaithirayanon, K., Vivithanaporn, P., Siangcham, T., Jattujan, P., Poomtong, T., Nobsathian, S., Sobhon, P., 2016. Extract of the sea cucumber, *Holothuria scabra*, induces apoptosis in human glioblastoma cell lines. Funct. Foods Health Dis. 6, 452–468.

Sant, G., 1995. Marine invertebrates of the South Pacific. In: Species in Danger, Traffic International. Cambridge University Press, Cambridge, UK, pp. 1–85.

Sant, G., 2006. CITES and sea cucumbers. In: B. A.W (Ed.), Proceedings of the CITES Workshop on the Conservation of Sea Cucumbers in the Families Holothuriidae and Stichopodidae, NOAA Technical Memorandum NMFS-OPR-34, Silver Spring, Maryland, USA.

Sari, P.A., 2020. Determination of nano-collagen quality from sea cucumber *Holothuria scabra*. In: IOP Conference Series: Earth and Environmental Science, vol. 430. IOP Publishing, p. 012005.

Sarma, N.S., Anjaneyulu, A.S.R., Rao, C.B., Venkateswarlu, Y., 1987. Triterpene glycosides and aglycones of sea cucumbers holothuria-atra and holothuria-scabra holothurioidea. Indian J. Chem. - B Org. Med. Chem. 26, 715–721.
Satyamurti, S.T., 1976. The Echinodermata in the collection of the Madras Government Museum. In: Bulletin of the Madras Government Museum of Natural History, New Series, vol. 7, pp. 1–284.
Schagerström, E., 2003. A Study of Juvenile *Holothuria scabra* (Jaeger) (Echinodermata) on Different Types of Sediment at Bolinao, Pangasinan. Degree Project Work, University of Kalmar, Sweden, Philippines.
Schoppe, S., 2000. Sea cucumber fishery in the Philippines. SPC Beche-de-mer Inf. Bull. 13, 10–12.
Selenka, E., 1867. Beiträge zur Anatomie und Systematik der Holothurien. W. Engelmann.
Selvin, J., Lipton, A., 2004. Antifouling activity of bioactive substances extracted from *Holothuria scabra*. Hydrobiologia 513, 251–253.
Sembiring, S.B.M., Hutapea, J.H., 2019. Gonad development of sandfish (*Holothuria scabra*) fed with vitamin E-supplemented diet. Indones. Aquac. J. 14, 47–54.
Sembiring, S.B.M., Darmansyah, X., Arfah, R., Sugama, K., 1996. The effect of feeding frequencies on the growth and survival rates of larvae of sea cucumber *Holothuria scabra*. Indones. Fish. Res. J. 2 (2), 42–47.
Sembiring, S.B.M., Sugama, K., Suastika, I.M., Makatutu, D., Jufri, 2004. Technical Guidelines of Sea Cucumber, Holothuria scabra Breeding Technology. Central Research Institute for Aquaculture. Marine and Fisheries Research Agency. Department of Marine and Fisheries, Jakarta.
Sembiring, S.B.M., Wardana, I.K., Haryanti., 2015. Sea cucumber, *Holothuria scabra* breeding technology. In: Recommendation Technology of Marine and Fisheries. Marine and Fisheries Research and Development Agency. Ministry of Marine Affair and Fisheries, Jakarta, pp. 187–200.
Sembiring, S.B.M., Wardana, I.K., Haryanti, X., 2016. Performance of sea cucumber, *Holothuria scabra* juvenile from different sources of broodstock. Aquac. Res. J. 11 (2), 147–155.
Sembiring, S.B.M., Wardana, I.K., Giri, N.A., Haryanti, H., 2017. Keragaan Rematurasi Gonad Induk Teripang Pasir, *Holothuria scabra* Dengan Pemberian Jenis Pakan Berbeda. J. Ris. Akuakultur 12, 147–159.
Sembiring, S.B.M., Wibawa, G.S., Giri, I.N.A., Hutapea, J.H., 2018. Reproduction and larval rearing of sandfish (*Holothuria scabra*). Mar. Res. Indones. 43, 11–17.
Sembiring, S.B.M., Wibawa, G.S., Hutapea, J.H., Giri, I.N.A., 2019. Effect of salinity on the survival, growth and immunity rate of juvenile sea cucumbers (*Holothuria scabra*). In: BIOTROPIA-The Southeast Asian Journal of Tropical Biology, vol. 26, pp. 163–171.
Semper, C., 1868. Holothurien Reisen im Archipel der Phillippinen. In: Wissenschaftliche Resultate. Weisbaden, vol. 2, pp. 1–288.
Senff, P., Blanc, P.-P., Slater, M., Kunzmann, A., 2020. Low-technology recirculating aquaculture system integrating milkfish *Chanos chanos*, sea cucumber *Holothuria scabra* and sea purslane *Sesuvium portulacastrum*. Aquac. Environ. Interact. 12, 471–484.
Senff, P., Elba, B., Kunzmann, A., Gillis, L.G., Robinson, G., 2022. Carbon supplementation promotes assimilation of aquaculture waste by the sea cucumber *Holothuria scabra*: evidence from stable isotope analysis. Aquaculture 547, 737295.
Serang, A.M., Rahantoknam, S.P.T., Tomatala, P., 2016. Effect of different stocking densities on growth and survival rates of sea cucumber *Holothuria scabra* seedlings. Aquac. Indones. 17, 30–34.
Setiyanto, R.N., 2012. Uji aktivitas antioksidan ekstrak air teripang pasir (*Holothuria scabra*) dengan metode 1, 1-difenil-2-pikrilhidrazil (DPPH) dan analisis kandungan kimianya. Universitas Muhammadiyah Surakrta.

Shelley, C.C., 1981. Aspects of the Distribution, Reproduction, Growth and 'Fishery' Potential of Holothurians (beche-de-mer) in the Papuan Coastal Lagoon. MSc Thesis, University of Papua New Guinea, Port Moresby, Papua New Guinea.

Shelley, C.C., 1985a. Growth of *Actinopyga echinites* and *Holothuria scabra* (Holothurioidea: Echinodermata) and their fisheries potential (as beche-de-mer) in Papua New Guinea. In: Harmelin-Vivien, M., Salvat, B., La Croix, C., Gabrie, C., Toffart, J.L. (Eds.), Proceedings of the 5th International. Coral Reef Congress, *vol. 5*, pp. 297–302.

Shelley, C.C., 1985b. The potential for re-introduction of a beche-de-mer fishery in Torres Strait. In: Torres Strait Fisheries Seminar, 1985, Port Moresby. Australian Gov. Pub. Service, Canberra, Australia.

Shelley, C.C., Puig, P., 2004. Management of sea cucumbers in the Northern Territory, Australia, and current research to further improve understanding of the fishery. In: Lovatelli, A., Conand, C., Purcell, S.W., Uthicke, S., Hamel, J.-F., Mercier, A. (Eds.), Advances in Sea Cucumber Aquaculture and Management. FAO Fisheries Technical Paper No. 463, FAO, Rome, pp. 221–230.

Sheppard, C.R.C., Ateweberhan, M., Bowen, B.W., Carr, P., Chen, C.A., Clubbe, C., Craig, M.T., Ebinghaus, R., Eble, J., Fitzsimmons, N., 2012. Reefs and islands of the Chagos Archipelago, Indian Ocean: why it is the world's largest no-take marine protected area. Aquat. Conserv.: Mar. Freshw. Ecosyst. 22, 232–261.

Shiell, G., 2004. Field observations of juvenile sea cucumbers. SPC Beche-de-mer Inf. Bull. 20, 6–11.

Shiell, G.R., Uthicke, S., 2006. Reproduction of the commercial sea cucumber *Holothuria whitmaei* [Holothuroidea: Aspidochirotida] in the Indian and Pacific Ocean regions of Australia. Mar. Biol. 148, 973–986.

Sicuro, B., Piccinno, M., Gai, F., Cesarina, A.M., Danieli, A., Daprà, F., Mioletti, S., 2012. Food quality and safety of Mediterranean sea cucumbers *Holothuria tubulosa* and *Holothuria polii* in southern Adriatic Sea. Asian J. Anim. Vet. Adv. 7, 851–859.

Silver, H., 2020. Histology of the autotomy region in the esophagus of *Holothuria scabra* Jaeger. In: Keegan, B.F., O'Connor, B.D.S (Eds.), Echinodermata: Proceedings of the Fifth International Echinoderm Conference. A.A. Balkema, Rotterdam, pp. 451–457.

Simoes, M., Knauer, J., Ghobadyan, F., Morovvati, H., Ghazvineh, L., Tavassolpour, E., Demeuldre, M., Eeckhaut, I., Morgan, A.D., Xing, K., 2012. Marking juvenile sandfish, *Holothuria scabra*, with the fluorochrome oxytetracycline—a preliminary report. SPC Beche-de-mer Inf. Bull. 32, 3–5.

Simon, C., San Martín, G., Robinson, G., 2014. Two new species of Syllis (Polychaeta: Syllidae) from South Africa, one of them viviparous, with remarks on larval development and vivipary. J. Mar. Biol. Assoc. U. K. 94, 729–746.

Sinsona, M.J., Juinio-Meñez, M.A., 2018. Effects of sediment enrichment with macroalgae, *Sargassum* spp., on the behavior, growth, and survival of juvenile sandfish, *Holothuria scabra*. Aquac. Rep. 12, 56–63.

Sinsona, M.J., Juinio-Meñez, M.A., 2019. Periphyton characteristics influence the growth and survival of *Holothuria scabra* early juveniles in an ocean nursery system. Aquacult. Res. 50, 2655–2665.

Sipahutar, D., Soeharmoko, K.D., 1989. Preliminary study on the sea cucumber (*Holothuria scabra*) in South-Bintan waters, Riau Archipelago. Jurnal Penelitian Budidaya Pantai 5, 13–18.

Sithisak, P., Pongtippatee, P., Withyachumnarnkul, B., 2013. Improving inland culture performance of juvenile sea cucumbers, *Holothuria scabra*, by co-culture with red tilapia. Songklanakarin J. Sci. Technol. 35, 501–505.

Sitoresmi, D.A.N., Pursetyo, K.T., 2020. The Effect of Concrete Tanks for the Breeding Technique of the Sand Sea Cucumber (*Holothuria scabra*). IOP Publishing, p. 012054.

Skewes, T., Dennis, D., Burridge, C., 2000. Survey of *Holothuria scabra* (sandfish) on Warrior Reef, Torres Strait, January, 2000: CSIRO Division of Marine Research Final Report.

Sloan, N.A., 1984. In: Echinorderm Fisheries of the World: A Review. Echinodermata (Proceedings of the Fifth International Echinoderm Conference). AA Balkema Publishers, Rotterdam, Netherlands.

Sluiter, C.P., 1901. Die Holothurien der Siboga-Expedition. In: Siboga Expeditie, vol. 44, pp. 1–142.

Smiley, S., 1988. The dynamics of oogenesis and the annual ovarian cycle of *Stichopus californicus*. Biol. Bull. 175, 79–93.

Smiley, S., McEuen, F.S., Chaffee, C., Krishnan, S., 1991. Echinodermata: Holothuroidea. In: Giese, A.C., Pearse, J.S., Pearse, V.B. (Eds.), Reproduction of Marine Invertebrates. vol. VI. The Boxwood Press, Pacific Grove, California, pp. 663–750.

Sommerville, W.S., 1993. Marketing of beche-de-mer. SPC Beche-de-mer Inf. Bull. 5, 2–7.

Soota, T., Mukhopadhyay, S., Samanta, T., 1983. On some holothurians from the Andaman and Nicobar Islands. Rec. Zool. Surv. India 80, 507–524.

South Pacific Commission, 1994. Beche-de-mer of the South Pacific Islands: A Handbook for Fishermen. South Pacific Commission, Noumea, New Caledonia.

South Pacific Commission, 1995. Holothuries et beche-de-mer dans le Pacifique Tropical, un manuel à l'intention des pêcheurs. South Pacific Commission, Noumea, New Caledonia.

Springhall, J.A., Dingle, J.G., 1967. Growth and pathology of chicks fed beche-de-mer meal. Aust. Vet. J. 43, 298–303.

Sroyraya, M., Hanna, P.J., Siangcham, T., Tinikul, R., Jattujan, P., Poomtong, T., Sobhon, P., 2017. Nutritional components of the sea cucumber *Holothuria scabra*. Funct. Foods Health Dis. 7, 168–181.

Stephenson, W., Endean, R., Bennett, I., 1958. An ecological survey of the marine fauna of Low Isles, Queensland. Mar. Freshw. Res. 9, 261–318.

Stewart, B., 1993. Evidence for a marked decline of beche-de-mer populations in the Suva and Beqa areas of Fiji, and a preliminary description of a method of identifying beche-de-mer individuals based on characteristic body wrinkles. In: Technical Reports of Marine Studies, University of the South Pacific, vol. 1, pp. 1–20.

Stonik, V.A., Ponomarenko, L.P., Makarieva, T.N., Boguslavsky, V.M., Dmitrenok, A.S., Fedorov, S.N., Strobikin, S.A., 1998. Free sterol compositions from the sea cucumbers *Pseudostichopus trachus*, *Holothuria* (*Microtele*) *nobilis*, *Holothuria scabra*, *Trochostoma orientale* and *Bathyplotes natans*. Comp. Biochem. Physiol. B Biochem. Mol. Biol. 120, 337–347.

Stott, R.S.H., Hepburn, H.R., Joffe, I., Heffron, J.J.A., 1974. The mechanical defensive mechanism of a sea cucumber. S. Afr. J. Sci. 70, 46–48.

Strehlow, H.V., 2004. Economics and management strategies for restocking sandfish in Vietnam. Naga 27, 36–40.

Sugama, K., Giri, N.A., Zairin, M., 2019. Aspek biologi dan budidaya teripang pasir, *Holothuria scabra*. AMaFRaD PRESS, Jakarta.

Suphamungmee, W., Nontunha, N., Chaichotranunt, S., Tinikul, Y., Poomtong, T., Sobhon, P., 2018. Electron microscopy reveals the ultrastructure and the chromatin organization of *Holothuria scabra* male germ cells. Songklanakarin J. Sci. Technol. 40, 321–328.

Suwansa-Ard, S., Chaiyamoon, A., Talarovicova, A., Tinikul, R., Tinikul, Y., Poomtong, T., Elphick, M.R., Cummins, S.F., Sobhon, P., 2018. Transcriptomic discovery and comparative analysis of neuropeptide precursors in sea cucumbers (Holothuroidea). Peptides 99, 231–240.

Taboada, M.C., Gonzalez, M., Rodriguez, E., 2003. Value and effects on digestive enzymes and serum lipids of the marine invertebrate *Holothuria forskali*. Nutr. Res. 23, 1661–1670.

Takano, T., Kano, Y., 2014. Molecular phylogenetic investigations of the relationships of the echinoderm-parasite family Eulimidae within Hypsogastropoda (Mollusca). Mol. Phylogenet. Evol. 79, 258–269.

Tan Tiu, A., 1981a. The intertidal holothurian fauna (Echinodermata: Holothuroidea) of Mactan and the neighboring islands, Central Philippines. Philipp. Sci. 18, 45–119.

Tan Tiu, A., 1981b. Systematics of Intertidal Sea Cucumbers (Echinodermata: Holothuroidea) of Mactan Island and Vicinity. MSc Thesis, University of San Carlos, Cebu City, Philippine.

Tangestani, M., Kunzmann, A., 2019. Isolation and characterization of bacteria from the lesion of juvenile sea cucumber *Holothuria scabra* (Jaeger, 1938) with symptom of skin ulceration disease. Iran. J. Fish. Sci. 18, 915–923.

Tangko, A.M., Madeali, M.I., Ratnawati, E., Danakusumah, E., Suwardi., 1993a. Polyculture of sea cucumber, *Holothuria scabra* and seaweed, *Gracilaria* sp. in Luki waters, Kolaka Regency, Southeast Sulawesi. In: Prosiding Seminar Hasil Penelitian, vol. 11, pp. 91–94.

Tangko, A.M., Rachmansyah, M.I., Madeali, S.T., Ismail, A., 1993b. Polyculture of sea cucumber, *Holothuria scabra* and seaweed, *Eucheuma* sp. in Sanisani Bay waters, Kolaka Regency. In: Prosiding Seminar Hasil Penelitian, vol. 11, pp. 85–89.

Taylor, A., 2016. A modified method for processing fluorescently marked sea cucumber ossicles. SPC Beche-de-mer Inf. Bull. 36, 54–57.

Taylor, A.L., Nowland, S.J., Hearnden, M.N., Hair, C.A., Fleming, A.E., 2016. Sea ranching release techniques for cultured sea cucumber *Holothuria scabra* (Echinodermata: Holothuroidea) juveniles within the high-energy marine environments of northern Australia. Aquaculture 465, 109–116.

Taylor-Moore, N., 1994. Bêche-de-mer Fishery Management Arrangements. Department of Primary Industries and Queensland Fish Management Authority, Brisbane, Queensland, Australia.

Tebchalerm, S., 1984. Edible Invertebrates in the Southern Part of Thailand. MSc Thesis, Chulalongkorn University, Thailand.

Thakur, N.L., Mainkar, S.P., Pandit, R.A., Indap, M.M., 2004. Mosquito larvicidal potential of some extracts obtained from the marine organisms-prawn and sea cucumber. Ind. J. Mar. Sci. 33, 303–306.

Théel, H., 1886. In: Report on the Scientific Results of the Voyage of HMS Challenger (Zoology) (Ed.), Report on the Holothurioidea, Part 2. vol. 14, pp. 1–290.

Thongbuakaew, T., Suwansa-Ard, S., Chaiyamoon, A., Cummins, S.F., Sobhon, P., 2021. Sex steroids and steroidogenesis-related genes in the sea cucumber, *Holothuria scabra* and their potential role in gonad maturation. Sci. Rep. 11, 1–16.

Tikader, B.K., Das, A.K., 1985. Glimpses of Animal Life of Andaman & Nicobar Islands. Zoological Survey of India, Calcutta, India.

Tlusty, M.F., Hardy, R., Cross, S.F., 2011. Limiting size of fish fillets at the center of the plate improves the sustainability of aquaculture production. Sustainability 3, 957–964.

To, A.W.L., Shea, S.K.H., Conand, C., 2018. Trade patterns of beche-de-mer at the global hub for trade and consumption—an update for Hong Kong. SPC Beche-de-mer Inf. Bull. 38, 64–67.

Todinanahary, G.G.B., Fohy, N., Andriatsialonana, C., Eeckhaut, I., 2018. Potential influence of phytoplankton density on doliolaria and pentactula appearances in sandfish hatcheries. SPC Beche-de-mer Inf. Bull. 38, 60–63.

Tomatala, P., Rahantoknam, M.A., 2017. Implementasi Budidaya Teripang di Desa Madwaer, Maluku Tenggara. JPPM 2, 135–140.

Tomatala, P., Letsoin, P.P., Kadmaer, E.M.Y., 2018a. The effectiveness of pen–culture construction for cultivation of sea cucumber. Jurnal Akuakultur Indonesia 17, 26–33.

Tomatala, P., Letsoin, P.P., Kadmaer, E.M.Y., 2018b. Potential development of marine culture in the Coastal of the Village of Rat, Southeast Maluku regency. Int. J. Biosci. 13, 401–407.

Tomatala, P., Letsoin, P.P., Ompi, M., Kadmaer, E.M.Y., 2018c. Size Identification of the adult sea cucumber, *Holothuria scabra*, in the coastal Labetawi Village, Tual Town of Maluku Province. In: Proceeding The 3rd International Conference on Operations Research, pp. 56–60.

Tomatala, P., Letsoin, P.P., Kadmaer, E.M.Y., 2019. Effectiveness of rearing sandfish, *Holothuria scabra* and seaweed, *Gracilaria* sp. with the polyculture system. Jurnal Ilmiah PLATAX 7, 266–273.

Tomatala, P., Letsoin, P.P., Kadmaer, E.M.Y., 2020. The nursery technique of juvenile sandfish, *Holothuri scabra*. Jurnal Ilmiah PLATAX 8, 89–94.

Toral-Granda, V., 2007. Facts on sea cucumber fisheries worldwide. SPC Beche-de-mer Inf. Bull. 25, 39–41.

Toral-Granda, V., Lovatelli, A., Vasconcellos, M., 2008. Sea Cucumbers: A Global Review of Fisheries and Trade. FAO Fisheries and Aquaculture Technical Paper 516, FAO, Rome.

Tortonese, E., 1979. Echinoderms collected along the eastern shore of the Red Sea (Saudi Arabia). *Atti Societa Italiana di Scienze Naturale*. In: Museo Sivico d'lstoria Naturale di Milano, vol. 120, pp. 314–319.

Tran, N.L., Ngo, V.S., 1963. Les holothuries de la Baie de Nhatrang. In: Annales de la Faculté des Sciences de Saigon, vol. 1963, pp. 237–248.

Tresnati, J., Yasir, I., Aprianto, R., Yanti, A., Bestari, A.D., Tuwo, A., 2019. Predators effects on mortality of sandfish *Holothuria scabra* cultured in multitrophic system. In: Journal of Physics: Conference Series, IOP Publishing, p. 022026.

Tresnati, J., Yasir, I., Aprianto, R., Yanti, A., Bestari, A.D., Tuwo, A., 2020. Early maturity of Sandfish *Holothuria scabra* offers more prospective broodstock supply of a commercially important aquaculture species. In: IOP Conference Series: Earth and Environmental Science, vol. 564. IOP Publishing, p. 012089.

Tresnati, J., Yasir, I., Aprianto, R., Yanti, A., Bestari, A., Tuwo, A., 2021. Peanut crab *Pinnotheres halingi*, a symbiotic commensal or parasite of Sandfish *Holothuria scabra*? In: IOP Conference Series: Earth and Environmental Science. IOP Publishing, p. 012004.

Trondillo, M.J.F., Yaptenco, K.F., Peralta, E.K., Carpio, E.V., 2018. Improved quality of dried Philippine sandfish (*Holothuria scabra*) using a combination of hot-air drying and solar drying. Philipp. J. Agric. Biosyst. Eng. 14, 45–53.

Tsiresy, G., Pascal, B., Plotieau, T., 2011. An assessment of *Holothuria scabra* growth in marine micro-farms in southwestern Madagascar. SPC Beche-de-mer Inf. Bull. 31, 17–22.

Tuapattinaja, M.A., Pattikawa, J.A., Natan, Y., 2014. Community structure of echinoderms at Tanjung Tiram, inner Ambon bay, Indonesia. Aquac. Aquar. Conserv. Legis. 7, 351–356.

Turner, L., 2016. Reproductive Conditioning and Spawning of the Sea Cucumber *Holothuria scabra* for Hatchery Production. PhD thesis, University of the Sunshine Coast, Queensland.

Tuwo, A., 1999. Reproductive cycle of the holothurian *Holothuria scabra* in Saugi island, Spermonde Archipelago, Southwest Sulawesi, Indonesia. SPC Beche-de-mer Inf. Bull. 11, 9–12.

Tuwo, A., 2004. Status of sea cucumber fisheries and farming in Indonesia. In: Lovatelli, A., Conand, C., Purcell, S., Uthicke, S., Hamel, J.-F., Mercier, A. (Eds.), Advances in Sea Cucumber Aquaculture and Management. FAO Fisheries Technical Paper No. 463, FAO, Rome, pp. 49–55.

Tuwo, A., Conand, C., 1992. Developments in beche-de-mer production in Indonesia during the last decade. SPC Beche-de-mer Inf. Bull. 4, 2–3.

Tuwo, A., Tresnati, J., 2015. Sea Cucumber Farming in Southeast Asia (Malaysia, Philippines, Indonesia, Vietnam). In: Brown, N., Eddy, S. (Eds.), Echinoderm Aquaculture. John Wiley & Sons, Inc, pp. 331–352.

Tuwo, A., Tresnati, J., Saharuddin, 2012. Analysis of Growth of Sandfish *Holothuria scabra* Cultured at Different Cultivated Habitat. pp. 17–21.

Tuwo, A., Yasir, I., Tresnati, J., Aprianto, R., Yanti, A., Bestari, A.D., Nakajima, M., 2019. Evisceration Rate of Sandfish *Holothuria scabra* During Transportation. IOP Publishing, p. 012039.

Tuwo, A., Yasir, I., Aprianto, R., Yanti, A., Bestari, A.D., Tresnati, J., 2020. Low salinity reduces survival rate of a commercially important sea cucumber (sandfish: *Holothuria scabra*). In: IOP Conference Series: Earth and Environmental Science. vol. 564. IOP Publishing, p. 012088.

Tuwo, A., Yasir, I., Bestari, A.D., Yanti, A., Aprianto, R., Tresnati, J., 2021. High salinity effect on mortality of sandfish *Holothuria scabra* (Jaeger, 1833). In: IOP Conference Series: Earth and Environmental Science, vol. 763. IOP Publishing, p. 012019.

Uthicke, S., 2001. Interactions between sediment-feeders and microalgae on coral reefs: grazing losses versus production enhancement. Mar. Ecol. Prog. Ser. 210, 125–138.

Uthicke, S., 2004. Overfishing of Holothurians: Lessons From the Great Barrier Reef. In: Lovatelli, A., Conand, C., Purcell, S.W., Uthicke, S., Hamel, J.-F., Mericer, A. (Eds.), Advances in Sea Cucumber Aquaculture and Management. FAO Fisheries Technical Paper No. 463, FAO, Rome, pp. 163–172.

Uthicke, S., Benzie, J.A.H., 1998. Improving the Conservation Management of the Commercial Sea Cucumber *Holothuria scabra* (Sandfish). Australian Institute of Marine Science, Townsville, Australia. Report project WRCP-019.

Uthicke, S., Benzie, J.A.H., 1999. Allozyme variation as a tool for beche de mer fisheries management: a study on *Holothuria scabra* (sandfish). SPC Beche-de-mer Inf. Bull. 12, 18–23.

Uthicke, S., Benzie, J.A.H., 2001. Restricted gene flow between *Holothuria scabra* (Echinodermata: Holothuroidea) populations along the north-east coast of Australia and the Solomon Islands. Mar. Ecol. Prog. Ser. 216, 109–117.

Uthicke, S., Karez, R., 1999. Sediment patch selectivity in tropical sea cucumbers (Holothurioidea: Aspidochirotida) analysed with multiple choice experiments. J. Exp. Mar. Biol. Ecol. 236, 69–87.

Uthicke, S., Purcell, S., 2004. Preservation of genetic diversity in restocking of the sea cucumber *Holothuria scabra* investigated by allozyme electrophoresis. Can. J. Fish. Aquat. Sci. 61, 519–528.

Uthicke, S., Benzie, J.A.H., Ballment, E., 1998. Genetic structure of fissiparous populations of *Holothuria* (*Halodeima*) *atra* on the Great Barrier Reef. Mar. Biol. 132, 141–151.

Uthicke, S., Purcell, S., Blockmans, B., 2005. Natural hybridization does not dissolve species boundaries in commercially important sea cucumbers. Biol. J. Linn. Soc. 85, 261–270.

Uthicke, S., Schaffelke, B., Byrne, M., 2009. A boom-bust phylum? Ecological and evolutionary consequences of density variations in echinoderms. Ecol. Monogr. 79, 3–24.

Vail, L.L., 1989. Trepang Resource Surveys: Melville Island, Gove Harbour, Croker Island. Northern Territory Museum of Arts and Sciences.

Vaitilingon, D., Smit, S., Watson, G., Miller, T., Alattas, S., Hock, K.O., Zainoddin, J., Zaidnuddin, I., Azhar, H., 2016. Production de juvéniles d'holothuries en écloserie en Malaisie: la recherche et le développement au service d'un projet pilote de production d'holothuries de sable *Holothuria scabra*. SPC Beche-de-mer Inf. Bull. 36, 67–75.

Valayudhan, P., Santhanam, R., 1990. Fish by-products of commerce. Fish. Chimes 9, 44–47.

Van Dyck, S., Caulier, G., Todesco, M., Gerbaux, P., Fournier, I., Wisztorski, M., Flammang, P., 2011. The triterpene glycosides of *Holothuria forskali*: usefulness and efficiency as a chemical defense mechanism against predatory fish. J. Exp. Biol. 214, 1347–1356.

Van Eys, S., 1986. The international market for sea cucumber. Infofish Marketing Digest 5, 41–44.
Van Eys, S., 1987. The Market for Sea Cucumber From the Pacific Islands. pp. 207–221 (Chapter 11).
Van Thanh, N., Dang, N.H., Van Kiem, P., Cuong, N.X., Huong, H.T., Van Minh, C., 2006. A new triterpene glycoside from the sea cucumber *Holothuria scabra* collected in Vietnam. ASEAN J. Sci. Technol. Dev. 23, 253–259.
VandenSpiegel, D., Ovaere, A., Massin, C., 1992. On the association between the crab *Hapalonotus reticulatus* (Crustacea, Brachyura, Eumedonidae) and the sea cucumber *Holothuria* (*Metriatyla*) *scabra* (Echinodermata, Holothuridae). In: Bulletin de l'Institut Royal des Sciences Naturelles de Belgique Biologie, vol. 62, pp. 167–177.
Vergara, W., Rodríguez, A., 2016. Nutritional composition of sea cucumber *Isostichopus* sp. Nat. Resour. 7, 130.
Vuki, V.C., Viala, F., 1991. Shrinkage and weight loss of nine commercial species of holothurians from Fijian waters. SPC Fish. Newsl. 51, 27–29.
Wahyuni, N.I., As'ad, S., Massi, M.N., 2020. The effect of *Holothuria scabra* on the level of docosahexaenoic acid in breast milk with preterm birth. Enferm. Clin. 30, 366–370.
Wainiya, W., 1988. On the taxonomy of commercial sea-cucumbers from Prachuap Khiri Khan-Surat Thani Provinces. Marine Fisheries Division, Department of Fisheries, Thailand, pp. 1–28. *Technical Paper No. 112531 Marine Fisheries Laboratory*.
Wang, J., Han, H., Chen, X., Yi, Y., Sun, H., 2014. Cytotoxic and apoptosis-inducing activity of triterpene glycosides from *Holothuria scabra* and *Cucumaria frondosa* against HepG2 cells. Mar. Drugs 12, 4274–4290.
Wargasetia, T., Widodo, N., 2018. PO-408 Methanol Extract of *Holothuria scabra* Inhibits Cell Growth and Induces Apoptosis in T47D Breast Cancer Cells. BMJ Publishing Group Limited.
Wargasetia, T.L., Ratnawati, H., Widodo, N., 2020. Anticancer potential of holothurin A, holothurin B, and holothurin B3 from the sea cucumber *Holothuria scabra*. In: AIP Conference Proceedings. AIP Publishing LLC, p. 040084.
Watanabe, S., Zarate, J.M., Lebata-Ramos, M.J.H., Nievales, M.F.J., 2012a. Evaluation of nutritional condition of juvenile sandfish (*Holothuria scabra*). In: Hair, C.A., Pickering, T.D., Mills, D.J. (Eds.), Asia–Pacific Tropical Sea Cucumber Aquaculture. ACIAR Proceedings, vol. 136. Australian Centre for International Agricultural Research, Canberra, pp. 50–56.
Watanabe, S., Kodama, M., Zarate, J.M., Lebata-Ramos, M.J.H., Nievales, M.F.J., 2012b. Ability of sandfish (*Holothuria scabra*) to utilise organic matter in black tiger shrimp ponds. In: Proceedings Asia – Pacific Tropical Sea Cucumber Aquaculture. Australian Centre for International Agricultural Research, *Canberra*, pp. 113–120.
Watanabe, S., Zarate, J.M., Sumbing, J.G., Lebata-Ramos, M.J.H., Nievales, M.F., 2012c. Size measurement and nutritional condition evaluation methods in sandfish (*Holothuria scabra* Jaeger). Aquacult. Res. 43, 940–948.
Watanabe, S., Kodama, M., Sumbing, J.G., Lebata-Ramos, M.J.H., 2013. Diet-tissue stable isotopic fractionation of tropical sea cucumber, *Holothuria scabra*. Jpn. Agric. Res. Q. 47, 127–134.
Watanabe, S., Sumbing, J.G., Lebata-Ramos, M.J.H., 2014. Growth pattern of the tropical sea cucumber, *Holothuria scabra*, under captivity. Jpn. Agric. Res. Q. 48, 457–464.
Watanabe, S., Kodama, M., Orozco, Z.G.A., Sumbing, J.G., Novilla, S.R.M., Lebata-Ramos, M., Hazel, J., 2015. Estimation of energy budget of sea cucumber, *Holothuria scabra*, in integrated multi-trophic aquaculture. Aquaculture Department, Southeast Asian Fisheries Development Center, pp. 307–308.

Weerasingha, R., Ajith Kumara, P.A.D., Fahim, M.S.M., Athukorala, D.A., 2015. Utilization of different types of fish silage on the growth of juvenile sea cucumber (*Holothuria scabra*) (sandfish). In: Proceedings o f the National Aquatic Resources Research and Development Agency (NARA), pp. 32–36.

Wiedemeyer, W.L., 1992. Feeding behaviour of two tropical holothurians, *Holothuria (Metriatyla) scabra* (Jager 1833) and *H. (Halodeima) atra* (Jager 1833), from Okinawa, Japan. In: Proceedings of the Seventh International Coral Reef Symposium, pp. 863–870.

Wolkenhauer, S.-M., 2008a. Impact of Removal—A Case Study on the Ecological Role of the Commercially Important Sea Cucumber *Holothuria scabra* (Echinodermata: Holothuroidea). University of Rostock, Moreton Bay, Australia. PhD Thesis.

Wolkenhauer, S.-M., 2008b. Burying and feeding activity of adult *Holothuria scabra* (Echinodermata: Holothuroidea) in a controlled environment. SPC Beche-de-mer Inf. Bull. 27, 25–28.

Wolkenhauer, S.-M., Uthicke, S., Burridge, C., Skewes, T., Pitcher, R., 2010. The ecological role of *Holtohuria scabra* (Echinodermata: Holothuroidea) within subtropical seagrass beds. J. Mar. Biol. Assoc. U. K. 90, 215–223.

Xia, J., Ren, C., Yu, Z., Wu, X., Qian, J., Hu, C., 2016. Complete mitochondrial genome of the sandfish *Holothuria scabra* (Holothuroidea, Holothuriidae). In: Mitochondrial DNA Part A, vol. 27, pp. 4174–4175.

Yaghmour, F., Whittington-Jones, B., 2018. First record of *Holothuria (Metriatyla) scabra* Jaeger, 1833 (Echinodermata: Holothuroidea) from the coastal waters of the United Arab Emirates. PeerJ 6, e5555.

Yahyavi, M., Afkhami, M., Javadi, A., Ehsanpour, M., Khazaali, A., Mokhlesi, A., 2012. Fatty acid composition in two sea cucumber species, *Holothuria scabra* and *Holothuria leucospilata* from Qeshm Island (Persian Gulf). Afr. J. Biotechnol. 11, 2862–2868.

Yamada, A., Tamori, M., Iketani, T., Oiwa, K., Motokawa, T., 2010. A novel stiffening factor inducing the stiffest state of holothurian catch connective tissue. J. Exp. Biol. 213, 3416–3422.

Yamanouchi, T., 1939. Ecological and physiological studies on the holothurians in the coral reef of Palao Islands. Palao Trop. Biol. Stud. 4, 603–636.

Yamanouchi, T., 1956. The daily activity rhythms of the holothurians in the coral reef of the Palao Islands. In: Publication of the Seto Marine Biology Laboratory Animal Handbooks, 5, pp. 347–362.

Yan, B., Li, L., Yi, Y., Tang, H., Sun, P., Zhang, S., 2005. Study on bioactive triterpene glycosides of sea cucumber *Holothuria scabra*. Jaeger Acad. J. Second Mil. Med. Univ. 26 (6), 626–631.

Yan, J., Wang, S., Zhou, Y., Zhu, Y., Feng, Y., 2012. Studies on embryonic and larval development of sea cucumber (*Holothuria scabra*). J. Fish. China 36, 1073–1080.

Yang, H., Hamel, J.-F., Mercier, A., 2015a. The Sea Cucumber Apostichopus japonicus: History, Biology and Aquaculture. Academic Press, London.

Yang, X.M., Wu, M.C., Zhang, L., Li, J.B., Chen, F.Y., Pan, Y., 2015b. Annual change of gonadal development in *Holothuria scabra* from Hainan Island. J. South. Agric. 46, 1117–1122.

Yang, L., Wang, Y., Yang, S., Lv, Z., 2018. Separation, purification, structures and anticoagulant activities of fucosylated chondroitin sulfates from *Holothuria scabra*. Int. J. Biol. Macromol. 108, 710–718.

Yanti, A., Tresnati, J., Yasir, I., Rahmani, P.Y., Aprianto, R., Tuwo, A., 2020. Size at the maturity of sea cucumber *Holothuria scabra*. Is it an overfishing sign in Wallacea Region? In: IOP Conference Series: Earth and Environmental Science, vol. 473. IOP Publishing, p. 012056.

Yao, X.M., Wang, J., Bei, R.-B., Xing, S.-L., 2006. Impact of pH on the digestive enzymes of the sea cucumber *Holothuria scabra*. Nat. Sci. J. Hainan Univ. 4.

Yao, X.M., Wang, H.Y., Xing, S.L., He, X.Y., 2007. The effects of temperature and water quality on feeding and growth of *Holothuria scabra*. Fish. Sci. Liaoning 26, 292–295.

Yaptenco, K.F., 2017. Moisture sorption isotherms and mathematical model selection for dried sandfish (*Holothuria scabra*). Agric. Eng. Int. CIGR J. 19, 176–186.

Yaptenco, K.F., Pangan, R.S., Pardua, S., Lapitan, E.S., 2013. Annual Report for Project 3: Enhancing the Philippine Sea Cucumber Industry Through Appropriate Processing Technology. DOST-PCAARRD/Institute of Agricultural Engineering University of the Philippines, Los Baños.

Yenti, F., Apriandi, A., Suhandana, M., 2019. PEMANFAATAN TERIPANG PASIR (*Holothuria scabra*) SEBAGAI MINUMAN FUNGSIONAL—Utilization of Sea Cucumber (*Holothuria Scabra*) as a Functional Drink. Marinade 2, 23–29.

Yin-Geng, W., Chun-Yun, Z., Xiao-Jun, R., Jie-Jun, C., Cheng-Yin, S., Hui-ling, S., Jing-Ping, Y., 2005. Diseases of cultured sea cucumber, *Apostichopus japonicus*, in China. In: FAO Fisheries Technical Paper, vol. 463, pp. 297–310.

Yunizal, T., Nasran, S., Murdinah., 1997. Effect of icing of raw material on the quality of dried smoked sand sea cucumber (*Holothuria scabra*). In: James, D.G. (Ed.), Summary Report of and Papers Presented at the Tenth Session of the Working Party of Fish Technology and Marketing, Colombo, Sri Lanka, 4-7 June 1996.

Yurasakpong, L., Apisawetakan, S., Pranweerapaiboon, K., Sobhon, P., Chaithirayanon, K., 2021. *Holothuria scabra* extract induces cell apoptosis and suppresses warburg effect by down-regulating Akt/mTOR/HIF-1 Axis in MDA-MB-231 breast cancer cells. Nutr. Cancer 73 (10), 1964–1975.

Yussuf, Y.S., Yahya, S.A., 2020. Size-distribution and length-weight relationship of a deep-water population of *Holothuria scabra* (Jaeger, 1833) in Zanzibar, Tanzania. West. Indian Ocean J. Mar. Sci. 19, 113–121.

Yusuf, S., Tuwo, A., Tresnati, J., Moore, A.M., Conand, C., 2017. Teripangfishing activities at Barang Lompo Island, Sulawesi, Indonesia: an update 20 years after a visit in 1996. 37, 99–102.

Zamora, L.N., Yuan, X., Carton, A.G., Slater, M.J., 2018. Role of deposit-feeding sea cucumbers in integrated multitrophic aquaculture: progress, problems, potential and future challenges. Rev. Aquac. 10, 57–74.

Zarate, J., Niwa, K., Watanabe, S., 2012. The Relationship Between Nutritional Stress and Digestive Enzyme Activities in Sea Cucumber *Holothuria scabra*. JIRCAS Working Report, pp. 97–105.

Zhang, Q., Liu, Y.H., 1998. Aquaculture Techniques of Sea Cucumber and Sea Urchin. The Press of Ocean University of Qingdao, *Qingdao*. 119 pp.

Zhao, Y., Ma, H., Zhang, W., Ai, Q., Mai, K., Xu, W., Wang, X., Liufu, Z., 2011. Effects of dietary β-glucan on the growth, immune responses and resistance of sea cucumber, *Apostichopus japonicus* against *Vibrio splendidus* infection. Aquaculture 315, 269–274.

Zhao, Y., Zhang, W., Xu, W., Mai, K., Zhang, Y., Liufu, Z., 2012. Effects of potential probiotic *Bacillus subtilis* T13 on growth, immunity and disease resistance against Vibrio splendidus infection in juvenile sea cucumber *Apostichopus japonicus*. Fish Shellfish Immunol. 32, 750–755.

Zhao, S., Yuan, C., Wan, L., Chen, W., Liu, J., 2017. Effect of COS on phagocytic activity of coelomic cell of tropical sea cucumber (*Holothuria scabra*). Gen. Appl. Biol. 36, 3237–3240.

Zoutendijk, D.A., 1989. Bêche de Mer. Rori of the Cook Islands.—Ministry of Marines Resources, Rarotonga, Cook Islands. In: Resource Profile, vol. 6, pp. 1–17.

FSC
www.fsc.org
MIX
Papier | Fördert
gute Waldnutzung
FSC® C083411

Kolibri 360 GmbH
Ferdinand-Jühlke-Straße 7
99095 Erfurt, Deutschland
produktsicherheit@kolibri360.de